INTRODUCTION TO
PLANT POPULATION
BIOLOGY

For our children: Daniel Silvertown
and Henry and Leo Lovett Doust

INTRODUCTION TO
PLANT POPULATION
BIOLOGY

Jonathan W. Silvertown

Department of Biology, The Open University,
Milton Keynes, UK

AND

Jonathan Lovett Doust

Department of Biological Sciences
University of Windsor
Ontario, Canada

b

Blackwell
Science

© 1993 by
J.W. Silvertown & J. Lovett Doust
Published by
Blackwell Science Ltd
Editorial Offices:
Osney Mead, Oxford OX2 0EL
25 John Street, London WC1N 2BL
23 Ainslie Place, Edinburgh EH3 6AJ
238 Main Street, Cambridge
    Massachusetts 02142, USA
54 University Street, Carlton
    Victoria 3053, Australia

Other Editorial Offices:
Arnette Blackwell SA
1, rue de Lille
75007 Paris
France

Blackwell Wissenschafts-Verlag GmbH
Kurfürstendamm 57
10707 Berlin
Germany

Blackwell MZV
Feldgasse 13
A-1238 Wien
Austria

First published 1982
by Longman Scientific & Technical
Reprinted 1984
Second edition 1987
**Third edition published 1993
by Blackwell Science Ltd**
Reprinted 1995

Set by Setrite Typesetters, Hong Kong
Printed and bound in Great Britain
at the University Press, Cambridge

DISTRIBUTORS

Marston Book Services Ltd
PO Box 87
Oxford OX2 0DT
(*Orders*: Tel: 01865 791155
            Fax: 01865 791927
            Telex: 837515)

USA
Blackwell Science, Inc.
238 Main Street, Cambridge, MA 02142
(*Orders*: Tel: 800 215-1000
            617 876-7000
            Fax: 617 492-5263)

Canada
Oxford University Press
70 Wynford Drive Don Mills
Ontario M3C 1J9
(*Orders*: Tel: 416 441-2941)

Australia
Blackwell Science Pty Ltd
54 University Street Carlton, Victoria 3053
(*Orders*: Tel: 03 347-5552)

A catalogue record for this title
is available from the British Library

ISBN 0-632-02973-0

Library of Congress
Cataloging-in-Publication Data

Silvertown, Jonathan W.
    Introduction to plant population biology/
Jonathan W. Silvertown,
Jonathan Lovett Doust. — 3rd ed.
        p.   cm.
    Rev. ed. of: Introduction to
plant population ecology. 2nd ed. 1987.
    Includes bibliographical references (p.   )
    and index.
    ISBN 0-632-02973-0
    1. Plant populations. 2. Vegetation dynamics.
    I. Lovett Doust, Jonathan.
II. Silvertown, Jonathan W.
Introduction to plant population ecology.
III. Title.
QK910.S54   1993
581.5248—dc20

Contents

Acknowledgements

We thank the following friends and colleagues for their comments on drafts of various chapters: Spencer Barrett, James Bullock, Deborah Charlesworth, Pam Dale, Tom de Jong, Tim Holtsford, Subodh Jain, Paul Keddy, Andrew Lack, David Lloyd, Nick Waser, Andrew Watkinson, Jake Weiner and Mark Westoby. Richard Law and J. Bastow Wilson deserve particular thanks for working their way through the entire book. These acknowledgements should not be taken to imply agreement with what we have written: responsibility for errors of omission and commission remains ours. Niki Koenig drew further vignettes of plant species, which have leavened the figures since the first edition of *Introduction to Plant Population Ecology*.

We are particularly grateful to Rissa de la Paz and to Lesley Lovett Doust for their help and support throughout.

We are grateful to the following for permission to reproduce copyright material. Academic Press for Fig. 4.6 (White 1985; Fig. 1), Fig. 5.16b (New 1958; Fig. 1) and Fig. 6.7b (Sarukhan *et al.* 1985; Fig. 2); Academic Press and the authors for Fig. 6.2 (Frissell 1973; Fig. 2), Fig. 8.7 (Jain 1976; Fig. 8) and Table 2.2 (Wyatt 1983; Table 1); American Society of Plant Taxonomists and the authors for Fig. 9.3 (Bayer & Stebbins 1987; Figs 4 & 5); *Annals of Applied Biology* and the authors for Fig. 4.16c (Burdon & Chilvers 1975; Fig. 1a); Annual Reviews, Inc., for Fig. 8.10a reproduced with permission from the *Annual Review of Ecology and Systematics*, Vol. 19, © 1988 by Annual Reviews, Inc. (Ashton 1988; Fig. 2); Association for Tropical Biology for Fig. 6.6a (Brokaw 1982; Fig. 1); *Australian Journal of Ecology* for Fig. 6.4 (Auld 1987; Fig. 6); British Ecological Society for Fig. 2.4b (McGraw & Antonovics 1983; Figs 1 & 4d), Fig. 4.4 (Kays & Harper 1974), Fig. 4.7 (Ford 1975), Fig. 4.8 (Mohler *et al.* 1978), Fig. 4.17 (Keddy 1981), Fig. 5.1b (Sarukhan & Harper 1973), Fig. 5.5 (Hassell *et al.* 1976; Fig. 2), Fig. 5.7 (Watkinson 1990; Fig. 6b), Fig. 5.8a (Sarukhan 1974), Fig. 5.8b (Pavone & Reader 1982; Fig. 2), Fig. 5.9 (Thompson & Grime 1979), Fig. 5.10 (Thompson 1986), Fig. 5.13 (Roberts & Feast 1973; Fig. 1), Fig. 5.15 (Harper *et al.* 1965), Fig. 6.1 (Agren & Eriksson 1990; Figs 1d, 3d & 5d), Fig. 8.3b (Goldberg & Fleetwood 1987; Fig. 1), Fig. 8.5a (Antonovics & Fowler 1985), Fig. 8.5b (Law & Watkinson 1987), Fig. 8.6 (Firbank & Watkinson 1985; Fig. 4a, b), Fig. 8.8 (Turkington & Harper 1979), Fig. 8.11 (Mahdi *et al.* 1989; Fig. 3) and Fig. 10.7 (Grime & Jeffrey 1965); Botanical Society of Japan for Fig. 6.7a (Kanzaki 1984; Fig. 2); Cambridge University Press and the authors for Fig. 1.1a (Huntley & Birks 1983; maps), Fig. 2.2b (Briggs & Walters 1984; Fig. 3.4a), Fig. 6.8 (Lieberman *et al.* 1985b; Fig. 5) and Fig. 8.12b (Fleming & Williams 1990; Fig. 2a); Carnegie Institution of Washington for Fig. 2.4a (Clausen *et al.* 1948; Fig. 1); CSIRO for Fig. 10.8 (Black 1958; Fig. 11); Ecological Society of America for Fig. 3.2c from 'Analysis of parentage for naturally established seedlings of *Chamaelirium luteum* (Liliaceae)' by T.R. Meagher & E. Thompson, *Ecology*, 1987, **68**, 803–812. Copyright © 1987 by ESA, reprinted by permission (Meagher & Thompson 1987; Fig. 4); Fig. 4.16a from 'A dynamic analysis of age in sugar maple seedlings' by J.M. Hett, *Ecology*, 1971, **52**, 1071–1074. Copyright © 1971 by ESA, reprinted by permission (Hett 1971); Fig. 5.14 from 'Ecophysiology of secondary dormancy in seeds of *Ambrosia artemisifolia*' by J.M. Baskin & C.C. Baskin, *Ecology*, 1980, **61**, 475–480. Copyright © 1980 by ESA, reprinted by permission (Baskin & Baskin 1980; Fig. 1); and Fig. 7.7 from 'An experimental study of competition among fugitive prairie plants' by W.J. Platt & I.M. Weiss, *Ecology*, 1985, **66**, 708–720. Copyright © 1985 by ESA, reprinted by permission (Platt & Weiss 1985; Fig. 5); Freeman & Co. for Fig. 9.2 from *Basic Concepts in Population, Quantitative, and Evolutionary Genetics* by James F. Crow. Copyright © 1986 by W.H. Freeman and Company, reprinted with permission (Crow 1986; Fig. 7.5); Gauthier-Villars for Fig. 6.6b (Faille *et al.* 1984; Fig. 6); Genetics Society of America for Fig. 2.5 (Mitchell-Olds & Bergelson 1990b; Fig. 1); Gustav Fischer Verlag for Fig. 5.11 (van der Reest & Rogaar 1988; Fig. 1); HarperCollins for Fig. 6.5b from Fig. 12.2 from *Community Ecology* by Jared Diamond & Ted J. Case. Copyright © 1986 by Harper & Row, Publishers, Inc., reprinted by permission of Harper-Collins Publishers, Inc. (Grubb 1986; Fig. 12.2); Harvard University Press for Fig. 10.2 reprinted by permission of the publisher from *Ecology and Evolution of Communities* edited by Martin L. Cody & Jared M. Diamond, Cambridge, Mass.: The Belknap Press of Harvard University Press. Copyright © 1975 by the President and Fellows of Harvard College (Schaffer & Gadgil 1975; Fig. 1); *Heredity* for Fig. 2.3 (Zuberi & Gale 1976; Fig. 2) and Fig. 3.1 (Gliddon *et al.* 1987; Fig. 1); National Research Council of Canada for Fig. 9.1 (Eis *et al.* 1965; Fig. 3); *Nature* and the author for Fig. 5.2a reprinted with permission from *Nature*, **303**, 164–167. Copyright © 1983 Macmillan Maga-zines Limited (Bennett 1983; Fig. 2); Princeton University Press for Fig. 3.7 from Endler, J.A.; *Natural Selection in the Wild*. Copyright © 1986 by PUP, reprinted/reproduced by permission of Princeton University Press (Endler 1986; Fig. 1.3); Sinauer Publishers for Fig. 1.5 (Caswell 1989; Figs 4.1a, b & 4.3); Society for the Study of Evolution for Fig. 1.2 (Linhart 1988; Fig. 2), Fig. 2.7 (Schoen 1982), Fig. 3.5b (Levin & Kerster 1968; Fig. 3), Fig. 3.6 (Fenster 1991b; Fig. 4), Fig. 6.5a (Epling *et al.* 1960; Table 2), Fig. 9.4 (Bush & Smouse 1991; Fig. 3), Fig. 10.3 (Schemske 1984; Figs 1–3) and Fig. 10.4 (Law *et al.* 1977; Fig. 3); Springer-Verlag and the authors for Fig. 1.1e (Bergmann 1978; Fig. 2), Fig. 2.2a (Garbut & Bazzaz 1987; Fig. 1a), Fig. 3.5a (Rai & Jain 1982; Fig. 3), Fig. 4.9 (Mithen *et al.* 1984; Fig. 1), Fig. 4.12 (Burdon *et al.* 1984; Fig. 1c, d) and Fig. 5.4 (Symonides *et al.* 1986; Figs 1a & 3); Professor M.E. Turner for Fig. 3.3 (Turner *et al.* 1982; Fig. 5); University of Chicago Press for Fig. 3.2a, b (Meagher 1986; Fig. 6a, b), Fig. 3.5c (Levin & Kerster 1969b; Fig. 5), Fig. 5.14 (Baskin & Baskin 1983; Fig. 1), Fig. 7.3 (Alvarez-Buylla & García-Barrios 1991; Fig. 1), Fig. 8.10b (Werner & Platt 1976; Fig. 2), Fig. 8.12a (Janzen 1970; Fig. 1), Fig. 8.13 (Huston 1979; Fig. 2) and Fig. 10.6 (Lloyd 1987; Fig. 1); Verlag Eugen Ulmer for Fig. 1.1c, f (Ellenberg 1978; Fig. 155); and Professor R.W. Willey for Fig. 4.2 (Willey & Heath 1969; Figs 1c, d & 2a, d).

Lastly, we acknowledge the assistance of Longman UK in the reproduction of artwork.

Symbols and Terms
used in this Book

In choosing symbols we have tried to be consistent and to follow convention. Inevitably, some symbols have more than one meaning, but it should be obvious from the context which is the correct one in any particular case.

A Genetic neighbourhood area

a (i) The area necessary to achieve w_m.
(ii) $a_{i,j}$ is the element in column i, row j of a population projection matrix. The same terms are used for transitions between stages in stage-classified life history diagrams such as Fig. 1.5b

α, β Competition coefficients

B Number of births in a population

B Birth *rate*. $B = B/N_t$

$B_{i,0}$ Reciprocal of maximum plant weight (i.e. $1/W_m$), used in the reciprocal yield model (Eqn 8.8)

$B_{i,i}$ Measure of intraspecific competition in the reciprocal yield model (Eqn 8.8)

b (i) Exponent of the yield/density equation (Eqn 4.3) and the competition−density effect (Eqn 4.4). When the law of constant final yield applies, $b = 1$.
(ii) The exponent whose value determines the shape of the recruitment curve in a model of density-dependent population dynamics (Eqn 5.11).

c (i) Intercept on the vertical axis of the self-thinning rule (Eqn 7.5) relating w and density. (ii) The rate at which vacant sites in a metapopulation are colonized

D Number of deaths in a population

D Death *rate*. $D = D/N_t$

d A measure of inbreeding depression: $d = 1 - W_i$, where W_i is the relative fitness of inbred progeny

d_x Mortality between ages x and $x + 1$ in a life table (e.g. Table 1.1), i.e. $l_x - l_{x+1}$

E Number of emigrants from a population

E Emigration *rate*. $E = E/N_t$

G Fraction of a seed population that germinates

G_{ST} Genetic differentiation among populations $= (H_T - H_S)/H_T$

H An index of genetic diversity, equivalent to the probability that two randomly chosen alleles will be different

H_B Heritability in the broad sense $= V_g/V_p$

H_N Heritability in the narrow sense $= V_a/V_p$

H_S Genetic diversity within a population

H_T Total genetic diversity at the species level

I Number of immigrants in a population

I Immigration *rate*. $I = I/N_t$

k Exponent in the self-thinning rule (Eqn 4.5) relating yield and density

K_i The maximum sustainable density of a species in a mixture. Subscripts denote different species. $K = m^{-1}$

λ The annual (or finite) rate of population increase. $\ln \lambda = r$

l_x Proportion of individuals in a cohort surviving to age x

m (i) Migration rate (Eqn 3.2).
(ii) Reciprocal of the maximum value that density can achieve $(m = K^{-1})$, limited by self-thinning (Eqn 5.12)

m_x Fecundity of an individual of age x

n The *number* (as opposed to the *density*) of individuals in a population

N_e Effective population size

$N, N_t,$ N_i, N_x Population density. Subscripts denote time (t), different species (i) or age (x)

p (i) An allele frequency. (ii) The

proportion of sites in a metapopulation that are occupied

p_x Survival rate per individual alive at the start of an age interval in a life table (e.g. Table 1.1), i.e. $1 - (d_x/l_x)$

P_t The number of occupied sites in a metapopulation at time t

q An allele frequency

q_x Mortality rate per individual alive at the start of an age interval in a life table (e.g. Table 1.1), i.e. d_x/l_x

r Intrinsic rate of increase

R_0 Net reproductive rate, i.e. $N_{t+\tau}/N_t$

S Selfing rate

s (i) Selection coefficient. (ii) Mean seed production per plant in a population

s_m Maximum possible seed production by a plant in uncrowded conditions

T The total number of sites in a metapopulation $(V_t + P_t)$

t (i) A subscript denoting the time associated with a parameter, e.g. N_t is the population size at time t.
(ii) $\hat{t}$ equilibrium outcrossing rate

τ The length of a generation in years

V_a Additive genetic variance

V_e Environmental variance

V_g Genetic variance

V_p Phenotypic variance

V_t The number of vacant sites in a metapopulation at time t

V_x Reproductive value of an individual of age x (see Chapter 10)

W Darwinian fitness (its maximum value is 1)

W_i Fitness of inbred progeny

w Mean plant weight

w_m Maximum possible weight of a plant growing in uncrowded conditions

x The rate at which the occupied sites in a metapopulation become extinct

Y Total plant yield of a population or plot

Chapter 1
Introduction

1.1 PLANTS

Plants are too often taken for granted. This book you are holding was once a plant. The oxygen you are breathing was freed to the air by photosynthesis. You are probably wearing clothes at least partially made from plant fibres, and if you are not sitting in front of a cup of coffee we invite you to make one now, for we want your full attention! Plants are sometimes seen as dull because they 'do not move', they 'do not behave' and they seem altogether passive. This is all a gross illusion. As we shall see, plants are exciting if only you are perceptive enough to appreciate the subtleties of their peculiar ways. There is enough sex and death in the plant kingdom to fill a thousand and one nights of storytelling, but the Sheherazade of botany has still not made her appearance. Instead, we bid you to read on, and we will do our best to prepare the way for her.

This book is an introduction to the population biology of plants. It therefore takes slices out of population biology and out of plant science and, considered as separate topics, will leave much unsaid about both. What we have to offer is an insight into the fertile and fast-developing area where population biology and plant science overlap.

1.2 POPULATION BIOLOGY

A **population** is a collection of individuals belonging to the same species, living in the same area — the water hyacinth *Eichhornia crassipes* in a ditch, the grass *Lolium perenne* in a lawn, and the Norway spruce *Picea abies* in a forest are examples. This definition has two components, a genetic one (individuals belong to the same species) and a spatial one (individuals live in the same area), but populations are neither genetically nor spatially homogeneous. Populations have several kinds of structure. The **genetic structure** describes the patchiness of gene frequency and genotypes, and the **spatial structure** describes the variation in density within a population. Populations also have an **age structure** that describes the relative numbers of youngsters and oldsters, and a **size structure** that describes the relative numbers of large and small individuals. Population biology attempts to explain the origin of these different kinds of structure, to understand how they influence each other, and how and why they change with time. Changes in genetic structure with time are the subject of **evolution**; change in numbers with time is the subject of **population dynamics**. These are the two principal organizing themes of this book. We deal with the special characteristics of plants that affect their population biology in the second half of this chapter.

1.2.1 Demography and fitness

The essence of population biology is captured by a simple equation that relates the numbers per unit area of an organism N_t at some time t to the numbers N_{t-1} one year earlier:

$$N_t = N_{t-1} + B - D + I - E \qquad (1.1)$$

where B is the number of births, D the number of deaths and I and E are respectively immigrants into the population and emigrants from it. B, D, I and E are known as **demographic parameters** and are central to both population dynamics and evolution. The dynamics of a population may be summarized by the ratio N_t/N_{t-1}, which is called the **annual** or the **finite rate of increase**, and is given the Greek symbol λ (lambda). The balance between the two demographic parameters which increase N_t (B&I) and those which decrease it (D&E) determines whether the population remains stable ($N_t = N_{t-1}$ and λ = 1), increases ($N_t > N_{t-1}$ and λ > 1) or decreases ($N_t < N_{t-1}$ and λ < 1). Among other things, the values of B, D, I and E in natural plant populations are variously influenced by pollinators, herbivores, diseases, animals that disperse seeds, soil, climate, by the density of the population itself and by that of other plant species.

Furthermore, the influence of all of these factors on the demographic parameters usually has a genetic component, because some individuals in a population are more susceptible to disease than others, some are more distasteful to herbivores than others, and some are more tolerant of climatic extremes than others. The consequence of this is that one genotype may be favoured in one locality, and another somewhere else, and this may produce local differences in allele frequency that correlate with environmental conditions. For example, allele frequencies at an acid phosphatase (APH) locus in *Picea abies* correlate with altitude in the Seetaler Alps of Austria (Fig. 1.1e), and on a larger scale with latitude in North Europe (Fig. 1.1d). Polymorphism at this single locus appears to reflect a more general correlation between the genetic structure and geographical location of *P. abies* populations in Europe. Using 22 enzyme loci to characterize genotypes, Lagercrantz and Ryman (1990) found a similar geographical correlation stretching across populations sampled from most of the species' natural range. Note, however, that *correlations* between allele frequencies and the environment do not prove that the particular loci studied are the direct *cause* of ecological differences between genotypes.

Equation 1.1 can be applied separately to different genotypes to produce a finite rate of increase for each of them. For *P. abies* we could compare $λ_H$ for the high altitude genotype with $λ_L$ for trees with the low altitude genotype in different parts of the Seetaler Alps. At low altitudes we might expect $λ_L > λ_H$, and at high altitudes we might expect $λ_H > λ_L$. The finite rate of increase for each genotype can be used as a measure of its

Darwinian **fitness** (W). Fitness is a measure of the relative evolutionary advantage of one genotype over another under particular conditions. By convention, the genotype with the highest fitness has a value $W = 1$, and the fitness of other genotypes is given as a proportion of this. So, *if* the APH genotypes are adapted to the respective environments in which they are most frequent (and this has *not* yet been tested experimentally), at low altitudes the fitness of the low genotype W_L would be 1 and the fitness of the high genotype $W_H < 1$, while at high altitudes $W_H = 1$ and $W_L < 1$. The relative fitness of different genotypes is not alone in determining gene frequencies, because mating between different genotypes and pollen transport and seed dispersal between different areas can blur genetic differences between them. We will look at this in Chapter 3.

Demographic processes underlie the distribution of species as well as the distribution of genotypes within species. At a local scale, differences in B, D, I and E must explain why *P. abies* forms a distinct altitudinal belt in the Alps (Fig. 1.1c). Distributions can be investigated experimentally by sowing or planting the species along transects that cross the boundary of its natural range, to assess why boundaries lie where they do (Chapter 7). Climate changes, and in response, species' boundaries are, to varying degrees, dynamic. On a continental scale this is easily seen by comparing the historical spread of *P. abies* since the last glaciation (Fig. 1.1a) with its present geographical distribution (Fig. 1.1b). If rates of migration are slower than climatic change, boundaries may not reflect demographic limits but historical ones. This is true of the distribution and genetic structure of Norway spruce in Europe today.

1.2.2 Adaptation by natural selection

The genetic structure of a population is subject to a number of forces that may bring about evolutionary change. The two principal ones are **gene flow** and **natural selection**. Gene flow simply describes the changes in gene frequency brought about by the migration of individuals, their seeds and pollen. Special attention is given to natural selection because this is the *only* known process that can produce **adaptive** evolutionary change.

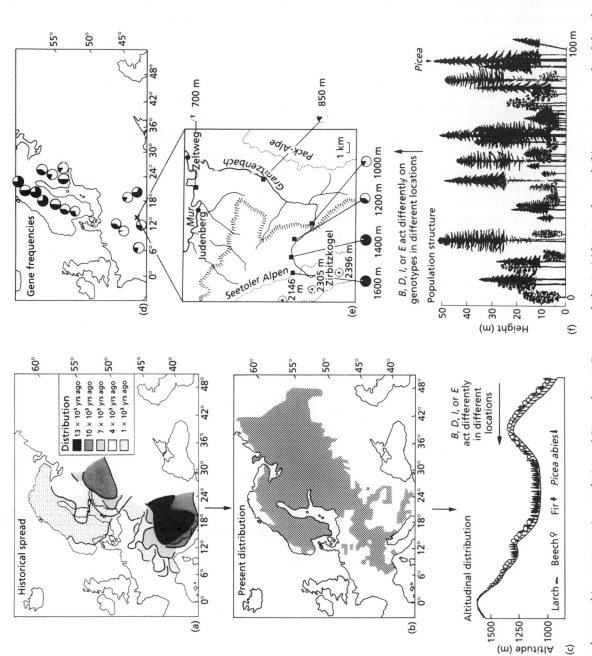

Fig. 1.1 How demographic parameters in populations of *Picea abies* in Europe underlie its (a) historical spread, (b) present geographical distribution, (c) local, altitudinal distribution, (d, e) APH allele frequencies and (f) population structure (from Silvertown 1987a).

This was the great, enduring insight in Charles Darwin's book *The Origin of Species*, first published in 1859. Three conditions are necessary for natural selection to operate:

1 Variation between individuals.

2 Inheritance of the variation.

3 Differences in fitness between variants.

Given these three conditions, genotypes with phenotypes that confer higher fitness will multiply faster than those with lower fitness and there is the potential for adaptive evolutionary change to take place. Whether change actually occurs or not may depend upon the level of migration of seeds and pollen from neighbouring areas where selection has a different direction or is absent. However, field observations have shown that natural selection may be quite strong enough to produce marked genetic differences between adjacent parts of the same population and adaptation to local conditions. For example, the annual *Veronica peregrina* grows in and around temporary pools, called vernal pools, that form in early spring time in the Central Valley of California. Though pools are only metres wide, plants in the centre are genetically different from those around

the periphery (Keeler 1978); they are more tolerant of flooding (Linhart & Baker 1973), and differ in a number of other ways that adapt them better to the conditions of intense intraspecific competition that occur in the centre than to competition with grasses that affects genotypes growing at the pool edge (Fig. 1.2) (Linhart 1988). We shall explore natural selection and other evolutionary forces in more detail in Chapter 3.

1.2.3 Life tables and age-dependence

The probability of an individual dying or giving birth is often related to its age, so the age structure of a population is likely to affect its future. The proportion of youngsters is an indication of the likely future, but this does depend upon the chances of youngsters surviving to adulthood. This information is conventionally summarized in a **life table** (Table 1.1).

Acacia suaveolens is a small shrub that grows in fire-prone arid habitats of Southwest Australia. Figure 1.3 shows a **survivorship curve** for *A. suaveolens*, based upon the data in the life table. The life table and survivorship curve for *A. suav-*

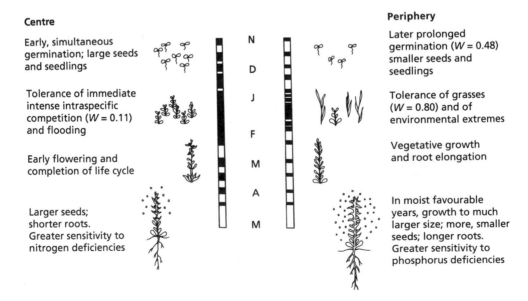

Centre

Early, simultaneous germination; large seeds and seedlings

Tolerance of immediate intense intraspecific competition ($W = 0.11$) and flooding

Early flowering and completion of life cycle

Larger seeds; shorter roots. Greater sensitivity to nitrogen deficiencies

N
D
J
F
M
A
M

Periphery

Later prolonged germination ($W = 0.48$) smaller seeds and seedlings

Tolerance of grasses ($W = 0.80$) and of environmental extremes

Vegetative growth and root elongation

In moist favourable years, growth to much larger size; more, smaller seeds; longer roots. Greater sensitivity to phosphorus deficiencies

Fig. 1.2 Genetically determined differences between plants of *Veronica peregrina* sampled from the centre and periphery of a vernal pool. Bars show the periods of adequate soil moisture in the two micro-habitats. *W* indicates the relative fitness, based upon seed production, of plants from the other micro-habitat when grown experimentally in the conditions indicated (after Linhart 1988).

Table 1.1 Life table and fecundity schedule for a population of the shrub *Acacia suaveolens*, in Australia. — indicates missing values (data from T. Auld & D. Morrison pers. comm.)

Age (years) x	Number N_x	Survival l_x	Mortality d_x	Mortality rate* q_x	Survival rate† p_x	Seeds/plant m_x	Reproductive value‡ V_x
0	1000	1	0.174	0.174	0.826	0	169.25
1	826	0.826	0.145	0.176	0.824	41	176.38
2	681	0.681	0.159	0.233	0.767	33	141.18
3	522	0.522	0.122	0.234	0.766	31	115.41
4	400	0.4	0.093	0.233	0.768	31	91.61
5	307	0.307	0.076	0.248	0.752	18	64.84
6	231	0.231	0.057	0.247	0.753	9	49.07
7	174	0.174	0.043	0.247	0.753	9	42.29
8	131	0.131	0.015	0.113	0.885	9	35.52
9	116	0.116	0.013	0.112	0.888	7	27.32
10	103	0.103	0.012	0.117	0.883	5	20.88
11	91	0.091	0.011	0.121	0.879	3	16.23
12	80	0.08	0.009	0.113	0.888	6	13.95
13	71	0.071	0.009	0.127	0.873	—	7.95
14	62	0.062	0.007	0.113	0.887	—	7.95
15	55	0.055	0.007	0.127	0.873	2	5.49
16	48	0.048	0.005	0.104	0.896	4	6.68
17	43	0.043	—	—	—	3	3.00

* Mortality rate $= d_x/l_x$
† Survival rate $= 1-(d_x/l_x)$
‡ See Chapter 10

eolens indicate that substantial mortality occurs throughout the lifespan of the population. In fact the mortality rates affecting this population of *A. suaveolens* are quite modest by comparison with many plant populations that have been studied, for which 90% mortality in the first year of life is not unusual. Such mortality may create alterations in the genetic structure of a population if some genotypes are more susceptible than others.

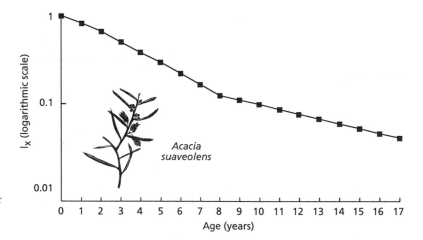

Fig. 1.3 A survivorship curve for *Acacia suaveolens*, based upon the data in Table 1.1.

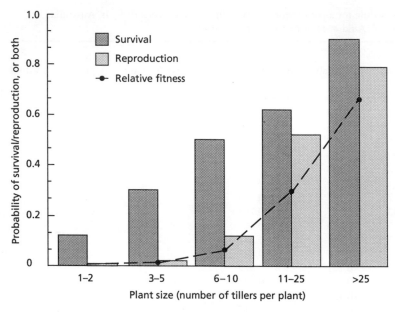

Fig. 1.4 Relationships between size and survival, reproduction and relative fitness (survival × reproduction) of the grass *Bouteloua rigidiseta* (data from Fowler 1986a).

Life tables were originally devised as a means of studying the human population, and were used by actuaries to calculate the risk of insuring the life of their clients (Hutchinson 1978). A person's age is of course not the only factor that influences their longevity, and neither is this so in plants. In fact, life tables have been imported into plant population biology following their use on animal populations, but in many ways they are inappropriate to the peculiarities of plants. The rate of growth in plants is highly dependent upon the local environment and consequently very variable, so that two genetically identical plants of the same age may be quite different in size. This is an example of **phenotypic plasticity**, a feature in which plants excel. Genetically identical plants may have quite different phenotypes if they have been exposed to different environments.

1.2.4 Life-cycle graphs and stage-dependence

Size is a major influence on the fate of individual plants and on their fecundity (Fig. 1.4). Because of phenotypic plasticity, size and age tend to be only loosely correlated. The dependence of plant fate on age is further weakened by the existence of dormant stages, notably seeds, but also tubers

and stunted seedlings, which may persist for many years before entering a phase of active growth. So, though it is clearly possible to draw up plant life tables and for some purposes it is useful to do so, this approach leaves much demographic variation in plant populations unaccounted for.

A simple alternative to the life table that uses age or plant **stage** to classify individuals in the population is the **life-cycle graph** (Hubbell & Werner 1979). Each age class or stage in the life cycle is represented by a node, and transitions between nodes are shown by arrows joining them (Fig. 1.5a, b). Appropriate rates of transition between stages and seed production are obtained from field studies. For example, Werner (1975) studied a population of the herb *Dipsacus sylvestris* in an old-field in Michigan, and classified plants in the stages: (1) first-year dormant seeds; (2) second-year dormant seeds; (3) small rosettes; (4) medium rosettes; (5) large rosettes; and (6) flowering plants. A life-cycle graph based on the annual rates of transition between these stages is shown in Fig. 1.5c. *Dipsacus sylvestris* dies after flowering, so the arrows from (6) to the other stages all represent numbers of individuals in those stages that were produced from seeds that were produced that year.

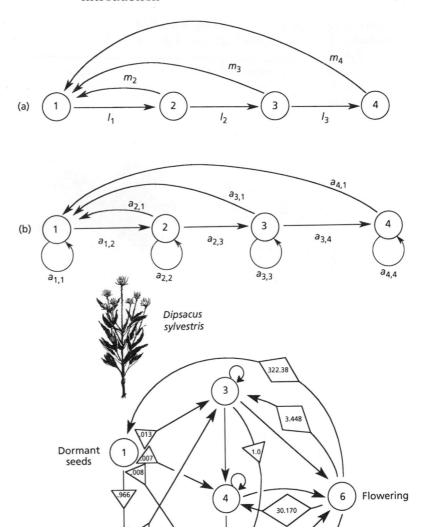

Fig. 1.5 Examples of life-cycle graphs. (a) An age-classified life cycle with four age classes. (b) A stage-classified life cycle with four stage classes. In both cases ① is the seed pool. l_x and m_x are, respectively, age-specific rates of survival and fecundity from the life table. In the stage-classified life cycle, fecundities are the values $a_{x,1}$; values $a_{x,x}$ are the rates of survival for plants that do not change stage class and values $a_{x,x+1}$ are the rates at which plants move up a stage class. (c) The life-cycle graph for a population of the herb *Dipsacus sylvestris*. Diamonds represent *numbers* (of seeds) making a transition, and triangles show *rates* of transition between nodes (after Caswell 1989).

1.3 SOME CONSEQUENCES OF BEING A PLANT

Two features shared by all plants deserve special mention because of their fundamental consequences for the population biology of these organisms: plants' growth and development from meristems, and the fact that they are sessile.

1.3.1 Meristems

Not all cells in a plant are capable of dividing. In

fact dividing cells occur in discrete regions called **meristems**, the chief of which are found in the apices of shoots, in the buds that sit in the axils of leaves, near the tips of roots and in the cambial layer beneath the surface of the stem in dicots and gymnosperms.

1.3.1.1 Structure and life history

The number and distribution of meristems on a plant, which of them develop and when, determine how a plant grows and its overall structure (Fig. 1.6). A bud that develops into a shoot usually

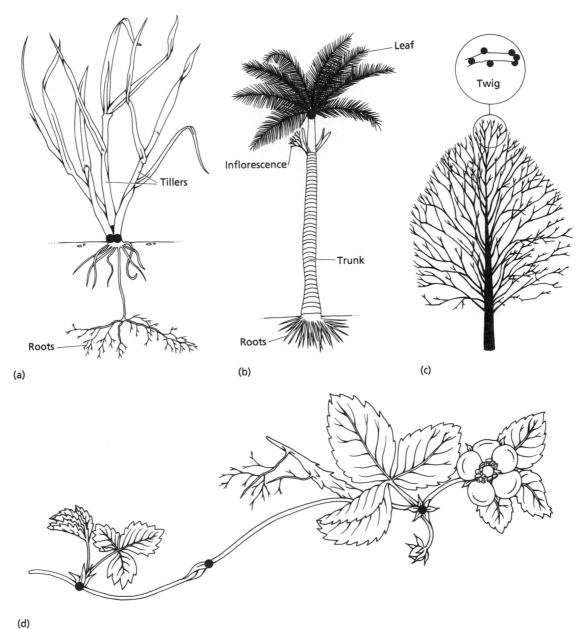

Fig. 1.6 Schematic diagrams showing plant construction and the distribution of shoot meristems, indicated by solid dots, in: (a) a grass, (b) a palm, (c) a dicot tree (*Acer saccharum*) and (d) a dicot clonal herb (*Fragaria vesca*).

multiplies the number of meristems on a plant because each shoot has its own meristems. This produces the **modular construction** typical of plants. A bud that develops into a flower or inflorescence consumes the meristem and ends its career, so the plant must use other meristems for vegetative development if it is to continue to grow and survive. For example, when the shoot of a grass, called a **tiller**, flowers its apical meristem differentiates, causing the death of the tiller after reproduction. If all an individual grass' tillers flower simultaneously, the whole plant will die.

The **life history** of a plant describes how long it typically lives, how long it usually takes to reach reproductive size, how often it reproduces and a number of other attributes that have consequences for demography and fitness. Because of the manner in which plants develop from meristems, there is a close relationship between the structure of a plant and its life history. Torstensson and Telenius (1986) found that the major difference in life history between the annual herb *Spergularia marina* and the related perennial *S. media* was due to a difference in how axillary meristems behaved. Axillary meristems were committed to flowering at the sixth or seventh node in the annual, but in the perennial these meristems produced shoots instead and flowering began much later in development.

As well as allowing plants to grow and branch vertically, meristems allow plants to proliferate and spread horizontally. This is **clonal growth** (Fig. 1.6d). The whole plant comprises the **genet**, which is defined as an individual arising from seed, however large and fragmented it may later become through clonal growth. A clonally produced part of a plant, with its own roots and potentially independent existence, is known as a **ramet**. The tillers produced from lateral meristems in grasses are an example. The branching pattern of growth produced by meristem activity results in a hierarchical structure: a genet is composed of ramets, ramets are composed of one or more branches, these bear inflorescences, inflorescences bear flowers, and flowers contain ovules and pollen. This hierarchical structure affects life history because the survival and reproduction of the genet depends upon the behaviour of ramets,

and ramets depend upon the behaviour of their parts (Silvertown 1989a). Flowering terminates the life of a meristem, so if all meristems on a branch produce flowers the branch will die, if all branches flower the ramet dies, and if all ramets flower the genet dies. Plants in which the genet dies after flowering are termed **semelparous** (or monocarpic), and those which can flower more than once are **iteroparous** (or polycarpic). An extreme example of semelparity occurs in some bamboos. The plant is clonal, forming additional, perennial ramets, but all of them flower at the same time and then the whole genet dies. In the giant semelparous bamboo *Phyllostachys bambusoides* the pre-reproductive period is 120 years! (Janzen 1976.)

There is a continuum of types of plant structure between trees whose branches all depend upon a common trunk and clonal plants in which the 'trunk' is reduced to no more than the crown of a rootstock and the 'branches' develop roots of their own. Both extremes can be seen by comparing *Salix pentandra*, a small tree, with *S. herbacea*, a dwarf creeping shrub (Fig. 1.7). There are similar contrasting examples among birches and dogwoods. Among clonal plants there is also great diversity in the method of spread, including stolons (above-ground creeping stems), rhizomes (below-ground creeping stems), tubers like the potato, bulbs such as the onion or corms as in crocus.

In clonal plants where the distance between ramets is short they present a 'phalanx'-like growth form (Fig. 1.7c), while in others where the distance is greater, there is a 'guerrilla' growth form, with shoots able to infiltrate between other plants (Fig. 1.7d) (Lovett Doust 1981a). Older parts of a clone generally supply younger parts with carbohydrates. Once a ramet has its own roots, it is usually self-sufficient in carbon, though water and minerals may still be imported from other parts of the clone if the connections between ramets persist (Marshall 1990). Plant species vary a good deal in how long the connections between ramets last. In white clover the stolons that connect ramets together are short-lived and the parts of a genet can wander far from each other, so that a single successful genet may be represented in many parts of a field. At the other extreme is

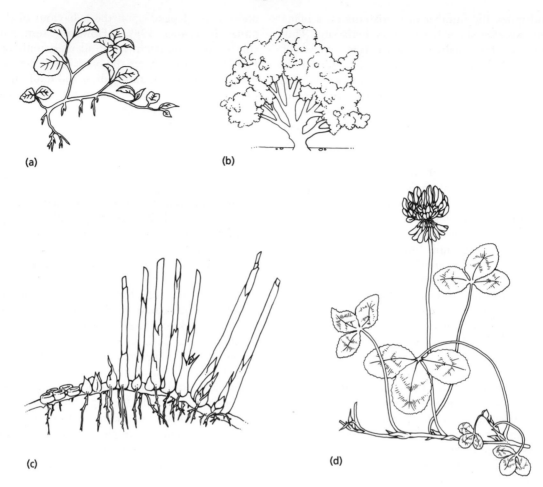

Fig. 1.7 The growth form of (a) *Salix herbacea*, (b) *S. pentandra*, (c) rhizome of *Cyperus alternifolius*, (d) *Trifolium repens*.

what is probably the biggest tree on Earth: a banyan *Ficus benghalensis* growing in Calcutta, India. It is a clone over 200 years old, with a thousand connected tree trunks covering more than 1.5 ha.

1.3.1.2 Genetics and evolution

Meristematic growth and modular construction have a number of genetic and evolutionary consequences for plants. First, the number of flowers a plant is capable of producing is limited by the number of meristems, so its size is likely to be a major determinant of its total reproductive success and fitness. Secondly, meristems are like

perpetually embryonic tissue — they are 'forever young', so it is possible for a very old and large genet to be entirely composed of much younger ramets that show no sign of senescence. This is at variance with a major body of evolutionary theory that has been developed only with unitary (i.e. non-modular) animals in mind (Rose 1991).

One view of evolution is that individual organisms are merely vehicles for their genes to replicate themselves (Dawkins 1976). Genes are immortal, passing down the generations from one mortal vessel to the next. This view is based upon a doctrine propounded by August Weismann (1893), a zoologist who worked mostly with insects, who concluded that an organism's cells are separated into a **germ line** that produces the gametes, and a **soma** (Greek for 'body') that constitutes the rest of the organism's tissues. Buss (1987) has described how, in the development of evolutionary theory, the modern synthesis of neo-Darwinism was predicated upon Weismann's doctrine. However, the separation of germ line and soma does not occur in plants, or for that matter in protists, fungi, corals or bryozoans. Weismann's doctrine does not apply to plants because virtually every meristem is capable of producing gametes. Clonal plants then, which proliferate indefinitely from their meristems, are potentially as immortal as their genes. Huge clones in which the genet is inferred to be thousands of years old are known in many species. Probably the record is held by the creosote bush *Larrea tridentata* which has some clones in the Southwest USA that, judged by their size and rate of growth, may be 11 000 years old (Vasek 1980). This is as ancient as the deserts of the American Southwest themselves.

Since meristems occur at the apices of all shoots and roots, their number in a large plant may be tremendous. The apical meristem of a shoot tip is made up of cells that have already divided mitotically many times giving rise to all of the cells of the stem, leaves, buds, flowers and fruits. During these repeated mitoses it is always possible that a cell in the apical meristem or one of the buds may mutate. This mutant cell may give rise to large numbers of descendant cells within the shoot, and indeed entire shoots may have a distinct genotype through this kind of **somatic mutation**. Unless somatic mutants are inviable, such genetic variability within individuals should increase over time and be particularly evident in long-lived species, such as trees and clonally-spreading plants, especially those species in which there is a single apical cell (Klekowski 1988). Somatic mutations are responsible for the origin of most varieties of vegetatively propagated crops including bananas, potatoes and sugar cane. Similarly, the origins of many varieties of fruit trees (e.g. nectarine from peach, which involved genetic change at a single gene locus) are due to spontaneous somatic mutations. Mutations that arise in the apical cell of a meristem will be present in all subsequent growth, and will be incorporated in any gametes that are produced. In contrast to this, mutations arising in a multicellular meristem will form **chimeras** or genetic mosaics, a mixture of cellular genotypes that may or may not show up in gametes depending on the developmental role of specific cells. Mosaics in plants are well illustrated in streaky coloured petals or foliage, although streaky petals (e.g. in cultivated tulips) may also be due to viral infections or the presence of transposons in a lineage of cells.

1.3.1.3 Behaviour

Textbooks of ethology are strangely reluctant to define 'behaviour', though ethologists clearly know what they mean by the word and some mistakenly believe plants to be incapable of any activities that deserve the term. To encompass all the *animal* activities that ethologists study, behaviour must be defined broadly, for example as a response to some event or change in the organism's environment (Silvertown & Gordon 1989). Most plants are capable of some kind of movement, for example in the orientation of leaves, and *Mimosa pudica*, the 'sensitive plant', is able to fold up its leaves, exposing a spiny stem, when touched by a herbivore. However, rapid movement is rare and the chief method of behavioural response in plants depends upon their modular pattern of growth and the ability to alter the size, type and location of new organs to match an environmental change. Thus, for example, the tropical rainforest liana *Ipomoea phillomega* produces several types of shoot, depending upon the nature of the light environment. In shaded conditions, stolons with long internodes and rudimentary leaves extend rapidly over the ground (Fig. 1.8). When a gap is reached, twining shoots with large leaves are produced and these ascend to the tree canopy where the liana forms a crown of its own (Peñalosa 1983).

The parallel between the behaviour of *I. phillomega* and animal foraging is difficult to resist. Such behaviour is especially common in climbers which, because of their extensive shoot systems, traverse several types of micro-environment during growth and often have different types of leaf or shoot to match. Perhaps the simplest

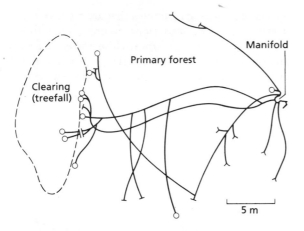

Fig. 1.8 A map of the shoot system of *Ipomoea phillomega* on the floor of a tropical rainforest in Veracruz, Mexico. This plant originated at the 'manifold' which has an ascending shoot and a crown in the canopy. Ascending shoots with crowns are represented by circles. Stolons that have lost their tips end in a 'T', and those that are still growing are shown with a 'Y' (from Peñalosa 1983).

behaviour of which virtually all plants are capable is etiolation in response to shade. Most plant responses to shade from other plants are mediated by the substance **phytochrome** which has two forms that interconvert with each other. The Pr form absorbs red light and is then converted to the Pfr form which absorbs far-red light. The equilibrium between the two forms is determined by the *ratio* of red to far-red light. Leaves absorb red and blue light (this is why they are green), but are relatively transparent to far-red so a plant can detect the presence of leaf shade by the effect this has on the Pr : Pfr ratio. In fact the system is so sensitive that plants can detect the quality of reflected light and respond to the presence of neighbours that do not shade them directly (Ballaré *et al.* 1990; Novoplansky *et al.* 1990). Using the phytochrome system, seeds of many species can discriminate between shade caused by other plants and darkness caused by burial. They will germinate in the dark but not when shaded by leaves (Silvertown 1981). The study of plant behaviour is still in its infancy, but it promises to be a field full of surprises (Silvertown & Gordon 1989; Lovett Doust 1990).

1.3.2 Sessileness

1.3.2.1 *Defence, dormancy and dispersal*

The fact that plants are rooted to the spot has a number of important consequences. Plants are unable to escape their herbivorous enemies by flight, but they are often packed with toxins that make them unpalatable. Others resist grazing with spines or, as grasses and sedges do, by hiding their meristems in the very base of the plant. In unfavourable seasons of the year plants cannot migrate, but they have a range of protective strategies that include leaf deciduousness, seed dormancy and retreating to underground storage organs from which they may regrow in the next season. Some movement is of course possible, but creeping stems and roots travel too slowly to form any effective means of escape from most hazards. However, by forming an extensive network stolons and roots can reduce the risk that local damage will kill the entire plant, and it has been suggested that this is a major advantage of the clonal growth habit (Eriksson & Jerling 1990).

Though plants are sessile, their genes are not, and these are transported at two different stages in organs equipped with a variety of dispersal aids. Some seeds have wings, parachutes, are surrounded by a fruit that attracts animal dispersal agents, or are sometimes even propelled explosively from the fruit as in the squirting cucumber *Ecballium elaterium*. Pollen grains are produced in vast numbers, and wind-dispersed grains are light and sometimes, as in pines, have buoyancy aids attached. Animal pollinated plants have colourful, fragrant flowers that attract insects, bats and birds as pollinators that are often rewarded with nectar. It is worth remembering that the fruit and flowers so enjoyed by humans originate from organs evolved by sessile organisms that rely on animals for transport.

1.3.2.2 *Mates and neighbours*

Because they are sessile, most of the interactions between plants are with their immediate neighbours. If below-ground resources such as water or nutrients limit plant growth, competing neighbours will chiefly interact by the depletion of resources in the common pool. However, when plants grow in a closed stand and light limits growth, a plant that is taller than its neighbours will not only receive a disproportionate share, but it is also likely to *suppress* the neighbours in its shade. As a result of this inherent asymmetry among plants competing for light, even a slight difference in height or canopy area can be decisive in determining which plant wins. This introduces a further complication, because which plant is initially the largest among its neighbours often depends upon which one germinated first. The early plant catches the light. So, because plants are sessile, not only *where* a plant is but also precisely *when* it appeared there may be a matter of life and death.

Despite the local character of plant–plant interactions, studies of plant demography often ignore the spatial structure of populations and concentrate on changes in average density or in age or stage structure. In Chapter 4 we shall see whether this simplification affects the accuracy of population dynamics models. In contrast to plant demography, population genetics has explicitly dealt with the spatial structure of populations from the early days of the subject. This is because, despite the agency of wind and animals in moving pollen, most plants mate with their neighbours and are frequently surrounded by relatives, creating genetic patchiness within populations. The genetic and spatial structures of populations interact with one another. Spatial patchiness in the density of populations reinforces the genetic isolation of patches, and this can lead to genetic divergence.

It is arguable that a well-rounded view of population biology that merges demography and genetics into a single subject can only emerge when each branch of the subject pays sufficient attention to the variables that are important in the other (Silvertown 1991b). At present, genetic models rarely incorporate density as a variable and population dynamics models tend to ignore spatial as well as genetic structure. There are good reasons for these differences — for one thing, models with too many variables tend to be too complicated for clear interpretation, and their behaviour too sensitive to parameter values and assumptions to be of much use. A central

aim of this book is to bring the genetic and ecological branches of plant population biology closer together. In some cases we have only been able to lay the two approaches side by side; the reader must judge if anything more has happened between these covers!

1.4 SUMMARY

A **population** is a collection of individuals belonging to the same species, living in the same area. Population biology aims to explain **spatial structure**, **age structure** and **size structure** in terms of the four basic **demographic parameters** which measure birth, death, immigration and emigration rates. When these rates affect genotypes differently they may produce **genetic structure**. Changes in genetic structure with time are the subject of **evolution**; changes in numbers with time are the subject of **population dynamics**. The dynamics of a population may be summarized by the ratio N_t/N_{t-1}, which is called the **annual** or the **finite rate of increase** (λ). The finite rate of increase for a genotype can be used as a measure of its Darwinian **fitness** (W), which is a measure of the relative evolutionary advantage of one genotype over another. The principal forces that bring about evolutionary change are **natural selection** and **gene flow**, but only the former can produce **adaptive** evolutionary change. Natural selection can only operate when there is variation between individuals in heritable characters that affect fitness.

Birth and death rates are influenced by the age and size of individuals. Plant size is particularly sensitive to local environmental conditions and shows great **phenotypic plasticity**. A **life table** may be used to describe the age-dependence of demographic parameters and a **life-cycle graph** to describe a stage- (or size-) classified population. The **life history** of plants tends to be closely tied up with their **modular construction** which arises from **meristematic** development. **Clonal growth** is common in plants. The individually rooted parts of a clone are called **ramets**, and the whole clone is called a **genet**. Unlike in unitary animals where gametes and somatic tissues derive from separate cell lines, in plants both cell types derive from meristems. This has the important consequence that clonal plants may be potentially immortal.

Modular growth and the ability to produce structures that match a change in the environment give plants a **behavioural** repertoire that enables them to forage for light and avoid competitors.

Because plants are unable to escape their animal enemies, they often rely on **chemical defences**. Seeds and pollen transport plants' genes, but plants still usually end up mating with their **neighbours**.

Chapter 2
Variation and its Inheritance

2.1 INTRODUCTION

Offspring tend to be *like* their parents, but not quite. The two most important foundation stones of modern biology, Darwinian evolution and Mendelian genetics, both arose when their authors broke from the prevailing view that variation within populations was of no real importance, and instead attempted to explain the significance of the fact that offspring are *not quite* like their parents.

Darwin, in propounding his theory of evolution by natural selection, realized that variation among individuals in a population could be the raw material for evolutionary change if the differences were inherited. He demonstrated in breeding experiments that variation could appear between one generation and the next, and that traits missing in one generation could reappear in subsequent crosses. Darwin's work on inheritance in plants is little known today because he never figured out how inheritance worked, but with the benefit of hindsight, we can now make sense of his results in terms of Mendelian genetics. Darwin crossed a peloric snapdragon (*Antirrhinum majus*) which has narrow tubular flowers, with pollen of the normal asymmetrical form, and the latter reciprocally with peloric pollen (Fig. 2.1). Among the many seedlings produced, none was peloric. He then allowed these crossed plants to self-fertilize, and out of 127 seedlings, 88 proved to be like normal snapdragons, two were in an intermediate condition and 37 were perfectly peloric (Allan 1977). Although Darwin went on to carry out several such experiments with cabbages and antirrhinums, what we now recognize as typical 'Mendelian' ratios (approximately 3 : 1) did not catch his attention.

Mendel discovered the particulate nature of inheritance because he chose characters which happened each to be determined by a single gene locus, and he was able to recognize consistent ratios among the progeny of crosses. The bringing together of Darwinian evolution and Mendelian genetics in the 1930s created the neo-Darwinian synthesis.

This chapter is about heritable variation and the genetic structure of plant populations. In the next chapter we look at natural selection and plant ecological genetics. Most of these topics recur in various contexts throughout the rest of the book, particularly in chapters 9 and 10.

2.2 TYPES OF TRAIT

Plants make a speciality of morphological variation. We can divide variation into four types, depending upon whether a trait varies markedly *within* a genotype or *between* genotypes, and upon whether the variation is continuous or discontinuous (Fig. 2.2). An individual that changes its phenotype in response to the environment exhibits **phenotypic plasticity**. When a trait changes *continuously* in response to a range of environments, the set of phenotypes expressed by a single genotype across the environmental range is called a **reaction norm** (Fig. 2.2a). In contrast, phenotypic plasticity that produces *discontinuous* variation in a trait, for example between the floating and submerged leaves of *Ranunculus heterophyllus* (Fig. 2.2c), is known as **heteromorphism**.

Phenotypic variation *between* genotypes also occurs in two forms. The majority of phenotypic traits exhibit continuous variation, so that individuals differ from each other in a **quantitative** way (e.g. larger or smaller) (Fig. 2.2b). These

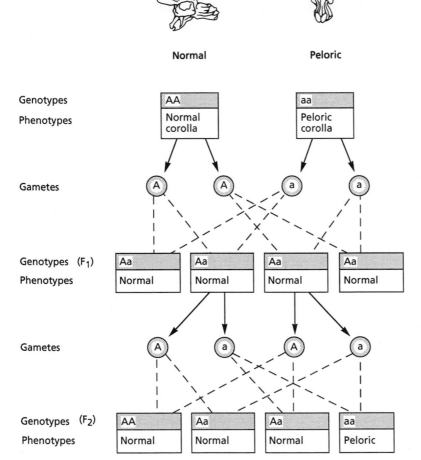

Fig. 2.1 Segregation of peloric snapdragon (*Antirrhinum majus*), a qualitative trait. Mendel realized that a 3 : 1 ratio in the F$_2$ generation of such a cross could be explained if the traits were inherited as particles, of which each plant had two, and that one character was dominant to the other. In this case the normal corolla is dominant to the peloric.

traits are described as **metric**, because they are *measured*. Phenotypic traits that vary in a discrete, all-or-none fashion between one individual and another (e.g. peloric or normal flowers) (Fig. 2.2d) are described as **qualitative** and are usually determined by only one or two genetic loci with two or more alleles each. Early geneticists thought that different modes of inheritance lay behind continuous and discontinuous traits, but today we recognize that continuously varying traits are generally influenced by alleles at several loci. Technically though 'quantitative' refers to the means by which these traits are described, not the number of genes involved. Qualitative traits are classified, rather than measured.

A trait that exhibits qualitative variation that has a simple genetic basis is described as **polymorphic** if variants are present at a frequency of 5% or more. Enzyme polymorphism has been particularly well studied (see Fig. 1.1). Different molecular forms of an enzyme catalysing the same reaction, termed **isozymes**, may be the product of different loci, or they may be produced by alternative alleles at the same locus, when they are called **allozymes**. Isozyme variants can be distinguished by the technique of **electrophoresis** which separates proteins on the basis of their different rates of migration through a gel matrix placed in an electric field. Most plant enzymes which have been studied have two to several isozymes (Gottlieb 1981). In plants, about 50% of isozyme loci are polymorphic and have

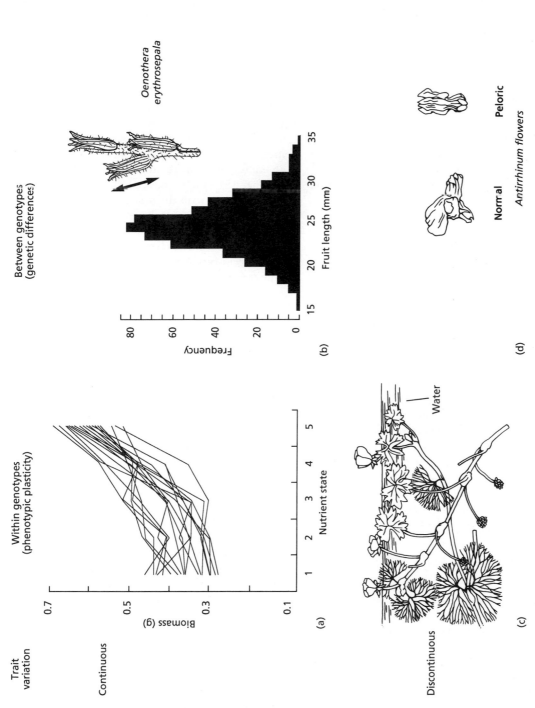

Fig. 2.2 Types of plant trait: (a) the reaction norm of plant size in response to a nutrient gradient for the progeny of 20 maternal families of *Abutilon theophrasti* (from Garbutt & Bazzaz 1987), (b) quantitative variation in length of fruit of the evening primrose *Oenothera erythrosepala* (from Briggs & Walters 1984), (c) heteromorphism in leaves of *Ranunculus heterophyllus*, (d) qualitative variation in normal versus peloric flowers of *Antirrhinum*.

different allozymes (Hamrick & Godt 1990). When discussing *Picea abies* in Chapter 1 we noted that correlations between allozyme frequencies and environment cannot be taken as evidence that the allozyme loci themselves are under selection. In fact the commonly accepted view is that molecular differences between allozymes at the same locus generally have no functional significance. Correlations between allozyme variation and environmental variables may be caused by linkage between these loci and others that *are* selectively important.

Enzymes are gene products, so isozyme and allozyme variation is phenotypic. At present, most of our information about genetic variation in populations comes from electrophoretic studies, but other techniques are also available for comparing genotypes directly. DNA sequencing is extremely laborious, and is not at present feasible for analysing variation within populations. However, DNA sequence polymorphisms can be detected at a more generalized level by the use of restriction enzymes which cut DNA molecules at sites of particular nucleotide sequences (the restriction sites of the enzymes), creating DNA fragments of different lengths. These show up as restriction fragment length polymorphisms (**RFLPs** or 'riflips') when the location of restriction sites differs between individuals. Genetic interpretation of RFLPs is often more complicated than for isozyme banding patterns. Allelic RFLPs are not commonly available, except where two inbred lines are crossed to get F_1 and F_2 generations. RFLP data are usually scored as presence or absence of a particular band, corresponding to a particular fragment length, without identifying the allele or locus related to that restriction site.

2.3 GENOTYPE AND PHENOTYPE

The genotype determines the *potential* of an organism to express a particular character in its phenotype, but its actual phenotypic expression always depends upon the environment to some degree. No organism can exist or develop without an environment, and most plant characters are phenotypically plastic to a greater or lesser extent. When phenotypic differences between genotypes vary between environments, there is a **genotype × environment interaction**. For example, Maddox and Cappuccino (1986) found that when plants were well watered there were significant differences in the resistance of plants to aphid infestation between four half-sibships of *Solidago altissima*, but when the plants were less well watered, differences between the families disappeared. Zuberi and Gale (1976) planted 20 inbred lines (genotypes) of the poppy *Papaver dubium* in 16 soil environments, differing in nutrient status. Eleven metric traits were measured, and all showed significant genotype × environment interactions. Interactions were so strong for one character, leaf number at 10 weeks, that genotype number 15, which produced the largest plants in the treatment with the best soil, produced almost the smallest plants on the poorest soil. Figure 2.3 shows a regression of the trait mean for each genotype grown in each environment on the mean for all genotypes in each environment. A genotype × environment interaction occurs when regression slopes differ significantly, because this tells us that genotypes are not responding in a similar way to environmental variation (Finlay & Wilkinson 1963). Genotype × environment interactions are particularly important for characters that affect fitness when regression lines cross each other, because this suggests that which genotype is favoured by selection will depend upon particular environmental conditions.

A simple way of separating the effects of environment from the effects of genotype on phenotypic variation is to transplant individuals or their offspring to a uniform environment in a common garden. So long as precautions have been taken to prevent the carry-over of environmental effects from the wild (for example in soil, or via non-genetic maternal effects) and plants are randomized, differences between plants grown in the common garden should be largely genetic. In an early study that is now justly famous, Clausen *et al.* (1948) demonstrated large genetic differences between populations of *Achillea lanulosa* along an altitudinal gradient in the Sierra Nevada of California, by raising plants in a common garden at Stanford from seed collected along the transect (Fig. 2.4a).

The common garden does not represent a natural habitat for any of the genotypes planted

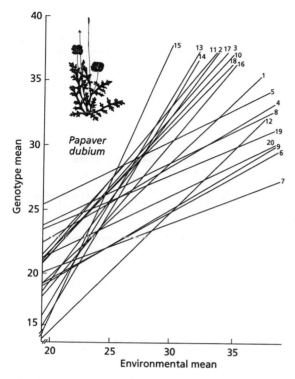

Fig. 2.3 Regressions of the mean number of leaves at 10 weeks for each of 20 genotypes of *Papaver dubium* grown in 16 environments, on the mean for all genotypes in each environment. Regression lines are numbered for genotypes 1–20 (from Zuberi & Gale 1976).

in it. Another approach, also taken by Clausen *et al.* (1948), is to use reciprocal transplants between gardens at different sites, but a still better one is to carry out reciprocal transplants between field sites and to include replants, where plants are sampled and returned to their native site, as controls (Fig. 2.4b).

2.3.1 Heritability and environmental variation

The variability of a particular phenotypic trait in a population, such as leaf length or seedling growth rate, can be measured by its variance, or what is termed its **phenotypic variance** (V_p) (Falconer 1989). The phenotypic variance of a character has a genetic component that is due to genetic differences between individuals, and an environmental component that is due to differ-

ences between individuals caused by their different local environments. The genetic component of V_p is called the **genetic variance** (V_g) and the environmental component is called the **environmental variance** (V_e). The simplest model of how the contributions of genotype and environment influence phenotypic variance assumes that their effects are additive:

$$V_p = V_g + V_e + 2COV_{g,e}$$

where the term $2COV_{g,e}$ is the **covariance** of genotype and environment and reflects any tendency for particular genotypes to be found in particular environments. The ratio of genetic to phenotypic variance V_g/V_p is a measure of the **heritability in the broad sense** of the trait in question (H_B).

The genetic variance V_g can be subdivided into a number of components, including the **additive genetic variance** V_a, the dominance variance V_d and the interaction (epistatic) variance V_i:

$$V_g = V_a + V_d + V_i$$

The dominance variance quantifies the effect that dominant alleles have on the phenotype, which in heterozygotes is disproportionate to allele frequency. The interaction variance quantifies effects on the phenotype when the influence of an allele at one locus depends upon the presence of an allele at another locus. For example, in white clover *Trifolium repens* two polymorphic loci affect the ability of plants to produce cyanide when their tissues are damaged by herbivores such as slugs. Two alleles at one locus determine the presence (allele *Ac*) or absence (allele *ac*) of a glucoside that releases hydrogen cyanide when it is broken down, and two alleles at the other locus determine the presence (allele *Li*) or absence (allele *li*) of an enzyme capable of catalysing the reaction. *Ac* and *Li* are both dominant. Of the four possible phenotypes, only *AcLi* is cyanogenic. This kind of interaction between loci is known as **epistasis**.

In crops, many of the traits associated with yield are non-additive because final yield is in large part a consequence of sequential events that each depend on developmental stages that precede them (e.g. Lovett Doust & Eaton 1982). The interaction of traits affecting fitness can

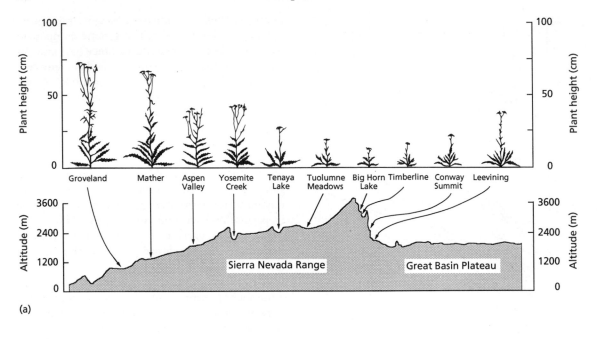

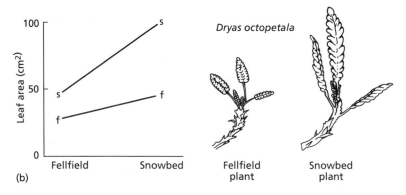

(b)

Fig. 2.4 (a) Results of a common garden experiment with *Achillea lanulosa* (after Clausen *et al.* 1948), (b) reciprocal transplant–replant experiment between fellfield (f) and snowbed (s) populations of *Dryas octopetala* in Alaska, showing the effects on leaf area (from McGraw & Antonovics 1983).

be quantified by calculating correlations among quantitative characters and applying a path analysis. For example, in a study of a population of the annual woodland herb *Impatiens capensis* in a forest gap, Mitchell-Olds and Bergelson (1990a, b) found that fitness was correlated with germination date, size in June, growth rate early and growth rate late in the season. A number of traits had stronger indirect than direct effects upon fitness (Fig. 2.5), though it so happened that there was no genetic variance for them.

The additive genetic variance has a special importance because the additive effects of genes contribute most of the response to selection, though V_a does also contain some effects due to dominance (see Falconer 1989). The ratio of the additive genetic variance to the total V_a/V_p is called the **narrow-sense heritability** (symbolized H_N), and this is the measure of heritability generally quoted. A simple method of calculating H_N is from the correlation in the trait of interest between parents and offspring. Experimental pro-

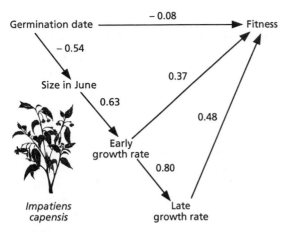

Fig. 2.5 Path diagram showing the correlations (partial regression coefficients) among traits in a population of *Impatiens capensis* (after Mitchell-Olds & Bergelson 1990b).

cedures and the technicalities of calculating heritability are dealt with by Falconer (1989) and Nyquist (1991).

In what Fisher (1958) called the **fundamental theorem of natural selection**, he proposed that 'the rate of increase in fitness of any organism is equal to its (additive) genetic variance (in fitness) at that time'. As Fisher himself fully realized, the fitness of an organism cannot increase indefinitely in any meaningful sense and so his fundamental theorem has given rise to some confusion and debate as to what he actually meant (Frank & Slatkin 1992). In reality Fisher's theorem implies that traits which are highly correlated with fitness should have low values of H_N, because their additive variance is rapidly removed by selection (Maynard Smith 1989). In a large survey of genetic variation in natural animal populations, Mousseau and Roff (1987) examined more than 1100 estimates of H_N for a wide array of traits. Life history traits associated with fitness had significantly lower heritabilities than morphological, behavioural and physiological traits. No comparable survey has yet been carried out for wild plants, but in domesticated plants the narrow-sense heritabilities of reproductive traits are typically much lower than those for traits less closely correlated with fitness (Istock 1983).

Note that even if the additive genetic variance V_a has a fixed value, heritability may vary from one environment to another and from one year to the next because it is a function of environmental variance and genotype × environment interactions. For example, Hurka and Benneweg (1979) measured the heritability of seed size in shepherd's purse *Capsella bursa-pastoris* and found values varying from 0.05 to 0.79 between populations. Price and Schluter (1991) argue that life history traits are so strongly affected by environmental variation that their typically low heritabilities may be due to high values of V_e, not low values of V_a.

2.3.2 Phenotypic plasticity

Phenotypic plasticity has two particularly important consequences for the fitness of plants. First, it allows plants to alter their phenotype in a behavioural manner to suit local conditions (Section 1.2), so affecting fitness and generating natural selection that may mould phenotypic plasticity. Second, phenotypic plasticity can partially decouple genotype and phenotype, giving phenotypically plastic characters low heritability and consequently preventing natural selection from operating on the underlying genes. The phenotypic plasticity of a trait may have a heritable component and can be inherited and selected independently of the trait itself (Bradshaw 1965; Schlichting & Levin 1986; Sultan 1987).

It has been suggested that, by permitting one genotype to adapt to many environments, phenotypic plasticity may substitute for adaptive genetic variation and that there should therefore be a negative correlation between the genetic variance V_g and the environmental variance V_e (Thoday 1953; Levins 1963). On the other hand, by decoupling phenotype and genotype, phenotypic plasticity may shelter genetic variation from selection, leading to zero or even positive correlation between V_g and V_e. Marshall and Jain (1968) grew seed collected from co-occurring populations of two wild oats *Avena fatua* and *A. barbata* in a series of 24 diverse environments and recorded nine quantitative characters, including the length of the primary tiller, the total number of tillers, number of fertile tillers and the number of spikelets on the primary panicle. *A. barbata* was more phenotypically plastic, but less genetically variable than *A. fatua*. Scheiner and Goodnight (1984) measured phenotypic plasticities (V_e), genetic variances (V_g) and broad-sense heritabilities of 12 traits in five populations of the grass *Danthonia spicata*. Individuals were clonally replicated and then grown in six different environments. Comparisons revealed highly significant differences among genets in amounts of phenotypic variance (V_p), but *no* significant differences in the amounts of genetic variance and no significant correlation between genetic variance and phenotypic plasticity. Because this study involved five populations and used clonal replicates it is a stronger test than the study of the wild oats that involved only two populations and used seedlings rather than ramets. It does not support the hypothesis that V_e and V_g are negatively correlated.

MacDonald and Chinnappa (1989) investigated morphological plasticity and allozyme variation in five populations of *Stellaria longipes*. Trait means were generally different between populations, but most of this morphological variation proved to be due to phenotypic plasticity. There were also interactions between genotype and environment affecting the plasticity of traits. As far as the relationship between morphological and allozyme variation is concerned, the direction and extent of morphological divergence between populations did not reflect differences in their allozyme phenotypes. MacDonald and Chinnappa (1989) concluded that, on the basis of what we know so far, plasticity and genetic variance may evolve independently of each other.

2.4 GENETIC VARIATION

Genetic variation can be analysed at a hierarchy of levels. It originates within the genome where mutations occur, it is amplified by sexual reproduction which generates genotypic diversity through recombination, and it is ultimately expressed as genetic variation within and among populations.

2.4.1 Sources of genetic novelty within the genome

Gene replication occurs with nearly perfect fidelity, time after time, cell division after cell division, and generation after generation. The rare, uncorrected mistakes are the ultimate source of polymorphic loci and all genetic variation. Hugo De Vries (1905) first proposed that sudden heritable changes, which he termed **mutations**, were responsible for inherited variation. The most usual mutation is the substitution of one base pair for another, but DNA can also be changed by the removal or insertion of a few base pairs or even large segments, as is the case for transposable elements (transposons). Mutations are rare and apparently random events, occurring for most loci at average rates of about one per $10^5 - 10^{10}$ cell generations. Copying errors during DNA replication are thought to cause a large fraction of spontaneous mutations, and these can be induced by ionizing radiation and environmental mutagens. Stadler (1928) first showed in plants (at the same time as Herman Muller made the point for fruit flies) that X-rays induce mutations. Stadler assessed the effects of mutations in barley, and estimated that only about one mutation in a thousand was beneficial to the plant in terms of doing something useful such as improving seed yield, or drought or pest resistance.

Clearly, the simplistic image of allele '*A*' mutating to allele '*a*' masks the fact that a gene may be 1000 or more base pairs in length, so there are an enormous number of possible base sequences for that gene. In many cases, a single base pair change may have no detectable effect on the phenotype. One striking property of plants is that they appear to be more tolerant than animals of deletions and the presence of extra or missing chromosomes. The reason for this resilience is not known. In general, duplications of DNA are less detrimental than deletions. Deletions of numbers of nucleotides in multiples other than three are particularly disruptive, because the message will have a frameshift change, and codons beyond the mutation may be read as nonsense.

Transposons, or 'jumping genes', are DNA sequences that may move from one part of the genome to another or even between the genomes of two associated, but unrelated, organisms. Although jumping genes were first discovered in maize, by Barbara McClintock in the 1940s, they have not been studied in many other plants (Saedler *et al.* 1983). However, there is general agreement that they occur in all organisms. A transposon can cause a change in a gene that shows up in the phenotype, but subsequently the transposon may move elsewhere, restoring the gene (and the phenotype) to its original state. At the moment we can only speculate as to how important transposons may be for plant evolution. Perhaps the different rates of mutation noticed for different genes are related to the number of target sites available for transposons, which appear to be associated with chromosome breakage 'hot spots'.

Another intriguing possibility is that transposons may move between unrelated species that are in frequent ecological association, such as hosts and parasites of specialized herbivores. For example, the genes promoting crown galls are transmitted from the bacterium *Agrobacterium tumifaciens* to the plants that they infect (Mettler *et al.* 1988). The *leg* haemoglobin gene in leguminous plants is remarkably similar to a gene for haemoglobin in invertebrates (Go 1981), suggesting that invertebrate genes may have become incorporated in the genome of eukaryotic plants. Harper *et al.* (1991) have speculated that the sudden appearance of roots in the fossil record of land plants might be explained by gene transfer from some soil microorganism such as the bacterium *Agrobacterium rhizogenes* which carries a plasmid that is able to induce root formation in stems. Alternatively, of course, some ancestor of *A. rhizogenes* may have picked up these genes from a plant in the first place. It must be emphasized that we do not yet know if such transfers occur in the wild and *permanently* transform plants.

2.4.2 Sex and the regulation of genetic variation

In essence the mechanism of sexual reproduction involves three steps: **meiosis** to form haploid cells, genetic **recombination** which occurs during meiosis, and later the **fusion of gametes** derived from the haploid cells to form a new zygote. The important genetic consequence of sexual reproduction is that it produces offspring having novel combinations of genes, and hence new genotypes. Novelty can arise at each of the three stages in the process. The first opportunity comes during the first meiotic division when crossing-over may occur between homologous chromosomes. The second is due to the independent assortment of chromosomes at meiosis: for any particular chromosome half the gametes will contain a copy derived from the mother and half will contain a copy derived from the father of the individual whose cells are dividing. The third source of novelty occurs with fusion of the gametes when these are derived from genetically different parents. Naturally enough, if the parents are closely related (the extreme case of this being where a homozygous plant is self-fertilized) the novelty produced at this stage may be reduced.

The **mating system** of a population is a model describing who mates with whom. Its most important aspect is the degree of genetic relatedness among mates, which determines the level of **inbreeding**. As we shall see later in this chapter and the next, inbreeding influences levels of genetic variation within and between families and genetic differentiation between populations. In the **mixed mating model**, mating systems are described in terms of the **selfing rate** S, which is the proportion of zygotes formed by self-fertilization, and the **outcrossing rate** $\hat{t}$, which is defined as $1 - S$. According to this model, variation in mating systems is continuous between complete **selfing** ($S = 1$) at one extreme and total outcrossing at the other ($\hat{t} = 1$). Many plants have **mixed mating systems** that lie somewhere between these two extremes, but not all mating systems can be adequately described with the mixed mating model because it ignores situations where inbreeding arises from mating between relatives (called **bi-parental inbreeding**) rather than from selfing. However, in populations where inbreeding is bi-parental this may be quantified by calculating the *equivalent* amount of selfing that would produce the same level of inbreeding, and S may be used to measure this. So, for example, a brother–sister mating where the siblings have

half their genes in common is equivalent to a
selfing rate of:

$$S = \frac{1}{2} \times \frac{1}{2} = 0.25$$

The mating system of a population is strongly
influenced by the sexual characteristics of the
plants in it. These characteristics define a plant's
sex habit and determine who is *able* to mate with
whom. Selection on these traits drives the
evolution of mating systems, which we shall
examine in some detail in Chapter 9. In sexual
plants and animals the differentiation of gametes
into two kinds, large (female) gametes that are
relatively few and stationary, and small (male)
gametes that are relatively numerous and mobile,
is called **oogamy**. This fundamental dichotomy
between gametes is the evolutionary basis of all
the subsequent sex differences that affect mating
systems (Maynard Smith 1978).

The existence of oogamy permits plants to
allocate their resources differently toward male
and female function. The **gender** of a plant denotes
the relative importance of contributions to its
fitness through ovule production (maternal
function) and pollen production (paternal function). The highly varied sex habits of flowering
plants are traditionally classified according to
three main features of the reproductive system
that affect which plants are able to mate with
each other.
1 The spatial separation of male and female
reproductive organs within and between plants.
2 The temporal separation of male and female
reproductive activity within flowers.
3 The presence or absence of self-incompatibility.
The variants under each of these headings and
the sex habits that they define are described in
Table 2.1.

Linnaeus (1737) provided some of the original
terminology to describe plant sex habits (hermaphrodite, monoecious, dioecious, etc.) and
later Darwin (1877) introduced additional terms
to describe more unusual arrangements (e.g. and-
romonoecious, gynodioecious). However, *most*
plant populations have only one gender class
combining maternal and paternal functions in
the same **cosexual** individual. The principal
direction of evolution of sex habits has been from

this condition (typically hermaphroditism or
monoecism) toward separate sexes (Lloyd 1982).

In most populations the sex habit is quite
readily recognized, however none of the categories
is absolute and the range of plant sexuality is
by nature continuous. Darwin (1877) himself
declared that 'the classification is artificial, and
the groups often pass into one another'. One
major problem with a typological approach to sex
habits is the existence of individuals of ambiguous
sex, as well as the occurrence of sex change which
has been recorded in a number of plant species
(Freeman *et al.* 1976). In dioecious flowering
plants, the two sexes can sometimes have varying
degrees of functional hermaphroditism. It is more
appropriate to regard gender as a continuous
variable (Lloyd 1980).

2.4.2.1 *Spatial separation of the sexes*

In most angiosperms, stamens and ovaries are
placed within the same flower — termed a **hermaphrodite** flower (Fig. 2.6a). In some plants,
male and female organs occur in separate flowers
with males formed toward the tips of the stems,
which enhances pollen dispersal, while females
are lower down. This sex habit is called **monoecy**.
Good examples of this are cucumber or corn
plants (Fig. 2.6b). The monoecious arrangement
of sex organs in plants exactly parallels 'hermaphroditism' in animals.

Complete separation of sexual organs into male
and female individuals is termed **dioecy**, and is
seen in, for example, willows, asparagus, cannabis
and spinach (Fig. 2.6c). The separation of the
sexes opens the way for the evolution of differences between them in secondary sexual characteristics such as the size of flower, and in some
dioecious species there is also a divergence in the
ecology of males and females (Chapter 9).

2.4.2.2 *Temporal separation of the sexes*

Temporal differences in the maturity of male
and female parts are often seen in species with
flowering spikes and composite flowers. In fox-
gloves *Digitalis purpurea* and fireweed *Epilobium
angustifolium*, flowers on the spike bloom from
the bottom up. Anthers ripen and shed pollen

Table 2.1 Three major features of flowering plant sexuality and their variants. Percentages refer to the sex habits occurring in a sample of 121 492 *angiosperm* species (about half of all species) compiled by Yampolsky and Yampolsky (1922). Gymnosperms are mostly monoecious or dioecious

Spatial separation of male and female reproductive organs (herkogamy)

One gender class (all plants in a population have the same gender)

HERMAPHRODITISM. (72%) There is only one kind of flower, containing organs of both sexes, e.g. most plants, including most lilies and most orchids

Heterostyly: two or three types of hermaphrodite plant with different style lengths occur in the same populations. Stamens are always longer or shorter than the style and a sporophytic incompatibility system is usually associated with the morphological differences

Distyly: two types of individual occur, one has flowers with a long style and short stamens ('pin') and the other has flowers with a short style and long stamens ('thrum'), e.g. Primulaceae, Linaceae, Turneraceae

Tristyly: three types of individual occur that bear only long-, only mid-, or only short-style flowers. In each type of flower, anthers occur in the two positions not occupied by styles. Crosses between stigmata and anthers at the same level (on different plants) are compatible, e.g. Lythraceae (including *Lythrum salicaria*), Oxalidaceae, Pontedariaceae (including *Eichhornia crassipes*)

MONOECISM. (5%) Separate male and female flowers occur on the same plant, e.g. *Typha, Zea, Quercus, Cucurbita*, most Pinaceae

ANDROMONOECISM. (1.7%) Plants bear hermaphrodite and male flowers, e.g. most Apiaceae

GYNOMONOECISM. (2.8%) Plants bear hermaphrodite and female flowers, e.g. many Asteraceae, including *Bellis* and *Solidago*

Two gender classes (two kinds of plant in a population)

DIOECISM. (4%) Plants bear only female or only male flowers, e.g. *Salix, Vallisneria, Elodea, Cannabis*

GYNODIOECISM. (7%) Plants bear either hermaphrodite or female flowers, e.g. many Lamiaceae including *Glechoma, Thymus*

ANDRODIOECISM. Plants bear either hermaphrodite or male flowers. Almost unknown, but reported in *Datisca glomerata* (Liston *et al.* 1990)

Temporal separation of male and female reproductive activity (dichogamy)

PROTANDRY. Anthers ripen and pollen is shed before stigmata become receptive, e.g. *Daucus, Ranunculus acris*

PROTOGYNY. Stigmata become receptive before anthers ripen and pollen is shed, e.g. *Magnolia, Brassica*

Presence or absence of self-incompatibility

Self-incompatibility

Polymorphism at one or more incompatibility loci. Pollen unable to fertilize ovules if pollen or the pollen parent (depending on the system) carry the same incompatibility alleles as the ovule parent. All systems are 'leaky' to some extent and allow a few 'illegitimate' fertilizations to take place

GAMETOPHYTIC SYSTEMS. The incompatibility type of the pollen grain is determined by the haploid genotype of the gamete itself. Common, e.g. in Fabaceae, Solanaceae, Poaceae, Rosaceae and Ranunculaceae

SPOROPHYTIC SYSTEMS. The incompatibility type of the pollen grain is determined by the diploid genotype of its parent (the sporophyte), e.g. Asteraceae, Brassicaceae, Lythraceae and Primulaceae

Self-compatibility

Plants lacking incompatibility loci. Most pollinations, including self-pollination, result in fruit set, e.g. Apiaceae

---

before stigmata are receptive (i.e. flowers are protandrous), so flowers go through an early male phase and then effectively become female, while younger flowers, higher up the spike, are just beginning their male phase. The older 'female' flowers are richer in nectar, which means that bees visit them first, depositing the pollen they have picked up elsewhere. The bee will then work its way up the flower spike until the nectar rewards no longer justify the effort, and will then move to the lowest flowers of another plant, effecting cross-pollination (Waser 1983; Meeuse

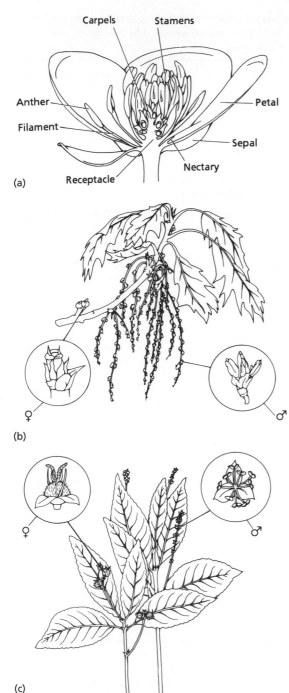

(a)

(b)

(c)

Fig. 2.6 (a) Cross-section through the hermaphrodite flower of the herb *Ranunculus repens*, (b) male and female flowers of the monoecious tree *Quercus rubra*, (c) male and female ramets of the dioecious woodland herb *Mercurialis perennis*.

& Morris 1984). This combination of protandry and bee foraging behaviour limits the amount of pollen transfer within a flower spike and facilitates cross-pollination.

In contrast, wind-pollinated plants such as plantain (*Plantago* spp.) have their female parts ripen first and therefore the uppermost, youngest flowers are functionally female while lower flowers on the spike are male. This increases the accessibility of female-phase flowers to wind-borne pollen, and diminishes the likelihood of self-pollination since pollen rarely travels up the spike.

Because sex habits are variable, and because the efficiency of most mechanisms that prevent certain matings and permit others depends upon the activities of pollinators, the actual selfing rate may vary considerably between local populations that superficially possess the same sex habit. For example, in studies of the perennial herb *Gilia achillefolia* (Schoen 1982) the degree of protandry (see Table 2.1) varied between populations and significantly affected selfing rates (Fig. 2.7).

2.4.2.3 *Self-incompatibility*

In many species of angiosperms, mating is regulated after *pollination* but before fertilization has taken place. To effect *fertilization*, a pollen grain that has reached a stigma has to germinate

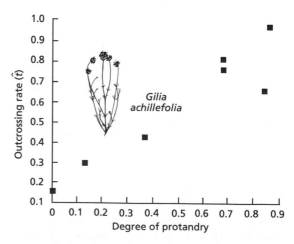

Fig. 2.7 Relationship of protandry to selfing rates in populations of *Gilia achillefolia* (from Schoen 1982).

and produce a pollen tube which grows through the tissues of the style, carrying the male nuclei to the ovary. Physiological reactions may selectively block this process on the stigma or in the style, allowing only non-self pollen to achieve fertilization.

Two major forms of self-incompatibility (SI) are generally recognized:

1 Gametophytic SI is the most common mechanism, found in about 10 families including Fabaceae, Solanaceae and Poaceae. The incompatibility types of pollen and stigma are determined by one, two or four genetic loci with many alleles, called S-alleles. If an S-allele carried by the haploid pollen grain matches any of the alleles in the diploid tissue of the pistil, it will be rejected.

2 Sporophytic SI occurs in Brassicaceae, Asteraceae and other families. The SI genotype of the pollen grain is expressed in its outer case or exine which is derived from its diploid parent. This therefore represents that sporophytic parent's genotype, rather than the haploid genotype of the gamete within. The incompatibility types of pollen and stigma are determined by one genetic locus with two alleles. Pollen tubes fail to penetrate stigmata with the same SI genotype as the pollen grain. Superimposed on this mechanism of biochemical discrimination, sporophytic SI can be further classified into **homomorphic** and **heteromorphic** systems on the basis of whether or not the mating types are morphologically alike. Most are homomorphic.

Two classes of heteromorphic incompatibility are known, **distyly** and **tristyly**, depending on whether there are two or three mating groups (Fig. 2.8; see Table 2.1). The mating groups usually differ in style length, anther height, pollen size and production, and aspects of incompatibility behaviour. In species having heteromorphic incompatibility, the frequencies of mating types can be readily observed. Surveys of distylous species show that in general the long- and the short-styled morphs are equally frequent. Where unequal morph frequencies prevail, it may be because of a host of external factors, including founder effects and differential rates of clonal growth, mating asymmetries among the various style morphs, differential rates of selfing owing to relaxation of self-incompatibility, or it may be due to an evolutionary breakdown of

heterostyly due to the absence of pollinators (S.C.H. Barrett 1988).

Recent research has shown that there is actually a much greater diversity of self-incompatibility mechanisms than was once thought, and that many of these do not fit neatly into the classic definitions of gametophytic and sporophytic SI (e.g. Seavey & Bawa 1986; Waser & Thornhill 1992). Sometimes self-pollen is able to fertilize ovules when it is the only kind in the pistil, but not when pollen from another individual is also present (Weller & Ornduff 1989).

2.4.2.4 Pollination syndromes

The manner in which pollen is transferred from the male to the female reproductive structures of flowers is an integral part of mating systems. The pollination syndromes of animal-pollinated flowers are suites of floral traits that lure, bribe or deceive pollinators, and variously attract bees, bats, birds, beetles, butterflies, moths and wasps. Each of these animal groups is susceptible to a different suite of floral characters (Table 2.2). The specificity of the relationship between plant and pollinator varies. Some evening primroses *Oenothera* spp. have a long, narrow corolla tube from which only hawkmoths can suck the nectar, pollinating the plant in the process. In contrast, the flat, open inflorescences of members of the Apiaceae such as wild carrot *Daucus carota* offer a kind of 'open house' letting most insect visitors serve as pollinators. Extreme specificity between plant and pollinator is relatively rare. In two quite separate, fascinating and famous examples, the larvae of the moths which pollinate *Yucca* spp. in North America and the larvae of the wasps that pollinate figs (*Ficus* spp.) in many parts of the world, feed within the developing fruit of their respective plants. Cross-pollination by the adult insects is not only essential to their host plants but also to the insects themselves because the fruit on which their larvae feed will not develop without it (Proctor & Yeo 1973).

Scarlet gilia *Ipomopsis aggregata* growing in Colorado is pollinated by humming-birds and hawkmoths, but despite this apparent non-specificity shows a remarkable adaptation to both its main pollinators. Paige and Whitham (1985) found that, at about the time when hum-

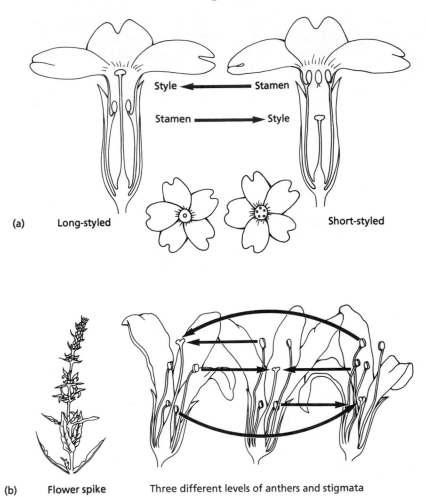

Fig. 2.8 Examples of the two types of heterostyly: (a) distyly in *Primula vulgaris*, (b) tristyly in *Lythrum salicaria*. Arrows indicate compatible matings.

ming-birds migrated from the plant's habitat and hawkmoths became the most important pollinator, individual plants and the population as a whole shifted from a darker to a lighter flower colour. When plants showing the shift were compared with plants that did not change flower colour, it was found that changing plants had a greater reproductive success, and that this was a direct function of their greater attractiveness to hawkmoths.

2.4.3 Patterns of genetic variation and population genetic structure

Genetic variation is the raw material of evolution. Its quantity and distribution are therefore of vital interest if we want to know how much potential for evolutionary change and adaptation exists in plant species and their constituent populations. **Genetic diversity** can be measured at a hierarchy of levels: differences between species, differences between populations within species, differences between parts of the same population, differences between individuals. It is even possible for there to be differences between

Table 2.2 Floral syndromes of animal-pollinated plants (from Wyatt 1983)

Pollinators	Anthesis	Colours	Odours	Flower shapes	Flower depth	Nectar guides	Rewards
Beetles	Day and night	Variable, usually dull	Strong, fruity or aminoid	Actinomorphic	Flat to bowl-shaped	None	Pollen or food bodies
Carrion and dung flies	Day and night	Purple-brown or greenish	Strong, often of decaying protein	Usually actinomorphic	None, or deep if traps involved	None	None
Syrphids and bee flies	Day and night	Variable	Variable	Usually actinomorphic	None to moderate	None	None or pollen or nectar
Bees	Day and night or diurnal	Variable but no pure red	Present, usually sweet	Actinomorphic or zygomorphic	None to moderate	Present	Nectar (41.6%) and pollen; open or concealed
Hawkmoths	Nocturnal or crepuscular	White or pale to green	Strong, usually sweet	Actinomorphic; held horizontal or pendant	Deep, narrow tube or spur	None	Ample nectar (22.1%); concealed
Small moths	Nocturnal or crepuscular	White or pale to green	Moderately strong, sweet	Actinomorphic; held horizontal or pendant	Moderately deep tube	None	Nectar; concealed
Butterflies	Day and night or diurnal	Bright red, yellow or blue	Moderately strong, sweet	Usually actinomorphic; upright	Deep, narrow tube or spur	Present	Nectar (22.8%); concealed
Birds	Diurnal	Bright red	None	Actinomorphic or zygomorphic	Deep, wide tube or spur	None	Ample nectar (25.4%); concealed
Bats	Nocturnal	Dull white or green	Strong, fermented	Actinomorphic or zygomorphic	Brush- or bowl-shaped	None	Ample nectar (18.9%) and ample pollen; open

parts of the same plant (Chapter 1). Ecological genetics, which is dealt with in Chapter 3, is the unfolding, in finer and finer detail, of the relationship between these levels of genetic structure and environment.

Genetic diversity H may be measured with the index:

$$H = 1 - \sum_{i=1}^{i=n} p_i^2 \qquad (2.1)$$

where p_i is the frequency of allele i in a total of n alleles. If the population is random mating, H is the probability that two alleles chosen at random are different, though H measures genetic diversity regardless of whether mating is random or not. H may be calculated for allozyme loci individually, for all the loci studied in a population (H_S) or for several populations of a species (H_T).

Hamrick and Godt (1990) compiled and analysed the results of over 650 electrophoretic surveys covering 480 plant species. This analysis provides some promising insights into the ecological and life history variables that influence genetic diversity, although it should be cautioned that the eight variables used only explained about a quarter of the total variance in genetic diversity at the species (H_T) or at the population (H_S) levels. At the species level, geographic range was the single most important variable: endemic plants had less than half the genetic diversity of the most widespread species. Life form was next most important, and among perennials, woody species had higher genetic diversity than herbs. Mating system and mode of seed dispersal were next and each accounted for about 17% of the explained variance, with a tendency for outcrossing species to be more diverse than selfing ones, and animal-dispersed species to have greater diversity than others. The same variables were important in explaining variation within populations, although their order of importance was different. Mating system was most important, then geographic range, life form, taxonomic status (gymnosperm/dicot/monocot) and mode of seed dispersal. If we wished to pick a species likely to have high genetic diversity it should be a widespread, long-lived, outcrossing (wind-pollinated) tree. Most gymnosperms fit this description, which is probably why they as a group had significantly higher genetic diversity than dicots (though not monocots).

In the sample as a whole, allozyme diversity was about 30% greater at the level of the species than at the level of the population. This indicates significant **genetic differentiation** between populations, which can be measured by the statistic G_{ST}:

$$G_{ST} = \frac{H_T - H_S}{H_T} \qquad (2.2)$$

In Hamrick and Godt's survey G_{ST} was greatest in species with a selfing mating system and those with an annual life form. One might expect populations of clonal plants to contain lower levels of genetic diversity than populations of non-clonal species because clonal genets can grow to very large size. However, Hamrick and Godt's survey showed no significant difference in H_S between clonal and non-clonal species. Ellstrand and Roose (1987) catalogued electrophoretic data on the genetic structure of populations of clonal plants in which the recruitment of new genets was thought to be rare or absent. They concluded that most genotypes are restricted to one or a few populations, and that widespread clones are the exception. The general pattern seems to be that populations are multi-clonal, and have intermediate levels of genetic diversity. Eriksson (1989) surveyed demographic studies of clonal plants and concluded that seedlings of these species occur in populations quite frequently. The recruitment of new genets to a stable population need not be very great to maintain its genetic diversity (Soane & Watkinson 1979).

The genetic structure of populations may change through ecological succession, particularly in clonal plants. In honey locust trees *Gleditsia triacanthos*, Schnabel and Hamrick (1990) found high genetic diversity within populations, and low, though significant differentiation between populations. They found significant spatial genetic structure in both adults and juveniles and concluded: (i) that even small amounts of clonal growth can cause large increases in the amount of genetic structure in populations; (ii) that spatial structure among juveniles at two sites most likely represents the presence of family groups; and (iii) the overall level of spatial genetic structure at both sites was somewhat greater in the juvenile classes than in the non-clonal adult classes. This latter conclusion fits with theoretical studies which imply that limited gene movement causes a steady increase in spatial structure from generation to generation.

Studying old-fields of differing ages, Maddox *et al.* (1989) examined both the physical connections and the allozyme phenotypes of ramets of the clonal plant, *Solidago altissima*. They found that the 1-year-old field contained genets composed of single ramets, a 5-year-old field contained many small contiguous clones of *S. altissima* with highly interconnected ramets, while in the oldest two fields ramets were highly intermixed, and ramets of the same genotype were not extensively interconnected. The studies of *Gleditsia* and *Solidago* are only two of many that show that the genetic structure of a population is dynamic. The forces which act upon and create genetic structure are ecological as well as genetic, and are addressed in the next chapter.

2.5 SUMMARY

Inherited variation is the raw material of evolution. Plants make a speciality of morphological variation, which can be classified according to whether it occurs within genotypes (**phenotypic plasticity**) or between genotypes, and whether it is continuous or discontinuous. Variation within genotypes is referred to as the **reaction norm** if it is continuous and **heteromorphism** if it is discontinuous. Variation between genotypes is **metric** (or quantitative) if it is continuous and **polymorphic** (or qualitative) if it is discontinuous. The actual phenotype expressed by an organism always depends upon both genes and environment, and when genotypes respond differently to different environments there is said to be a **genotype × environment interaction**.

Phenotypic variability is measured by its variance. **Phenotypic variance** may be partitioned into a genetic component called the **genetic variance** and an environmental component called the **environmental variance**. **Broad-sense heritability** is the ratio of the genetic to the phenotypic variance. **Narrow-sense heritability** is the ratio of the **additive genetic variance** to the phenotypic variance, and is the better indicator of the amount of heritable variation present in a population upon which natural selection can act. Fisher's **fundamental theorem of natural selection** suggests that narrow-sense heritability should be low for characters closely associated with fitness.

Phenotypic plasticity is an important character in its own right, and is subject to selection. **Genetic variation** ultimately arises from **gene mutation**, but it is amplified and maintained by **sexual reproduction** which can produce novel genotypes by recombination. The **mating system** of a population describes who mates with whom. Its most important aspect is the degree of genetic relatedness among mates, or the amount of **inbreeding**. The **mixed mating model** describes mating systems on the basis of the degree of **selfing**. The mating system in a population is strongly influenced by plants' **sex habit** which defines who is *able* to mate with whom. The **gender** of a plant denotes the relative importance of male and female function to reproductive success. Sex habits are traditionally classified according to three features: (i) the **spatial separation** of male and female reproductive organs; (ii) the **temporal separation** of male and female activity within flowers; and (iii) the presence of **self-incompatibility**. The manner in which pollen is transferred from the male to the female reproductive structures is an integral part of mating systems and is reflected in the **pollination syndromes** of flowers.

The amount of genetic variation can be quantified by measuring **genetic diversity** (H). This may be calculated for variation within populations (H_S) or between populations (H_T), and these measures may be used to calculate the degree of **genetic differentiation** (G_{ST}) between populations of a species. Surveys of genetic diversity have shown that the strongest influence on H_T is geographic range (widespread species have high H_T) and the strongest influence on H_S is mating system (selfers have low H_S). Genetic differentiation between populations is strongest in selfing species and annuals.

Chapter 3
Ecological Genetics

3.1 INTRODUCTION

Genetic diversity is a puzzle. Why is it that nearly half of the enzyme loci studied in plants are polymorphic? New alleles appear infrequently, and in any case most mutations being random must be deleterious and hence quickly removed by selection. Those few mutations that are beneficial should spread rapidly, displacing alternative alleles: so beneficial mutations are no more likely to increase genetic diversity than are deleterious ones. If these observations are correct, where does all the genetic variation found in plant populations come from, and how does it persist?

The solution to this puzzle takes us into the realms of population and evolutionary genetics. The issues have been argued about for several decades and are still controversial. On the one hand is the view held by Kimura (1983, 1991a, b) who believes that most allozyme variation is **selectively neutral**, and on the other the view held by an apparently dwindling band of geneticists that this variation is maintained by selection. A full treatment of the controversy is beyond the scope of this book, but since plant populations are especially genetically diverse and contain more variation than (at the moment) we can explain on the basis of natural selection alone, neutralism provides a useful starting point for exploring plant population genetics.

3.2 THE HARDY−WEINBERG LAW

If two alleles A and a at a single locus are selectively neutral, and individuals mate with each other at random with respect to their genotype, we can calculate the expected frequency of genotypes among offspring from the allele frequencies in the parental generation. If the frequencies of A and a are respectively p and q, the alleles will pair with each other in the frequencies $(p + q)^2$. By expanding this expression we get the genotype frequencies expected in the next generation, which sum to one:

$$(p + q)^2 = p^2 + 2pq + q^2 = 1$$

As shown in Table 3.1, p^2 is the frequency of AA; $2pq$ is the combined frequency of Aa and aA; and q^2 is the frequency of aa. This equation is called the **Hardy−Weinberg law**, and is a benchmark against which we can test for deviations from the assumptions on which the law is based. For any particular locus with two alleles present at known frequencies, a population is said to be at **Hardy−Weinberg equilibrium** if genotypes are present in the ratios given by the expansion of $(p + q)^2$. The Hardy−Weinberg law is based on the assumption that mating is random, but if it is not and there is some degree of inbreeding, there will be a loss of heterozygotes. For example, if inbreeding occurs through selfing and selfing is complete, the frequency of heterozygotes decreases by a half in each generation, because only half the progeny of each $Aa \times Aa$ mating are Aa.

Many plant populations are relatively short-lived, and may not have had sufficient time to

Table 3.1 The ratios of genotypes expected at Hardy−Weinberg equilibrium for two alleles A, a at a single locus

Allele (frequency)	$A(p)$	$a(q)$
$A(p)$	$AA(p^2)$	$Aa(pq)$
$a(q)$	$aA(pq)$	$aa(q^2)$

reach equilibrium; however a number of reports, particularly for populations of trees, show allozyme frequencies that *are* in Hardy–Weinberg equilibrium. For example, Bousquet *et al.* (1987) carried out electrophoretic surveys in 22 natural populations of green alder *Alnus crispa*, and found 16 allozyme loci for the 11 enzymes studied. Nine of the loci were polymorphic. For every polymorphic locus in each population, the observed genotype distribution was compared, and found to match that expected under Hardy–Weinberg equilibrium. The equilibrium state of these loci suggested random mating, and indeed the species is outbreeding with highly dispersive pollen and seeds.

A number of factors can produce deviations from Hardy–Weinberg equilibrium. These are:

1 mutation;
2 linkage disequilibrium;
3 non-random mating;
4 genetic subdivision of the population;
5 gene flow;
6 genetic drift;
7 natural selection.

Mutation rates are generally too low to have significant impact in the short run typical of the scale of demographic studies (Chapter 2), but we will look at the other factors in this chapter, concluding with natural selection.

3.2.1 Linkage disequilibrium

When alleles at two or more loci exist in combination at higher or lower frequencies than would be predicted from their individual frequencies, the alleles are said to be in **linkage disequilibrium**. Linkage disequilibrium can be from selection, from epistasis or from physical linkage between loci that occur near each other on the same chromosome. Consider the following example. Alternative alleles segregate at each of two independently assorting loci (A,a and B,b), and the gene frequencies of the respective alleles are $p(A)$, $q(a)$, $r(B)$ and $s(b)$. Considering each locus separately, the Hardy–Weinberg law tells us that genotype frequencies at equilibrium should be: $p^2(AA):2pq(Aa):q^2(aa)$ for locus A and $r^2(BB):2rs(Bb):s^2(bb)$ for locus B. Considered together, the proportions of the bi-factorial genotypes at

equilibrium are given by the expansion of the expression:

$$(p + q)^2(r + s)^2 = (p^2 + 2pq + q^2)(r^2 + 2rs + s^2)$$

This is equivalent to squaring the equilibrium frequencies of the gametes *AB, Ab, aB, ab*. If actual frequencies at the start deviate from Hardy–Weinberg equilibrium and there is incomplete linkage (because the loci in question are on different chromosomes or because there is crossing-over between their positions on the same chromosome), gamete frequencies will converge on the equilibrium values asymptotically. At the maximum, disequilibrium will decay at a rate of half the difference from linkage equilibrium each generation. Within four to five generations, less than 10% of the original linkage disequilibrium remains. For three genes the speed of approach to equilibrium is diminished, and it becomes slower still as more genes are involved. When the loci involved are linked, the closer the linkage the longer it will take for the gametic genotype frequencies to reach equilibrium. The speed of approach to equilibrium is inversely proportional to the closeness of the linkage.

A good example of disequilibrium in which favourable combinations of alleles at separate loci are held together by linkage is the suite of traits found in heterostyly in the primrose, *Primula vulgaris*. The 'pin' form of flower, with a long style and low-set anthers, is inherited as if it was recessive to the 'thrum' form, which has a short style and high-set anthers (see Fig. 2.8). The reciprocal locations of style and anthers in pin and thrum morphs enhances the likelihood that insects visiting both kinds of flower will deposit compatible pollen in the right place. At first glance the genetics behind the pin and thrum forms appears simple — it looks like a single locus where pin is homozygous recessive *ss* and thrum is heterozygous *Ss*. However, very occasionally plants are found with a long (pin-type) style *and* long (thrum-type) anthers in the same flower. These plants are called homostyles. A likely explanation for their occurrence is that the S-locus is in fact a supergene comprising a group of at least three tightly linked loci corresponding to: (i) G/g influencing three traits — style length, female incompatibility type and stigmatic papilla type; (ii) *P/p* influencing pollen size and male incompatibility type; and (iii) *A/a* affecting anther height (Richards 1986). The thrum morph has the genotype *GPA/gpa* and the pin is homozygous *gpa/gpa*. Because of linkage *GPA* tend to segregate as a group and are rarely broken up by crossing-over. Though very rare, the most frequent cross-over event is between *G* and the other two loci (Lewis & Jones 1992). Naturally, since only the thrum morphs are heterozygous, it is only in thrums that crossing-over can occur. Brown (1984) examined linkage disequilibrium in five plant species by looking at the levels of association among allozyme loci. He found a striking dichotomy between outbreeders and inbreeders. In outbreeders, associations among allozyme loci were low, whereas in the inbreeders associations were generally high because of the lack of recombination in selfers.

3.2.2 Non-random mating

Random mating, or **panmixis**, is a convenient theoretical assumption, but if the phenotype of a plant determines which other members of the population it is likely to mate with, mating is said to be **assortative**. Sometimes the bias may favour matings between similar individuals, termed **positive assortment**, and sometimes matings between unlike individuals will be favoured (as in the case of self-incompatibility); this is called **disassortative mating**. Positive assortment will tend to cause inbreeding, while disassortative mating will tend to produce outcrossing.

Pollinators may play a significant part in causing positive assortative mating by constantly selecting flowers of a particular kind (Waser 1986). The behaviour of pollinators with different flower preferences appears to be responsible for maintaining flower colour polymorphism by positive assortment in a number of species (Kay 1978, 1982). In sky pilot *Polemonium viscosum*, a perennial herb found in the Rocky Mountains, populations are polymorphic for flower scent: there are sweet-smelling flowers visited by bumblebees, and skunky-smelling ones visited by flies. The frequency of the bumblebee-pollinated morph increases with altitude and is correlated with the

relative abundance of flies and bees (Galen *et al.* 1987). Levin and Kerster (1973) found that bees pollinating purple loosestrife *Lythrum salicaria* tended to maintain their altitude when flying from one plant to another and that the correlation between the statures of mating plants (those that followed each other in a take-off and landing cycle for a single bee) was high (0.86). In an experimental population composed of two strains that differed in height, 60% of flights were intrastrain. However, when the taller plants were trimmed to be of the same height as the shorter strain, there was no assortative pollination by strain.

In contrast to the role of pollinators in positive assortative mating, disassortative mating is likely to depend on flowering phenology, a property of the plant itself. For example, strong herkogamy, dichogamy or self-incompatibility (see Table 2.1) may produce disassortment. Because it reduces inbreeding, disassortative mating is a mechanism that tends to maintain polymorphism. The pin/thrum polymorphism in heterostyle populations is maintained by disassortative mating between the two morphs.

3.2.3 Genetic subdivision of the population

For obvious reasons, the probability of two individuals mating tends to be an inverse function of the physical distance between them. As a consequence of this, and the limited dispersal of offspring from their parents, neighbours tend to be related, and the two processes together result in the genetic subdivision of populations. For any given locus, the probability of two alleles drawn at random being identical to one another is greater in part of a genetically structured population than in part of an unstructured one of the same area. In fact the probability in a structured population will be the same as that for an unstructured population of *smaller area*. This fact can be used to measure the degree of genetic structure. The **genetic neighbourhood area**, *A*, is defined as the area centred on an individual, within which there is an 86.5% probability of finding its parents. If seeds and pollen travel equally in all directions from parent plants, the genetic neighbourhood area is:

$$A = 4\pi\sigma^2 \qquad (3.1)$$

where σ^2 is the variance of the dispersal distance between parents and offspring. This formula was first devised by Wright (1943, 1951) and its important assumptions include that offspring per parent and dispersal distances are normally distributed, that the sex ratio is 1 : 1, and that the population is at equilibrium. There have been many refinements to take account of deviations from these assumptions (Crawford 1984a, b), but for present purposes we will just look at how σ^2 can be broken down into variances that can be measured, and how to allow for the effect of the outcrossing rate $\hat{t}$. Figure 3.1 shows that σ^2 is actually composed of a series of variances that represent dispersal via the clonal spread of the parents, pollen dispersal and seed dispersal. These movements of genes represent **gene flow**. The variances of these components of gene flow are independent, so their contribution to the total variance is additive:

$$\sigma^2 = \frac{\hat{t}}{2}\sigma^2_{\text{pollen}} + \sigma^2_{\text{clonal}} + \sigma^2_{\text{seed}} \qquad (3.2)$$

where $\hat{t}$ is the outcrossing rate (Gliddon *et al.* 1987). Note that pollen variance only makes half the contribution of clonal and seed variances because pollen grains are haploid. The actual values of the variances of the gene flow components shown in Fig. 3.1 are affected by plant life history, pollination and seed dispersal mechanisms and are discussed in the next section.

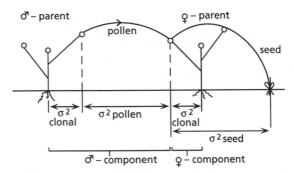

Fig. 3.1 Components of gene flow in plants with clonal spread (after Gliddon *et al.* 1987).

Another way of describing the degree of genetic structure in a population is to calculate the **effective population size**, N_e, which is the effective number of reproductive plants to be found within the genetic neighbourhood area:

$$N_e = A\frac{N}{2}(1 + \hat{t}) \qquad (3.3)$$

where N is the density of reproductive plants in the population and $\hat{t}$ the outcrossing rate (Crawford 1984a). N_e can be calculated from Eqns 3.1–3.3, or by genetic analysis (Husband & Barrett 1992). Some example values of N_e are given in Table 3.2. When N_e is equal to n, the total number of adult individuals in the population, mating is random and there is no genetic structure. In fact N_e is very variable, particularly in inbreeding species (Schoen & Brown 1991), but N_e is often much

Table 3.2 Effective population size N_e for some representative herbs and trees

Species	N_e	Source
HERBS		
Delphinium nelsonii	28	Waser & Price (1983)
Eichhornia paniculata	3–71	Husband & Barrett (1992)
Ipomopsis aggregata	69	Waser & Price (1983)
Lythospermum carolinense	5	Levin (1981)
Phlox pilosa	75–282	Levin (1981)
TREES		
Cedrus atlantica	207	Levin (1981)
Picea mariana	17	Barrett *et al.* (1987)
Pinus cembroides	11	Levin (1981)
Pseudotsuga taxifolia	26	Levin (1981)

smaller than total population size (Levin 1988). For an advantageous allele the rate of spread is an inverse function of N_e (Levin 1988), so ecological factors that lower N_e with respect to n increase the likelihood that different parts of the population will diverge in genetic composition.

In a population that is composed of genetically differentiated, discontinuous subpopulations, each partially isolated from the others, the amount of within-population genetic diversity (H_S) should be greater than in a homogeneous, randomly mating population of the same size. Outbreeding populations should have greater allelic diversity at the species level (H_T) and less differentiation among populations (lower G_{ST}) than self-fertilizing populations. All of these theoretical expectations are borne out by the patterns of genetic diversity found by Hamrick and Godt (1990) in their survey discussed in Chapter 2.

Another indicator that there may be genetic structure within a population is an apparent overall deficiency in the frequency of heterozygotes. This is likely to happen in two different ways: it can either be due to inbreeding or to the presence of two or more subpopulations. If each subpopulation is at Hardy–Weinberg equilibrium, but their equilibrium allele frequencies are different, a sample of genotypes taken indiscriminately from the whole population will provide genotype frequencies that *appear* to be out of Hardy–Weinberg equilibrium. This is termed the **Wahlund effect**. A symptom of the Wahlund effect is that individuals with similar genotype are clustered in space. Genetic variance due to the Wahlund effect can be distinguished from variance due to linkage disequilibrium by looking for these clusters (Brown & Feldman 1981).

3.2.4 Gene flow

New genes may enter a population by mutation (Chapter 2) or by gene flow through the dispersal of pollen, seeds or clonally produced ramets. Rates of gene flow can be measured in terms of the equivalent number of migrants moving between populations (of effective size N_e) per generation.

3.2.4.1. Pollen movement

Govindaraju (1988) surveyed data from studies of genetic differentiation in 115 plant species and used this information to calculate rates of gene flow in species with different pollination syndromes. In predominantly self-pollinated species gene flow averaged 0.83, in animal-pollinated species it averaged 1.211, while in wind-pollinated species it averaged 2.91.

It is not easy to determine who mates with whom in a plant population, because this depends upon the movements of insect pollinators and the vagaries of wind dispersal — to say nothing of the difficulty of tracking vast numbers of pollen grains, or dealing with the patchiness of natural populations. However, electrophoretic markers can be used for paternity analysis to distinguish among possible male parents and to say which *most probably* fathered a particular offspring. For each seed and each potential pollen donor, the likelihood of the donor being the male parent is compared with the likelihood of the father being a randomly chosen individual. This likelihood ratio is calculated for a number of loci and after the summed likelihood values for all potential male parents are compared, the individual having the greatest likelihood value is designated as the paternal parent of a given seed. Meagher (1986) used this method to determine the physical distances between mating pairs in a population of the lily *Chamaelirium luteum*, and found that there was a significant excess of matings between individuals less than 10 m apart, compared with expectations based on completely random mating (Fig. 3.2). Meagher's method is prone to overestimate distances between mates (Devlin *et al.* 1988; Brown 1989).

The behaviour of pollinators determines the distance pollen is carried, but pollinator behaviour is, in turn, influenced by plant traits and population properties. For example, Crawford (1984b) found a positive relationship between selfing rate and the total numbers of flowers per plant in a population of the herb *Malva moschata*, presumably caused by bees spending more time on plants with many flowers than on those with few. An inverse relationship between population density and neighbourhood area has been found

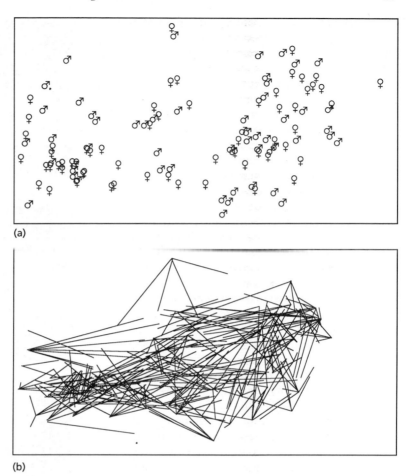

(a)

(b)

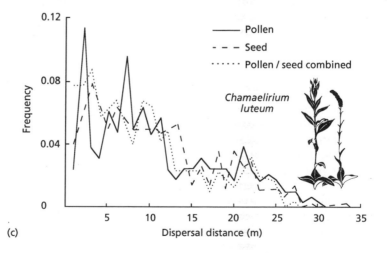

(c)

Fig. 3.2 Map of a population of *Chamaelirium luteum* showing: (a) the distribution of males and females, (b) lines connecting females with their mates in 381 natural crosses, and (c) the frequency distribution of pollination distances in the population (from Meagher 1986; Meagher & Thompson 1987).

in a number of other bee-pollinated species because flight distances were shorter in more dense stands (Levin & Kerster 1969a,b; Fenster 1991a).

In general, pollinators that routinely carry pollen over long distances will increase N_e and A, and decrease the probability of population differentiation, while restricted pollinator movements will have the opposite effect. Turner *et al.* (1982) carried out a computer simulation to see how pollination by nearest neighbour ($N_e = 4.4-5.2$) would affect the spatial distribution of genotypes in a hypothetical population of 10 000 annual plants that were outbreeding and polymorphic at a single, selectively neutral diallelic locus. Starting with the three genotypes (*AA, Aa, aa*) distributed randomly, marked patchiness developed within 100 generations, and by 600 generations heterozygotes were rare and the population was mostly composed of patches homozygous for one allele or the other (Fig. 3.3).

Ellstrand *et al.* (1989) used genetic markers in experimental and natural populations of wild radish, *Raphanus sativus*, to measure gene flow by pollen into small populations. In both types of population, considerable heterogeneity of gene flow receipt was found among populations. Even though experimental populations were within a few hundred metres of one another and were flowering at the same time, there was a 22-fold range in gene flow estimates. Between 0.2% and 4.5% of matings were between populations. Small natural populations in different years and locations showed an almost sixfold range in apparent gene flow (3.2–18.0% of mating occurred between populations). Kirkpatrick and Wilson (1988) found a range of 0–15% gene flow from experimental plots of cultivars of *Cucurbita pepo* to several single wild plants of *C. texana*. The variation in gene flow rates observed for *Raphanus, Cucurbita* and other species suggests that gene flow rates are likely to be particular to a population rather than a characteristic of species (Ellstrand *et al.* 1989).

Devlin and Ellstrand (1990) used paternity analysis to estimate gene flow by pollen entering two isolated natural populations of *Raphanus sativus*. At least 7% of the seeds in both populations were fathered by foreign pollen, even though there were no other radishes within 150 m

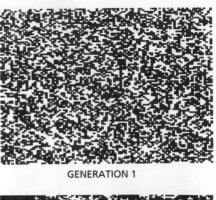

GENERATION 1

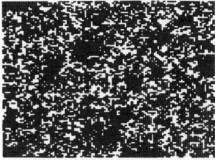

GENERATION 100

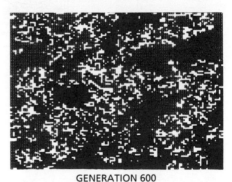

GENERATION 600

Fig. 3.3 Spatial distribution of three genotypes and development of genetic structure in a model population with nearest-neighbour pollination. White = *Aa*, grey = *aa*, black = *AA* (from Turner *et al.* 1982).

of the test populations. The principle behind this particular method of paternity analysis was that all potential local pollen parents were ruled out as sires of a particular seed before it was concluded that the pollen came from outside. This of course tends to underestimate gene flow because some foreign pollen may not be genetically distinguishable from the pollen produced by local flowers. On the other hand, in larger, less isolated populations gene flow from outside would probably be much less than 7%.

Rai and Jain (1982) measured gene flow in the wind-pollinated wild oat *Avena barbata* and found, not surprisingly, that gene flow and outcrossing rates were higher along transects orientated in some directions than others.

3.2.4.2 *Seed dispersal*

Seeds are dispersed by a number of agents: they may be water- or gravity-dispersed, explosive, winged or plumed, animal-ingested, or animal-attached (Fig. 3.4). Each mode of dispersal is likely to have a characteristic effect on population genetic structure, depending upon the mean and variance of dispersal distances, the shape of the dispersal curve and whether seeds are dispersed individually or with their kin.

Genetic analyses of plant populations indicate the presence of a great deal of spatial heterogeneity, and it seems quite likely that much of this is due to the limited distance of most seed dispersal, as well as inbreeding amongst related neighbours. In a review of the literature, Hamrick and Loveless (1986) found a significant correlation between genetic structure and seed dispersal mechanism. In plants which are wind-dispersed, most seeds do not move very far from their maternal parent (Fig. 3.5) (Levin & Kerster 1974), and in animal-dispersed species, seed vectors often carry away and deposit in their faeces groups of seeds from the same plant, if not from the same fruit (Howe & Smallwood 1982). Birds and rodents that cache seeds have a similar effect. Furnier *et al.* (1987) found that whitebark pine *Pinus albicaulis*, whose seeds are dispersed by Clark's nutcracker, occurred in family groups. A similar situation was found in limber pine *Pinus flexilis* by Schuster and Mitton (1991).

Several studies have attempted to estimate the relative contributions of every element in pollen and seed dispersal, including the fitness of dispersed seeds, into a quantification of gene flow. For example, in a natural population of the annual prairie legume *Chamaecrista fasciculata*, Fenster (1991a) found the outcrossing rate to be high ($\hat{t} = 0.8$), but gene dispersal to be so limited that the average A corresponded to a circle of radius 2.4 m containing about 100 individuals. The fitness of seed progeny, as manifested in seedling survival and fruit production, depended upon how far apart the parents were. The progeny of selfing had half the fitness of progeny with two different parents from the same neighbourhood, and seeds with parents more than one neighbourhood diameter apart had the highest fitness of all. Fenster (1991b) found that using these relative fitness values to weight the contribution of pollen dispersal to overall gene flow nearly doubled the total estimate (Fig. 3.6). Seed dispersal made a remarkably small contribution to gene flow.

Gliddon and Saleem (1985) estimated the three variance components of gene flow (Eqn 3.2) in two small populations of *Trifolium repens*, which has extensive clonal growth (Chapter 1). There were about 40 genotypes in each population, and because the species is self-incompatible, the outcrossing rates were high ($\hat{t} = 0.86$ to 0.95). Means of pollen and seed dispersal distances, and their variances, were remarkably similar, but the variance of clonal dispersal distance depended upon the size of plants in the population, and this depended upon their age. Using a computer simulation of clonal spread, Gliddon and Saleem (1985) found that clonal dispersal made a minor contribution to A in young populations, but nearly doubled A over the first 10 years. In their study populations N_e and n were similar, pointing to the absence of genetic structure.

Seed dispersal is often a two-stage process, involving primary dispersal that may be aided by animals or wind, and a secondary phase on the soil surface when water, wind, seed predators or ants move seeds still further. In several species of woodland violet *Viola* spp., seeds are expelled from dehiscent fruit and then they are collected from the ground by ants that carry them to their nests. There, ants remove a fatty appendage from the seed called the elaiosome and deposit the seed, still viable, on their refuse heap. Beattie and Culver (1979) estimated A from the variance of ballistic dispersal, secondary dispersal by ants and pollination in three species of *Viola* growing in woods in West Virginia. The three processes

(a) *Taraxacum officinale*

(b) *Arctium lappa*

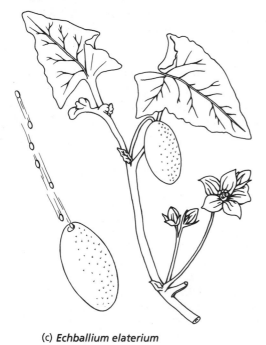

(c) *Echballium elaterium*

(d) *Fragaria vesca*

Fig. 3.4 Examples of modes of seed dispersal: (a) wind-dispersed seeds of *Taraxacum officinale*, (b) hooked, animal-dispersed seed head of *Arctium lappa*, (c) seeds explosively expelled from the fruits of *Echballium elaterium*, (d) fleshy, animal-dispersed 'berry' of *Fragaria vesca*.

accounted for 22%, 4% and 74% of the average neighbourhood area, respectively. In a study of another ant-transported herb *Corydalis aurea* growing in the Rocky Mountains, Hanzawa *et al.* (1988) compared the fitness of seeds transported by ants with that of seeds not carried away, and found that ant transport increased fitness by 38%. As Fenster (1991b) found in *Chamaecrista fas-*

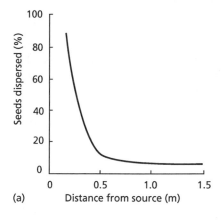

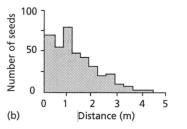

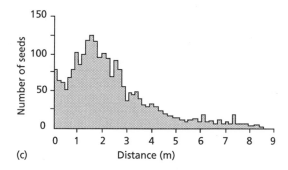

Fig. 3.5 Examples of seed dispersal curves: (a) seeds of *Avena barbata* (from Rai & Jain 1982), (b) *Phlox pilosa* (after Levin & Kerster 1968), and (c) *Liatris aspersa* (after Levin & Kerster 1969b).

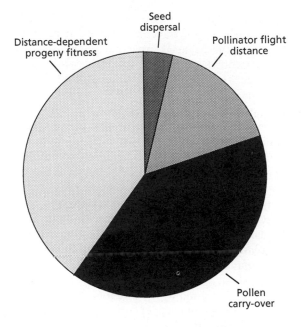

Fig. 3.6 The relative contributions of seed dispersal, pollinator flight distance, pollen carry-over and distance-dependent progeny fitness to gene flow in *Chamaecrista fasciculata* (from Fenster 1991b).

ciculata, and Waser and Price (1983, 1989) have also found in *Delphinium nelsonii* and *Ipomopsis aggregata*, it is likely that incorporating a measure of the fitness consequences or *quality* of dispersal as well as its distance would increase the estimate of the contribution that ants make to *A* in *Viola*.

Many species of violets are included among herb species that have two kinds of flower on the same plant: a **cleistogamous** one that never opens, but that produces self-fertilized seeds, and

chasmogamous flowers that open in the normal way and are pollinated by insects or wind. It is interesting that in all the species that display this syndrome there is a consistent correlation between poor seed dispersal of cleistogamous seeds and good dispersal of chasmogamous ones (Lord 1981). This is carried to an extreme in peanut grass *Amphicarpum purshii* and other **amphicarpic** species in which the cleistogamous seeds are produced from subterranean flowers (Cheplick 1987). The conclusion is inescapable that evolution in these plants seems to have favoured a strategy in which progeny genetically like their parents are kept at home while outcrossed progeny are dispersed (Schoen & Lloyd 1984).

There are some important genetic differences between pollen and seed dispersal. Not only does a seed carry twice the number of alleles of a haploid pollen grain, but it is also several steps nearer recruitment into the population than a pollen grain that, even after it has accomplished fertilization, still has hazards to face before its genes are successfully dispersed in a seed. Though most seeds do not travel far from their parents, the tail of the dispersal curve may be very distant, and way off the scale of graphs such as those in Fig. 3.5. The fate of the occasional seed that is carried a long way can be highly significant. A few seeds carried over a great distance can found a new population, which of course pollen grains cannot.

If a new site is colonized by a limited number of seeds that are genetically atypical of the source population simply due to sampling error, there may be a strong **founder effect**, giving rise to a genetic differentiation between populations. This kind of colonization event, or a reduction in plant population size from some other cause that forces the population through a **bottleneck**, may also bring about a significant decrease in genetic diversity. Lack and Kay (1988) found evidence of both these phenomena in British populations of the self-pollinating herb *Polygala vulgaris*. Populations in long-established grassland habitats displayed significant genetic variation, but a number of roadside and sand-dune populations of the species were monomorphic. It seems likely that populations in these more disturbed habitats were each founded by just one or a very few individuals. Schwaegerle and Schaal (1979) found particularly low genetic diversity in a large population of the pitcher plant *Sarracenia purpurea* at a bog in Ohio, where it had originated from the deliberate introduction of a single individual some 8–15 generations before their study. Taggart *et al.* (1990) found that genetic diversity in populations of the same species that had been introduced in small numbers to bogs in Ireland was also low, but within the range recorded for native North American populations. Many natural populations of *S. purpurea* are highly isolated from each other, and it is possible that for this reason founder effects are important in this species generally.

The actual loss of genetic diversity when a population passes through a bottleneck can depend upon how the population behaves after the event (Nei *et al.* 1975). If the population expands again quite rapidly after the bottleneck, genetic diversity may be little affected, even though rare alleles may be lost in the bottleneck.

This is because rare alleles make a negligible contribution to the index of genetic diversity H, as it is calculated (Eqn 2.1), and none at all if their frequency is less than 5%. Barrett and Kohn (1991) discuss the significance of bottlenecks further, particularly as they affect the conservation of rare species, and they list other examples of plant populations whose low levels of genetic diversity may be attributed to this cause.

3.2.5 Genetic drift

When effective population size is very small, or alleles are rare, the transmission of alleles from generation to generation is prone to sampling error that may cause changes in frequency in a random direction. This is described as **genetic drift**, and it may cause genetic differentiation between isolated populations or subpopulations in the absence of selection. Drift is likely to be much more important in plants than in animals, owing to the typically small values of N_e in plant populations (see Table 3.2). Wright (1948) showed that very low rates of gene flow are sufficient to prevent genetic drift from occurring. If the number of populations is reasonably large and the mutation rate is reasonably small, the amount of genetic differentiation between populations that will arise due to drift, as measured by G_{ST} (Chapter 2), is:

$$G_{ST} = \frac{1}{1 + 4N_e m} \qquad (3.4)$$

where m is the proportion of the effective population N_e that are migrants and $N_e m$ is the number of individuals migrating into a population per generation. For genetic drift to create significant genetic differences between populations, the value of $N_e m$ must be much less than one. In other words, a single migrant seed travelling between populations and establishing each generation is sufficient to prevent them diverging through drift. However, there is one proviso. The model on which this conclusion is based assumes that migrants travel at random between all populations, but if populations are arranged linearly in space, like stepping-stones for migrants, Eqn 3.4 underestimates the genetic divergence that may arise due to drift.

There is evidence that drift does play a role in some herb populations, though it is usually difficult to rule out other causes of genetic differentiation. In populations of *Impatiens pallida* and *I. capensis*, both of which produce cleistogamous as well as chasmogamous flowers, Schoen and Latta (1989) found small-scale spatial structure in the frequencies of three qualitative traits. Flower and stem phenotypes were clustered in patches of consistent size that did not appear to be related to any environmental heterogeneity on the same scale. This suggested that the genetic structure might have been a consequence of limited gene dispersal, rather than the formation of locally adapted patches. Figure 3.3 illustrates how this can come about.

Lythrum salicaria and *Decodon verticillatus* are two tristylous herbs (see Fig. 2.8) that in some populations lack one of the flower morphs. Eckert and Barrett (1992) found this to be much more common in Canadian populations of *L. salicaria* where the plant is an alien invasive weed of wetlands, than among native populations of the species in Europe. Most populations lacking a morph were small in size with n only 3–50. In the related species *D. verticillatus*, which is native to North America, geographically marginal populations more often lacked a morph than did populations at the centre of the species' range. Rare flower morphs of heteromorphic species have a frequency-dependent advantage (see Section 3.3.3), so one might expect selection to *prevent* populations from losing a morph through genetic drift. However, Eckert and Barrett (1992) used a computer simulation model to compare the opposing effects of selection and drift and found that drift and founder effects could be strong enough to cause some populations, particularly small ones, to lose a flower morph.

3.3 NATURAL SELECTION

Natural selection is the only known mechanism that can produce adaptive evolutionary change. How much of the genetic variation in plant populations can be accounted for by adaptation to local conditions, and how much is maintained by the other mechanisms we have discussed, operating on selectively neutral alleles?

There is no simple answer to this question, but what is clear is that natural selection is a ubiquitous force whose operation is frequently implied by the close correspondence between genotype and environment as, for example, in *Veronica peregrina* in vernal pools or *Achillea lanulosa* in the Californian Sierra (see chapters 1 & 2).

As we saw in Chapter 2, Clausen *et al.* (1948) were able to demonstrate genetic differences between altitudinal populations of *Achillea lanulosa* by transplant experiments in common gardens at different altitudes. If natural selection is responsible for the genetic differences between populations, then native genotypes should have a higher fitness than alien ones and this difference should be revealed in transplant/replant experiments. The fitness, W_A, of an alien can be estimated from the simple ratio:

$$W_A = \frac{\text{performance of alien}}{\text{performance of natives}}$$

The intensity of natural selection against the alien can be measured by the **selection coefficient** s, which is defined as $s = 1 - W_A$. Based upon survival percentages in the experiments of Clausen *et al.* (1948), the selection coefficient against *A. lanulosa* from Mather grown at Timberline was 0.57, and against plants from Timberline grown at Mather was 0.26 (see Fig. 2.4a).

In a population where there is phenotypic variation for some metric trait that is thought to have an effect upon fitness, the potential strength of selection upon the trait may be measured by a regression of trait values on some component of fitness such as survival or seed production. This measures **phenotypic selection**. A significant relationship between, say, leaf area and fitness satisfies two of the necessary conditions for natural selection to occur (variation between individuals and differences in fitness between variants), but the variation must also be *heritable* if phenotypic selection is to be translated into natural selection (Chapter 1). For metric traits that are correlated with each other, multiple regression can be used to identify which trait or set of traits is selectively most important (Lande & Arnold 1983), with due regard for the statistical assumptions of the method (Mitchell-Olds &

Shaw 1987), and the need to test any conclusions experimentally (Wade & Kalisz 1990).

3.3.1 Allele frequencies and selection

As a simple example of how selection can alter allele frequencies, we will look at an experiment by Ennos (1981) who sowed seeds of cyanogenic and acyanogenic genotypes of *Trifolium repens* in a field, and scored allele frequencies among the survivors after 6 months. Most of the seeds sown possessed the *Ac* allele that produces a cyanogenic glucoside, but some seeds had genotypes with (*Li*) and some without (*li*) the ability to produce the enzyme linamarase that releases cyanide (Chapter 2). For present purposes we can concentrate on the *Li* locus and ignore *Ac*. Because of dominance, genotypes *LiLi* and *Lili* had the same phenotype and fitness, but Ennos found that there was selection against *lili* genotypes (*s* = 0.3) due to poorer survival.

How does this degree of selection against the recessive homozygote alter the frequency of the *li* allele over a single generation? So long as mating remains random, the Hardy–Weinberg equation still works as a description of how genotype frequencies are derived from allele frequencies, and Table 3.3 shows how selection against *lili* can be incorporated into the calculation. The proportion of the three genotypes remaining after selection is $1 - sq^2$, because:

$$p^2 + 2pq + q^2(1 - s) = p^2 + 2pq + q^2 - sq^2$$
$$= 1 - sq^2$$

Notice that there is no selection against hetero-zygotes in Table 3.3, so these form a reservoir of the *li* allele that is sheltered from selection and *li* will continue to segregate, producing some *lili* genotypes even after many generations of strong selection against them. Since many deleterious alleles are recessive, **shelter from selection** in heterozygotes is an important mechanism that preserves genetic diversity.

3.3.2 Types of selection

Natural selection can alter the frequency distribution of phenotypes in three basic ways, depending upon whether it operates against one tail of the distribution, both tails of the distribution or the mode (Fig. 3.7).

3.3.2.1 *Directional selection*

Selection against *lili* in Ennos' experiment with *T. repens* is an example of directional selection on a qualitative character. For metric traits, directional selection occurs when fitness is positively or negatively correlated with the trait value, favouring individuals at one end of the distribution of phenotypes and causing the mode of the character distribution to evolve a higher (or lower) value (Fig. 3.7a). For example, in a study by Fowler (1986a), fitness in the grass *Bouteloua rigidiseta* was correlated with size (see Fig. 1.4), so there was directional selection on this character. If there were additive genetic variance for plant size, this phenotypic selection would be translated into selection on the underlying genotypes. The strength and sign of directional selection on a

Table 3.3 Selection against the recessive homozygote: one locus with two alleles, *Li* and *li*

	Genotype			Total
	LiLi	*Lili*	*lili*	
Initial genotypic frequencies	p^2	$2pq$	q^2	1
Fitness (*W*)	1	1	$1 - s$	
Ratio after selection	p^2	$2pq$	$q^2(1 - s)$	$1 - sq^2$
Genotypic frequencies after selection	$\dfrac{p^2}{1 - sq^2}$	$\dfrac{2pq}{1 - sq^2}$	$\dfrac{q^2(1 - s)}{1 - sq^2}$	1

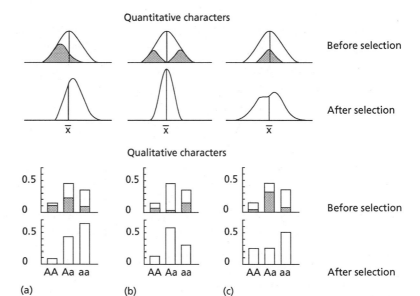

Fig. 3.7 Three modes of selection operating on quantitative and qualitative characters: (a) directional selection, (b) stabilizing selection, and (c) disruptive selection. Open areas are genotype frequencies, filled areas are genotypes selected against (from Endler 1986).

phenotypic character can be estimated from the regression (or partial regression) coefficient in a linear regression of trait values on fitness or one of its components (survival or reproduction).

3.3.2.2 Stabilizing selection

Stabilizing selection preserves the *status quo* because the individuals in the tails of the frequency distribution have lower fitness than individuals near the mode (Fig. 3.7b). Examples of this type of selection are rare (Endler 1986), no doubt because stabilizing selection has often already removed any additive genetic variance for the characters on which its action would be strongest. Stabilizing selection can be identified using the regression method by fitting a quadratic model (Lande & Arnold 1983). Using this approach, Widén (1991) detected stabilizing selection for flowering time in the grassland perennial *Senecio integrifolius*. The precise nature of phenotypic selection varied from year to year, and in one season it was entirely directional, favouring most the earliest plants to flower (Fig. 3.8). This environmental variability probably explains why Widén found that the population preserved some genetic variation for flowering time. Genotypes at a selective disadvantage in one year would be at a selective advantage in another.

One would expect stabilizing selection to operate most strongly, and consequently genetic variation to be reduced to the absolute minimum, in parts of the genome crucial to fitness. This is borne out by the extreme conservatism of the genes responsible for photosynthesis, which are carried in the chloroplast's own DNA. Complete sequencing of this DNA from tobacco, rice and a bryophyte has shown that they all possess the same genes, but these genes are uniquely absent from beechdrops *Epifagus virginiana*, which is a heterotrophic plant parasitic on beech trees. By contrast, chloroplast DNA in the hemiparasite *Striga asiatica* still has photosynthetic genes (Ranker 1991). It appears that only when photosynthesis becomes totally redundant is stabilizing selection on the underlying genes relaxed sufficiently to allow them to vary.

Stabilizing selection on qualitative characters with two alleles favours the heterozygote over either homozygote (Fig. 3.7b). This is known as **heterozygote advantage** and, unlike stabilizing selection on metric characters, has the effect of preserving genetic variation at the locus in question. It was once thought that this mechanism might be an important one in maintaining genetic diversity but, quite apart from certain theoretical problems (Lewontin 1974), it also appears to be too rare a phenomenon to make any major contri-

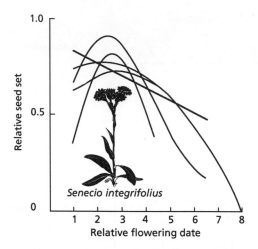

Fig. 3.8 Relationship of seed set (a component of fitness) to flowering time in a population of *Senecio integrifolius* in Sweden. Each line represents a different year. Straight lines indicate directional selection and curved lines indicate stabilizing selection (after Widén 1991).

that did occur between them would have produced progeny that were heterozygous at many loci too. It therefore seems likely that many more than three loci were responsible for the better survival associated with heterozygosity at the loci that were actually scored.

3.3.2.3 Disruptive selection

Disruptive selection occurs when individuals in the mode of the frequency distribution have lower fitness than individuals in both tails, causing the distribution to become bimodal (see Fig. 3.7c). Many examples of this have been found in plant populations occupying heterogeneous habitats where, despite countervailing gene flow, plants become adapted to local conditions because selection coefficients against non-adapted genotypes are very high. In 14 transplant experiments he surveyed in the literature, Waser (1992) found that the average selection coefficient against aliens was greater than 0.5, and that the distance separating subpopulations was often less than 10 m. Many of these populations straddled the boundaries of sites polluted with heavy metals (Bradshaw & McNeilly 1981), but other ecological variables that have been found to cause disruptive selection include herbivory (Schemske 1984; Polley & Detling 1990; Mopper *et al.* 1991), soil salinity (Antlfinger 1981), soil moisture (Farris 1987) and soil type (Nevo *et al.* 1988). Some

bution. Clegg and Allard (1973) reported heterozygote advantage at three electrophoretic loci in a wild population of *Avena barbata*, due to heterozygote seedlings surviving to adulthood up to 3.25 times better than homozygotes. The outcrossing rate in this population was extremely low ($\hat{t} = 0.02$), so most plants would have been homozygous at many loci and those few crosses

Table 3.4 The fitness of alien genotypes (W_A) relative to natives in some reciprocal transplant experiments between sites

Species	Mean W_A	Measure of fitness	Habitat	Source
ANNUALS				
Bromus tectorum	~1	λ	Steppe and forest	Rice & Mack (1991)
Impatiens pallida	0.65	l_x	Forest	Schemske (1984)
Phlox drummondii	0.57	λ	Grassland	Schmidt & Levin (1985)
Salicornia europaea agg.	0.32	λ	Salt marsh	Davy & Smith (1988)
PERENNIALS				
Anthoxanthum odoratum	1	m_x	Grassland	Plantenkamp (1990)
Dryas octopetala	0.53	l_x	Tundra	McGraw & Antonovics (1983)
Delphinium nelsonii	0.67	λ	Meadow	Waser & Price (1985)
Polemonium viscosum	0.17	l_x	Alpine tundra	Galen *et al.* (1991)
Ranunculus repens	0.57	Ramet production	Grassland/woodland	Lovett Doust (1981b)

examples are shown in Table 3.4. Notice that aliens are not always selected against.

3.3.3 Selection of variable direction

As we saw in Widén's study of flowering time in *S. integrifolius*, selection may vary in its strength or direction from one year to the next, and this is liable to preserve genetic variation for the traits in question. There are many other examples of selection varying in time. As plants colonize a site, conditions change and hence so may selective forces. Linhart (1988) detected genetic differences between cohorts within a ponderosa pine (*Pinus ponderosa*) population invading a grassland. Older trees, which would have been the first to colonize, had significantly different genetic constitutions from saplings established later. They differed at a peroxidase locus which shows high frequencies of allele *Per*-2 in cool conditions and allele *Per*-3 in warm conditions. The early arrivals, which had to get established in the relatively hot grassland, showed comparatively high frequencies of the *Per*-3 allele. In contrast, the saplings and seedlings, though they were the progeny of the early arrivals, had to establish under the cooler, closed canopy of the older trees and had a higher frequency of *Per*-2. Older age classes were also more heterozygous, suggesting that heterozygosity favoured survival.

Frequency-dependent selection is the most important kind of variable selection because it preserves genetic variation by relaxing selection against phenotypes that become rare, and increasing its toll of phenotypes that become too common. A simple example occurs in self-incompatible (SI) mating systems where the fecundity of an SI morph depends upon the frequency of compatible morphs. When a morph is rare it has many potential mates, when it is common it has fewer because so many individuals are like itself. The advantage this confers on rare SI morphs explains why one might expect selection to hinder the loss of morphs in tristylous species (see Section 3.2.5). If there are just two morphs, their equilibrium frequency is 50:50, because any deviation from this gives an advantage to the rarer one. Fisher (1958) argued that this is

the evolutionary explanation of 50:50 sex ratios (Chapter 9), and it also applies to the 50:50 ratio of pin and thrum flower morphs in populations of distylous plants such as *Primula vulgaris* (Chapter 2).

Interactions between plants and their parasites, particularly fungal pathogens, often involve frequency-dependent selection of a different kind. Flor (1956) discovered that flax *Linum usitatissimum* had 26 loci polymorphic for resistance or susceptibility to the flax rust *Melampsora lini*, and that there were the same number of loci in the pathogen population, which was polymorphic for virulence/avirulence against different flax genotypes. Whether a particular rust genotype is able to infect and multiply on a particular flax genotype depends upon the interaction in this **gene-for-gene** relationship (Table 3.5). This kind of relationship generates frequency-dependent selection upon host genotypes because when any particular genotype becomes common, it provides a pool of susceptible plants for those specific pathogen genotypes able to infect it. Other, resistant host genotypes are then favoured until their increase is limited by a similar frequency-dependent attack from their specific pathogens. In a situation like this one might expect a host genotype with dominant alleles at all resistance

Table 3.5 Genetic interactions in the gene-for-gene model for a plant host and a diploid pathogen, each with one locus. In the host, resistance *R* is dominant to susceptibility *r*, and in the pathogen, avirulence *V* is dominant to virulence *v*. No infection (−) occurs in any combinations of host and pathogen genotype that includes at least one *V* allele and one *R* allele. Infection is possible (+) for other combinations. When more than one locus is involved, a resistance at one locus will generally override susceptibility at another (after Burdon 1987)

		Host genotype		
		RR	*Rr*	*rr*
Pathogen genotype	*VV*	−	−	+
	Vv	−	−	+
	vv	+	+	+

loci to come out on top, but there appears to be a **cost of resistance** which lowers the fitness of genotypes with 'unnecessary' alleles for resistance to pathogen genotypes present at low frequency. Among pathogens there is a corresponding cost of virulence, so the gene frequencies of host and pathogen drive each other through cycles of boom and bust.

Gene-for-gene relationships have been demonstrated or postulated in at least 27 different host/pathogen systems affecting crop plants, but only in one wild system so far (Flor 1971; Burdon 1987; Burdon & Jarosz 1988). This may simply reflect the relative lack of knowledge about disease in wild plants, or it may imply that plant breeding for disease resistance has shaped gene-for-gene relationships out of natural systems that have a more metric basis. Frequency-dependence will occur in either case, and does not depend upon the dominance relationships (Table 3.5) typical of the gene-for-gene model (J.A. Barrett 1988). We shall see in Chapter 9 that frequency-dependent selection by pathogens may play a very important role in the evolution and maintenance of sexual reproduction.

3.4 THE GENETIC BASIS OF ECOLOGICAL AMPLITUDE

We have seen in this chapter how genetic structure is influenced by ecological processes, and how natural selection operating on additive genetic variance can lead to the adaptation of plants to their environment. So, the environment influences the genetic composition of populations, but how far does the genetic composition of a population influence the ecological abundance and distribution of a species, or what might be termed its **ecological amplitude**? To what extent does a species' ecological and geographical range depend upon its genetic diversity?

Because biological species are, by definition, genetically distinct from each other and because species also have distinct geographical and habitat distributions, we do know that genotypic differences at the species level have ecological effects. Although this is self-evidently true, the corollary that species' distributions have genetic limits is difficult to demonstrate. Circumstantial evidence

that genetic change can extend the range of a plant's distribution is seen in species, such as *Achillea millefolium* or *Rumex acetosella*, that include both diploid and derived polyploid populations. The polyploids are often more widely distributed than the ancestral diploids (Briggs & Walters 1984). Of course, if we do not know what the distribution limits of the diploid were before the polyploid appeared, such patterns could represent the displacement of diploids by polyploids, rather than an extension of the species' range caused by the appearance of a new genotype.

A clear example in which polyploidy was involved in range extension was the spread of a new genotype of cord grass *Spartina*, which appeared in Southampton, England in around 1890 following spontaneous hybridization between a native and an introduced species of *Spartina* there (Gray *et al.* 1990, 1991). A new species *Spartina anglica*, derived from the hybrid, colonized a coastal zone of mudflats where no *Spartina* was previously able to grow and it quickly spread around the coast of the British Isles.

A limited number of experimental studies have attempted to dissect a species' ecological niche or its ecological range along an experimental gradient, into genetic and phenotypic components. In essence, this is asking how far the reaction norms of different genotypes overlap. In separate experiments with *Abutilon theophrasti* (Garbutt & Bazzaz 1987) and *Phlox paniculata* (Garbutt *et al.* 1985) that were designed to answer this question by growing plants on experimental gradients of light, nutrients or moisture, there was in general a great deal of overlap between genotypes on all gradients. Using seed or flower number to measure plants' response to a light gradient, about 20% of the niche width of the population as a whole was due to genotype in both species. Genotype made a similar contribution to the niche breadth of *P. paniculata* on a moisture gradient, but no contribution at all to the response of *A. theophrasti* on a nutrient gradient.

In a similar study using *Phlox drummondii*, Schwaegerle and Bazzaz (1987) compared the reaction norms of plants, taken from nine different populations in Texas, along experimental gradients of moisture, light and nutrients. Reaction norms differed between populations, but the differences did not match variation in soil conditions among source sites. If the experimental gradients successfully mimicked field conditions, this suggests that differences in reaction norm were not adaptive and so it is difficult to see how they would have contributed very greatly to the ecological amplitude of the species.

Such as they are, these gradient experiments do not support very strongly the view that ecological amplitude has a large genetic component. The problem may be that simple, artificial gradients are not very relevant to the environmental variation that shapes reaction norms in field populations. By contrast, transplant experiments between field sites that show that genotypes common in one place are less fit in another do support the idea that genetic variation favours broad distribution, as well as the converse — that broad distribution increases genetic diversity. If this is true, then the corollary should also be true that species with limited genetic variation have limited distributions. Endemic plant species do typically have low levels of genetic variation, though it is not possible to say whether this is a cause or a consequence of their rarity: it could be both. However, there are also widespread, colonizing species such as *Echinochloa oryzoides* that are genetically depauperate.

The gap between population genetics and population ecology is still a significant one, and it forces us to make a deliberate conceptual jump between this chapter and the next. But the leap should not be blind, and it is worth noting a major difference between population genetics and population ecology before we reach the farther side. Genetics is mainly concerned with the *composition* of populations, so genotypes and alleles are usually measured as *frequencies*. In population ecology we are more interested in total *numbers*, and *density* is the usual measure. In the final section of this chapter we have briefly reviewed the evidence that genetic composition affects the numerical abundance of plants and, as you have seen, the evidence is largely circumstantial. That is the gap to cross.

3.5 SUMMARY

The large amount of genetic variation found in plant populations can only be explained if it is maintained by **natural selection**, or if some significant proportion of polymorphisms are **selectively neutral**. The **Hardy–Weinberg law** gives the frequencies of genotypes to be expected for a locus with two selectively neutral alleles. Deviations from **Hardy–Weinberg equilibrium** may theoretically be caused by **mutation**, or by linkage disequilibrium, non-random mating, genetic subdivision of the population, gene flow, genetic drift or natural selection. **Linkage disequilibrium** exists when alleles at two or more loci occur in combination at higher or lower frequencies than is to be expected from their individual frequencies. Physical linkage is one of the many possible causes of this. **Non-random mating** occurs when the phenotype of a plant determines with which other members of the population it may mate. Mating between like phenotypes is called **positive assortment** and is likely to increase inbreeding. Mating between unlike individuals is **disassortative mating** and is likely to reduce inbreeding.

Genetic subdivision of the population results when mating occurs among neighbours. The **genetic neighbourhood area** (A) is defined as the area centred on an individual, within which there is an 86.5% probability (one standard deviation from the mean) of finding its parents. The genetic neighbourhood area may be calculated from the variance of seed and pollen dispersal distances. **Effective population size** (N_e) is the average number of reproductive plants to be found within the genetic neighbourhood area and can be calculated from A, or by genetic analysis. **Gene flow** in plant populations is often restricted. This gives rise to genetic structure when N_e is significantly smaller than actual population size (n). If two or more subpopulations with different equilibrium allele frequencies are sampled as a single unit, genotype frequencies may *appear* to depart from Hardy–Weinberg equilibrium when they do not. This is termed the **Wahlund effect**.

Pollen movement is a major determinant of gene flow and is sensitive to, among other factors, pollinator foraging behaviour, plant density and wind direction. **Seed dispersal** is also influenced by the mode of animal dispersal. Most pollen and seeds do not travel far from their source, but the occasional dispersal event may carry a very few a long way. Dispersal may carry pollen or seeds to sites that vary in quality, so parental fitness is affected by where seeds and pollen are deposited as well as by the distance and quantity of dispersal. The genetic composition of populations founded by a limited number of individuals may exhibit a **founder effect**. When N_e is very small or alleles are rare the transmission of alleles from generation to generation is prone to sampling error that may cause **genetic drift**.

Only **natural selection** can produce adaptive evolutionary change. **Selection coefficients** may be calculated from transplant/replant exper-iments between sites. **Phenotypic selection** on a trait may be detected from the regression of trait value on fitness or one of its components. Three types of selection are distinguished: **directional selection** favours one tail of the frequency distri-bution of phenotypes, **stabilizing selection** favours the mode against the tails of the distribution, and **disruptive selection** favours both tails of the distribution against the mode. Selection may also vary in its direction with time, either because fitnesses change with environmental conditions or because fitnesses are **frequency-dependent**. One might expect the ecological and geographical range of species, or their **ecological amplitude**, to be influenced by genetic diversity, but convincing evidence for this is still lacking.

Chapter 4
Intraspecific Interactions

4.1 INTRODUCTION

Anyone who has grown a bed of vegetables, or has neglected one and turned around to find it filled with weeds, does not need to be told that plants will fill any available growing space. This observation is the basis of most plant population models. The underlying assumption is that the primary resource limiting plant growth is space. In arid or very cold habitats this may be false, but in most other situations it is a near enough approximation to the truth, because a plant can only acquire the essentials of life — light, water and nutrients — if it occupies the space to capture them. Space appears in population and genetic models in two guises, as an explicit statement of the area over which a plant is able to influence or be influenced by others (the ecological and the genetic neighbourhoods), and as density — the number of plants per unit area. We touch upon the relative merits of density and neighbourhood area as explanatory variables later in this chapter.

The simplest populations in which to study interactions between plants are monocultures, where differences between individuals are reduced to those found within the phenotypic range of one species. As we will see, this still leaves plenty of scope for differences between individuals, because plants are so phenotypically plastic. To reduce still further the complexities this creates, experiments are often carried out on monocultures of even-aged plants. Since plantation forestry and much of agriculture and horticulture operate with even-aged monocultures (so long as weeds can be kept out!), such populations are of practical as well as of theoretical interest. In fact, as we shall see, plant population biologists have learnt a great deal from experiments with crops.

4.2 YIELD AND DENSITY

As plants grow they occupy more and more space, and sooner or later the gaps between plants are filled and they begin to interfere with each other's access to resources. The total yield (weight of plant material per unit area) then approaches a ceiling whose level is determined by the species of plant and its growing conditions but which, above a threshold density, is *independent of plant density*. When the final yield of a crop is plotted against the density of plants sown we usually get a curve like that shown in Fig. 4.1a. The characteristic way in which this curve flattens off is referred to as the **law of constant final yield**. Remember that all plants need time to grow, so the full description of how yield changes with density should have a time dimension as well. If the crops planted at each density are followed through time the relationship between total yield and time follows a similar pattern to the yield/density curve (Fig. 4.1b). The higher the initial density of plants, the sooner a constant final yield is reached.

The asymptotic relationship between total yield Y and density N (Fig. 4.1a) can be described by a simple equation of the form:

$$Y = w_m N (1 + aN)^{-1} \qquad (4.1)$$

where w_m is the maximum potential biomass per plant and a is the area necessary to achieve w_m. At high densities $Y \propto w_m N (aN)^{-1}$ and density cancels out of the relationship to give final yield a constant value of $Y \propto w_m a^{-1}$. If final yield is constant it follows that mean yield *per plant* (w) must be inversely related to density (Fig. 4.1c). In fact: $w = Y/N$, so by dividing both sides of Eqn 4.1 by N we get an expression for the relation-

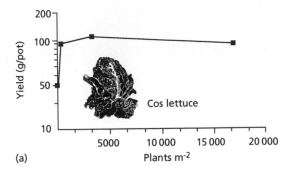

(a)

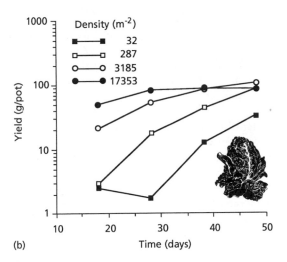

(b)

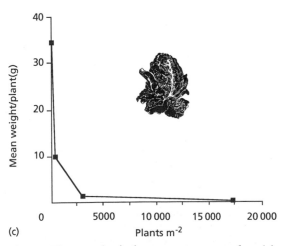

(c)

Fig. 4.1 The growth of a lettuce crop expressed as: (a) final yield from pots sown at different densities, (b) yield at different dates of harvest for different sowing densities, and (c) mean weight per plant at final harvest for different densities (from Scaife & Jones 1976).

ship between final yield per plant and density (Watkinson 1980):

$$w = w_m(1 + aN)^{-1} \qquad (4.2)$$

The curve of yield per plant corresponding to Eqn 4.2 has the hyperbolic shape shown in Fig. 4.1c.

The relationship between yield and density has been studied for many different crops, and while most conform to the law of constant final yield when the weight of the whole plant is measured (Fig. 4.2a, b, d), the yields of plant parts such as maize grain (Fig. 4.2c) or parsnip roots graded by size (Fig. 4.2d) may *decrease* at high density. Notice that the precise shape of the yield/density curve may vary with growing conditions (maize in Fig. 4.2c), and that the ceiling on the yield/density curve may vary from year to year (potatoes in Fig. 4.2b). Variation in the ceiling value of yield is due to variation in the maximum possible plant size w_m, but to allow for other *shapes* of yield/density curve we need to modify Eqns 4.1 & 4.2. This can be simply done by replacing the exponent -1 in the equations by a variable $-b$.

$$Y = w_m N(1 + aN)^{-b} \qquad (4.3)$$

$$w = w_m(1 + aN)^{-b} \qquad (4.4)$$

Equation 4.4 describes how mean plant weight (w) will vary between plots or populations at different densities at *a particular moment in time* (e.g. Fig. 4.1a). The relationship of w to density is called the **competition–density effect** or C–D effect (Firbank & Watkinson 1990).

The parameters w_m, a and b all vary during the growth of the crop (Watkinson 1984). When $b = 1$, these equations produce the curves appropriate for constant final yield (Fig. 4.1a, d). Biologically, b may be interpreted as the rate at which the effects of competition change with density (Vandermeer 1984). Figure 4.3 illustrates the effect of varying the value of b between 0.3 and 2 on total yield and on yield per plant. When $b = 1$, w is linearly related to the reciprocal of density ($w = Y/N$), but at lower and higher values of b this relationship is non-linear. Compared to when $b = 1$, at values of b greater than unity, yield per plant diminishes more as density rises, and at values less than unity yield per plant diminishes less as density

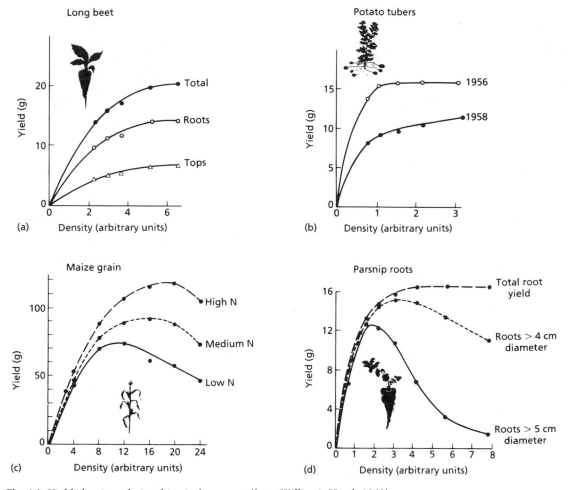

Fig. 4.2 Yield/density relationships in four crops (from Willey & Heath 1969).

rises. We will see when we come to Chapter 5 that these non-linear responses to density can have interesting effects upon plant population dynamics.

4.3 SELF-THINNING

The relationships between plant density and yield, described by Eqns 4.1–4.4, treat plant density N as a quantity that varies between populations, all recorded at the same time. This is appropriate for even-aged stands of plants that will be harvested or measured together, but it does not fully describe how w will vary *through time* in a population where N may change due to

mortality. This may be unimportant when dealing with annual crops deliberately sown at fairly low density by, say, a farmer who wants to avoid plant mortality and wastage of seed. But what of natural populations where there is no artificial limit upon the starting density, or perennial populations such as those of trees where the individuals start off very small and end up giants? If Eqn 4.4 is applied literally to *any* density, then an infinite density would produce infinitesimally small plants. Obviously this cannot happen, and at some point as density rises or crowding intensifies with plant growth, plants die reducing N with time. Because this decrease in N is due to crowding, it is described as **self-thinning**. Self-

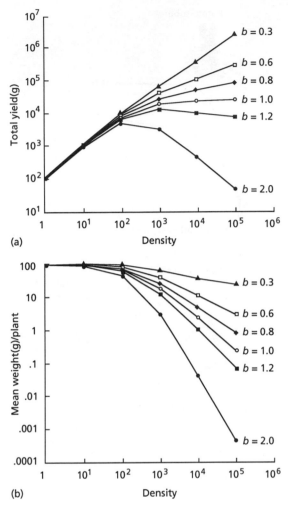

(a)

(b)

Fig. 4.3 Effect of varying the value of *b* when other variables are held constant in Eqns 4.3 & 4.4 upon: (a) total yield and (b) yield per plant.

thinking mortality is **density-dependent**. An example can be seen in Fig. 4.4 which follows the course of a number of populations through time.

Note that as time progresses *w* rises and then at a certain point *N* begins to fall. The fall in *N* happens soonest, and at lower values of *w*, in the most dense plots. Though often confused with it, self-thinning is quite distinct from the C–D effect. Self-thinning describes how density and yield change *in the same plot followed through time when density-dependent mortality is occurring*. It sets an upper boundary on the C–D effect

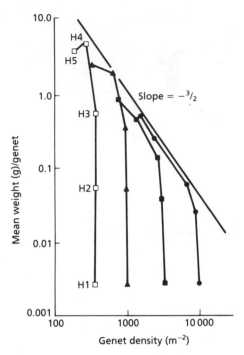

Fig. 4.4 Self-thinning in four populations of *Lolium perenne* planted at different densities, H1–H5 are replicates harvested at five successive intervals (from Kays & Harper 1974).

which describes the relationship at *one moment in time* between mean plant weight and density *across a range of plots*. Both of these characteristics can be seen in Fig. 4.4.

As mortality and growth proceed, the trajectory of populations on the self-thinning graph approaches a line of constant slope. There is an empirical relationship between mean plant weight *w* and surviving density *N* which is described by the equation:

$$w = cN^{-k} \tag{4.5}$$

The meaning of the terms in Eqn 4.5 is easier to see if we plot log *w* against log *N* (Fig. 4.5). Taking logs of both sides of the equation we get:

$$\log w = \log c - k \log N \tag{4.6}$$

We can see from this equation that −*k* is the slope of the self-thinning line in Fig. 4.5, and that log *c* is its intercept with the vertical axis. The term log *c* in Eqn 4.6 is a constant which has

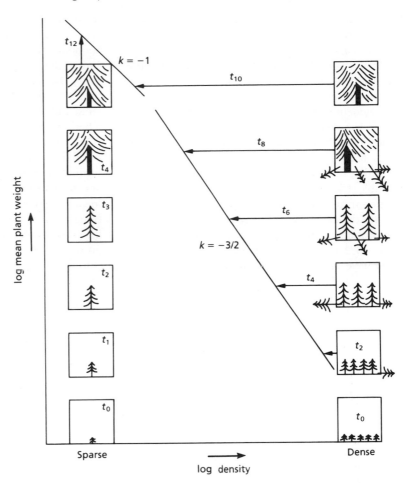

Fig. 4.5 The progress of a sparse and a dense tree population through time, illustrating the main features of the self-thinning process (Silvertown 1987a).

an empirical range of values between 3.5 and 5 (White 1985). Note that this is really quite wide, and that the range 3.5–5 on a log scale is nearly 10^5 when expressed on a linear scale. It is important to realize that mortality due to self-thinning often occurs *in the approach* to the thinning line, before it is actually reached (Osawa & Sugita 1989). The thinning line is a *boundary* and its slope (k) has a value of approximately 3/2 in many populations. Once the line described by Eqn 4.6 is reached by a population, it should follow it (e.g. Fig. 4.4). While following a line of slope $k = 3/2$, N is decreasing but w and total yield Y ($= Nw$) increase. Eventually the population hits a ceiling, and then the value of k changes to 1. From then on, Y is constant and any increase in mean plant size w is accompanied by

an equal and opposite decrease in plant density N, so that Nw does not change.

Yoda *et al.* (1963) originally proposed that the exponent in Eqn 4.5 had a constant value of $-3/2$ after they found that this relationship described self-thinning in experimental plots of a number of species, and since then the self-thinning rule has been known as the '**−3/2 power law**'. According to this law, self-thinning prevents populations crossing the $-3/2$ boundary.

In addition to the features of the self-thinning process already mentioned, Fig. 4.5 shows that dense populations reach the thinning line before sparse ones, and that it is possible for very sparse populations to reach maximum final yield (where $k = 1$) without any density-dependent mortality whatsoever. Below the self-thinning line the relationship between yield and density is governed by the competition–density effect discussed in Section 4.2.

Figure 4.6 shows in diagrammatic form the location of thinning lines for a whole range of monocultures of different species. Several things are evident from this graph. First, lines for species of enormously different size fall parallel to one another within a clearly defined band, and second, this band has a slope of approximately −3/2. Thirdly, of course, since this graph is plotted on double log axes it should be noted that these generalizations cannot be precise.

Are the deviations by particular species within the band shown in Fig. 4.6 significant? Critics of the −3/2 thinning rule have said that they are, and have questioned the universality of the rule

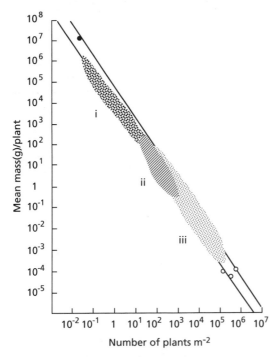

Fig. 4.6 Relationship between mean plant mass (w) and density (N) in monoculture populations of (i) trees, (ii) clonal perennial herbs, and (iii) annual and perennial herbs in experimental populations. Each shaded region includes many thinning lines, as well as points of maximum biomass (wN), such as those observed in field populations of *Sequoia sempervirens* (filled circle) and duckweed *Lemna* spp. (open circle). From White (1985).

(Zeide 1985, 1987; Weller 1987a, b, 1990), while others have pointed out that the rule describes a boundary to the behaviour of populations, not a course that all populations always follow (White 1985; Osawa & Sugita 1989). The existence of self-thinning and the inverse relationship between yield and density are not in doubt, but at issue is whether there is an 'ideal' thinning line of slope −3/2 that all populations would obey under standard conditions, whether there is some limited range of values for log c, and also whether deviations from these values have any biological meaning (Weller 1989). Answering these questions requires comparison of the values of k and log c in different populations. For these statistical purposes a slightly different formulation of the thinning rule, expressed in terms of total yield Y instead of mean plant weight w, is preferred (Westoby 1984; Weller 1987a; but see Prairie & Bird 1989). Since $Y = wN$, we can re-express Eqn 4.5 in terms of Y by multiplying both sides of the equation by N:

$$Y = cN^{-1/2} \qquad (4.7)$$

It is important to realize that this equation and the −3/2 power law are the same thing. Because the reader may come across Eqn 4.7 elsewhere we have given it here, but otherwise we use the original formulation of the power law which is easier to explain (see below).

In an excellent review of the evidence for many species, Lonsdale (1990) concluded that there are too many deviations from the expected value of $-k = -3/2$ for this value to be taken as a rule, but also that deviations from −3/2 are not easy to explain. It is notable that the deviations from −3/2 that Lonsdale (1990) found were fewer and slighter than those identified by critics of the rule, and that his deviant populations had thinning slopes *steeper* than −3/2. This is what is to be expected in thinning populations before they reach the −3/2 boundary (Osawa & Sugita 1989). We believe that a view consistent with all the present evidence is that, through time, self-thinning approaches a slope of −3/2, and that the intercept varies widely, depending on species and growing conditions (White 1985).

One reason for being reluctant to give up the idea that plants cannot transgress a boundary

.

of −3/2 is that Yoda *et al.* (1963) were able to provide a simple geometrical explanation for the rule. The weight of a plant (*w*) is proportional to the volume of space it occupies. This volume is measured by the *cube* of a linear dimension *l*³. Every plant volume sits upon an area, which is the *square* of a linear dimension *l*². When a plant fills the whole volume available to it in a given area, as it does in the crowded conditions where self-thinning occurs, then the ratio of weight (or volume) to area *cannot* exceed 3:2. This explanation fits well with Lonsdale's finding that when he took thinning lines for trees that were originally plotted using trunk size as a measure of *w* and replotted them using the canopy volume instead, slopes came nearer to −3/2. Burrows (1991) goes into the geometric explanation of the −3/2 thinning rule based on canopy volume in more detail than we have space for here, and argues that variation in the slope of the thinning line is to be expected.

If the geometrical explanation of the thinning rule is correct, then it should apply to mixtures of species as well as to monocultures. Though the experimental evidence available to test this is slight, White (1985) has found that tree mixtures and herb mixtures do fall within the boundaries that describe the thinning of monocultures in Fig. 4.6.

One group that might superficially appear to contravene the thinning rule is clonal plants, but stands of these are frequently composed of interconnected ramets. Genets can control rates of shoot production and the spatial distribution and density of their ramets, so as one might expect, densities are usually kept below the point that would cause mortality. When clonal plants do occasionally reach self-thinning densities their behaviour appears to be similar to that of non-clonal plants (Hutchings 1979; Pitelka 1984; White 1985).

4.4 SIZE VARIATION

Crowded populations typically begin life with a normal distribution of plant size, which quickly skews to an L-shaped distribution, with many small individuals and a few large ones (Fig. 4.7). When self-thinning occurs it is the smallest plants that are the first casualties. This of course contributes to the increase in mean plant weight (*w*) during self-thinning. In even-aged stands of trees where self-thinning continues over a long course of time, mortality may totally remove the smallest individuals, producing a more symmetrical size distribution once more (Fig. 4.8).

Why does size variability change in this way as plants grow? There are two simple alternative explanations. The first is based on the observation that plant growth is exponential during its early phase. If relative growth rate is normally distributed within a population, a rapid divergence can

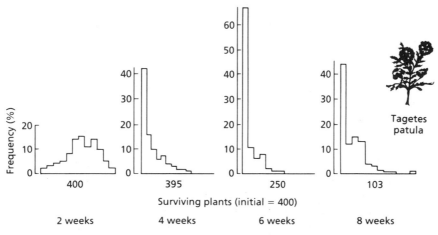

Fig. 4.7 The changing distribution of individual plant weight in a population of marigolds (*Tagetes patula*) with time. Frequencies are shown in 12 equal intervals in the range of weights present at each date (from Ford 1975).

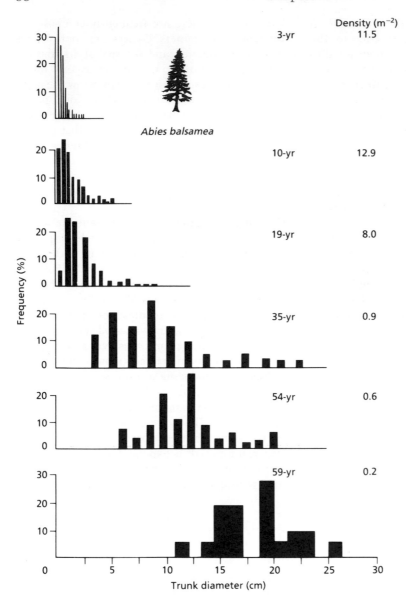

Density (m⁻²)

Fig. 4.8 Frequency distributions of trunk diameter for *Abies balsamea* stands divided into 12 equal intervals (from Mohler *et al.* 1978).

occur between plants with different growth rates, turning an originally normal size distribution into a log normal or L-shaped one.

An alternative explanation is that large plants suppress the growth of small ones. Of course it is possible that the truth lies somewhere between these two alternatives, but there is a simple observation which can tell us if one mechanism is more important than the other. If the change in plant size distribution with time is due to large plants suppressing small ones, it should not occur when plants are grown at low density. If, on the other hand, the change is due merely to exponential growth then it should occur in populations of all densities (Turner & Rabinowitz 1983).

To apply this test to the two hypotheses we need a measure of the size inequality in a plant population. The simplest effective one is the coefficient of variation (CV) of plant size, which is the standard deviation divided by the mean.

(This is highly correlated with another measure of inequality called the Gini Coefficient that is also often used (Weiner & Solbrig 1984).) In a survey of 16 experiments Weiner and Thomas (1986) found that in 14 size inequality increased with density. In the two tree species in their sample, *Abies balsamea* (Fig. 4.8) and *Pinus ponderosa*, size inequality fell once self-thinning took place because of the death of small individuals.

Clearly, size inequality in plant monocultures is strongly influenced by density and by intraspecific competition. A number of studies have thrown further light on the mechanism of this process by looking at the effects of neighbours on each other (Benjamin & Hardwick 1986). The effects of competitors on each other depend upon their relative size, but it is important to know precisely how competitive effects vary with size. When the effect of neighbours on each other is proportional to their relative sizes, competition is said to be **relative size symmetric**. When this kind of symmetric competition occurs, a plant weighing 100 g should reduce the growth of a 50 g neighbour by twice as much as the effect of the 50 g plant on the 100 g one. When the effect of neighbours on each other is disproportionate to their relative sizes, competition is said to be **asymmetric** (Weiner 1990). Firbank and Watkinson (1987) suggest that asymmetry is an essential part of the mechanism by which competition generates **size hierarchies**.

The extreme case of asymmetric competition is where competition is completely **one-sided** and the growth of a small plant is reduced by the presence of a larger neighbour, but the larger plant is unaffected by competition with the smaller. Cannell *et al.* (1984) found an example of one-sided asymmetry in plantations of sitka spruce *Picea sitchensis* and lodgepole pine *Pinus contorta*. Tall trees suppressed the height growth of shorter neighbours, but were themselves unaffected. However, height increment is not the most sensitive measure of growth for suppressed plants in dense stands because they etiolate. Using growth in stem volume rather than height as a measure of neighbour effects, Brand and Magnussen (1988) found competition in plantations of red pine *Pinus resinosa* growing in

South Ontario, Canada to be two-sided, though still asymmetric.

In what circumstances is plant competition asymmetric? Asymmetric competition is most likely to occur when light is the resource limiting plant growth, simply because shading of shorter plants by taller ones is one-sided. In a study with cultivated carrots Benjamin (1984) experimentally lowered the canopy of some of the plants in a population by pinning their normally erect leaves to the ground, while leaving other plants in the population with erect leaves. The root and shoot growth of pinned and unpinned plants in mixtures of the two treatments were compared with growth in populations with all plants pinned and with no plants pinned. Neither root nor shoot size at harvest was significantly affected by treatment when all plants in a treatment were treated the same way (pinned or unpinned), but in mixtures pinned plants did significantly worse than unpinned ones, and produced roots nearly 60% smaller than unpinned competitors. Asymmetry in competition for light was crucial in determining which plants grew large and which remained small.

A plant that, by being an exception, proves the rule that asymmetric shoot competition produces size inequality is the small annual saltmarsh herb *Salicornia europaea*. This plant has no leaves, but only a succulent stem and branches. When growing in dense stands of $10\,000\,\text{m}^{-2}$ or more it branches very little. By weeding out plants, Ellison (1987) thinned field populations to a range of densities at the beginning of the season and compared their size inequality at three harvests until the season's end. Size inequality was not related to density at any harvest, and there was no density-dependent mortality. *Salicornia*'s peculiar growth habit apparently minimizes the type of interference between shoots that generates asymmetric relationships between neighbours.

By contrast with competition for light, competition for soil resources is less likely to be one-sided or asymmetric. Weiner (1986) grew the climbing herb *Ipomoea tricolor* in experimental monocultures to compare the effects of root and shoot interference. Thanks to this species' twining habit he was able to separate the effects on size inequality of interference between roots and interference between shoots. Root competition was much more severe than shoot competition. Regardless of shoot competition, plants grown in pots where they shared root space were smaller than plants without root competition. Nevertheless, shoot competition and not root competition was the deciding factor in size inequality. There was significantly greater inequality among plants competing for light than those competing for soil resources. This is strong evidence that above-ground competition is asymmetric but below-ground competition may not be.

Although root competition has been frequently demonstrated, it has been suggested that its effects on small plants may be ameliorated by transfer of nutrients from larger plants through underground connections. This could either happen directly via root grafts, or indirectly via mycorrhizal connections. Root grafts are common in forest trees and intact neighbours can support the growth of stumps for several years after felling (Graham & Bormann 1966). The vast majority of vascular plants have mycorrhizal associations with fungi, and a large group of these — the endomycorrhizal fungi — tend to be non-specific in their host associations. The network of mycelia in the soil has been shown to channel mineral nutrients between plants of different size and different species, but it is not yet clear whether such transfers can cause a *net* transfer strong enough to affect population structure (Newman 1988; Eissenstat & Newman 1990). The population consequences of plant–plant interactions below ground are a subject ripe for investigation.

4.4.1 The influence of neighbours

It has been argued by Mack and Harper (1977) that, because plants interact with their neighbours, it is more appropriate to study interference in plant populations by the measurement of explicit spatial relationships than by measuring density, which is an average calculated over whole populations. Based as it is upon the undeniable fact that plants are rooted to the spot, there would appear to be a lot to be said for this argument. A common method used to quantify the space available to a plant surrounded by neighbours is to construct a map of the population and then to

divide the space into polygonal 'tiles' around each individual (Fig. 4.9).

In an experiment in which seeds of *Lapsana communis* were sown randomly Mithen *et al.* (1984) found that polygon area explained 60% of the variance in plant biomass, but the relationship between polygon area and plant size is more complicated than this would suggest. Franco and Harper (1988) planted seedlings of the annual herb *Kochia scoparia* in density gradients, but instead of the direct relationship between local density and plant size that one might expect, they found that plant size varied in wave fashion both along the gradients and across them. The largest plants were on the edge of the plot, these suppressed their immediate neighbours, and *their* neighbours were consequently that much bigger.

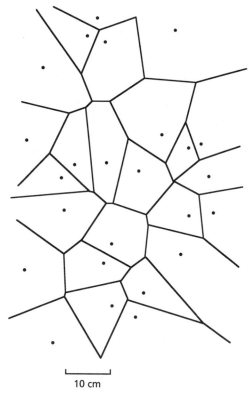

Fig. 4.9 A map of a plant population. Polygons were constructed around each individual by connecting all adjacent plants by a chord, bisecting each chord with a perpendicular line, and extending the perpendiculars until they met (from Mithen *et al.* 1984).

This effect rippled through the population, so that the size of a plant was always negatively correlated with that of its immediate neighbours and positively correlated with the size of neighbours one plant away. Such effects result from asymmetry in competition between neighbours and are accentuated by a density gradient, but they can occur along rows of plants sown at equal spacing too.

Waller (1981) mapped ramets of four *Viola* species in natural populations in New England and Canada, and found that the size of individuals was negatively correlated with the size and number of neighbours in only two out of 11 populations. This suggests that neighbours may have little influence on each other, but that conclusion would also be premature because asymmetric competition can cause small initial differences between competitors to become magnified, so loosening correlations between them. A small difference in the time of seedling emergence can be hugely magnified by asymmetric competition if seedlings are crowded. Also studying a species of violet in New England, Cook (1980) marked two cohorts of seedlings of *Viola blanda* that emerged 15 days apart in the same plots. The older plants remained consistently larger than the younger ones over the following 3 years and during an episode of severe mortality at the end of this period, older plants suffered significantly less mortality than plants only 15 days their junior in age. Mortality was size- rather than age-dependent. The fateful event that made some plants large and others small was the late emergence of one cohort in the midst of another 3 years previously.

The effect of **priority of emergence** on size and subsequent fate is an extremely common phenomenon that has been observed in a great many populations (Miller 1987). For example, emergence delays of only a day or two had an effect on the survival of the little sand-dune annual *Androsace septentrionalis* and 10 days made a difference to *Tragopogon heterospermus* (Fig. 4.10). Suggestive though these patterns are that older seedlings are suppressing younger ones, an experiment is needed to rule out completely the possibility that there is some other cause, such as a deterioration in the weather that disadvantages late-emerging seedlings. Weaver and Cavers (1979) carried out such an experiment in which they attempted to separate the effects of emergence *order* (who was first) from emergence *date* on the survival of successive cohorts of the dock *Rumex crispus*. Mortality was greatest in later cohorts, and emergence order was more important than emergence date in determining the contribution of each cohort to the final population.

The effect of emergence order on plant fate is largely due to the influence this has on plant size. This is greatly accentuated by cumulative differences in growth initiated by asymmetric competition between large plants and small ones. Later

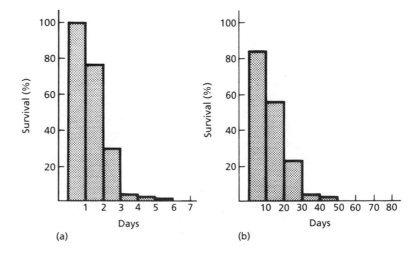

Fig. 4.10 Survival of seedlings from successive cohorts of: (a) *Androsace septentrionalis* emerging at daily intervals and (b) *Tragopogon heterospermus* emerging at 10-day intervals in natural populations on dunes in Poland (from Symonides 1977).

emerging seedlings are suppressed by interference from earlier ones, which have a greatly disproportionate advantage because competition for light is strongly asymmetric. Taken at face value, then, it would seem that the effects of neighbours should be detectable in the field, even where emergence order determines the size of individual plants, because it is size that really matters. However, things are not as simple as this. Although emergence order may *cause* size differences, asymmetry in competition for light makes the interaction between emergence order and size a strongly non-linear one.

Waller's (1981) finding that the size of neighbours was only rarely negatively correlated is not unusual. In experiments with the grass *Dactylis glomerata* sown in a random arrangement Ross and Harper (1972) found that emergence date accounted for 95% of the variance in final harvest weight of individuals, and that there was no consistent relationship whatsoever between the size of an individual and the size and number of its neighbours. They suspected that the effects of neighbours were obscured by being confounded with the effects of emergence time.

Using a computer simulation of neighbourhood competition in which competition from neighbours and emergence time were *known* to be the only variables affecting individual plant size, Firbank and Watkinson (1987) found that the variance in plant size explained by these two factors *decreased* as the asymmetry of competitive interactions and, in particular, the variance in emergence time *increased*. This confirms Ross and Harper's (1972) suspicion that the effect that neighbours have on each other is obscured by a non-linear interaction between emergence order and interference.

Although emergence time is of course important, asymmetric competition may also magnify other causes of initial difference in plant size. In experiments with velvetleaf *Abutilon theophrasti* sown in heterogeneous and homogeneous seed beds at three densities, Hartgerink and Bazzaz (1984) found that aspects of soil heterogeneity explained up to 76% of the variance in final seed output of plants, but they found no effect of neighbours, although there must have been interference between them. The effect of soil heterogeneity in this experiment was found to be greater than the effect of neighbourhood competition on plant size in experiments with the same species by Pacala and Silander (1987).

4.5 MORTALITY AND SPATIAL PATTERN

The neighbourhood approach has not yet produced many new insights that are unobtainable using density as a measure of plant crowding in monocultures. Pacala and Silander (1990), for example, found that neighbourhood models of *Abutilon theophrasti* were no more accurate predictors of this annual's population dynamics than were models based upon density. Although density does quite well as a predictor of how plants react in a crowd, the spatial pattern of a population is of some interest in its own right. We already know, from looking at size-frequency distributions, that density-dependent mortality is not random but that the grim reaper takes small plants in preference to large ones. Is this mortality spatially random, or are some small plants more susceptible than others? The spatial pattern of shrubs, such as the creosote bush *Larrea tridentata* growing in deserts, has been studied by many authors, interested in what it can tell them about intraspecific competition (e.g. Woodell *et al.* 1969; Phillips & MacMahon 1981; Wright 1982). The canopies of adjacent shrubs rarely meet, although their extensive root systems often do, so competition between neighbours, when it occurs, must be for below-ground resources — most probably water. The argument goes that the intensity of competition between individuals should be greatest between the closest neighbours. If this is intense enough, then individuals with close neighbours will have a higher mortality rate than individuals with more distant ones, and over time the distribution of individuals in the population will acquire a more regular spatial arrangement.

The crucial word in the last sentence is *more*. A once-off analysis of pattern can show regularity, but unless you know from what previous pattern this developed, any inference about the process that produced it is severely weakened, and a failure to find a regular pattern does not mean competition is absent. Ideally, spatial pattern should be measured before and after an episode of mortality. The distributions of young and old or live and dead plants in desert populations have been compared by a small number of authors who have usually found patterns that suggest competitive interactions. For example, Wright (1982) found that living plants of the perennial *Eriogonum inflatum* were more regularly spaced than a sample including dead individuals. Most studies have not made such a comparison. In a review of the extensive literature on plant competition and spacing in deserts, Fowler (1986b) found that only seven out of 33 species studied exhibited regular distributions, and in five of the seven, some populations were clumped or random. By contrast, it is noteworthy that many studies have found positive correlations between plant size and distance to nearest neighbour (Fowler 1986b).

As one might expect, self-thinning, which is driven by above-ground competition for light, alters the pattern of a population, tending to produce a more regular spatial distribution with time (Cannell *et al.* 1984; Mithen *et al.* 1984; Bonan 1988). Because this equalizes the space around surviving plants, the correlation between plant size and neighbourhood effects must become weaker, making it difficult or impossible to detect the influence of neighbours in retrospect if only living plants can be mapped. Kenkel (1988) compared the spatial distributions of living and dead trees in a natural stand of Jack pine *Pinus banksiana* at a site in Canada. The population was even-aged and had been recruited after a fire some 65 years before the study. Dead stumps of *P. banksiana* are slow to decay, so although there was no remaining clue as to the location of the first and smallest trees to die, there were still 916 stumps and standing dead trees to 459 living ones. The spatial distribution of the whole population, living and dead, was random but the distribution of survivors was highly regular and the distribution of dead trees was clumped. Surviving trees exerted a detectable influence on the fate of neighbours within a 3.5 m radius, which on average included six other trees. A polygonal tile (Fig. 4.9) has an average of six sides, suggesting that trees were competing only with immediately adjacent neighbours.

Another comparison of spatial patterns at two stages of the life cycle was made by Sterner *et al.* (1986) for four species of trees, two of them palms, in a tropical evergreen forest at La Selva, Costa Rica. By extreme contrast with the *P. banksiana* study site, La Selva is a species-rich forest and populations of the species studied contained individuals of all ages. In three of the four species adults were more regularly distributed than juveniles, suggesting that mortality was spatially non-random. A random pattern of juvenile mortality was applied to the map of living plants, and the adult spatial distribution was then compared with this simulation. This comparison showed that the spatial distribution of adults was more regular than that of randomly thinned juveniles in the same three species. Although all of the species in this study were abundant, every tree must have had neighbours of other species, greatly complicating the possible neighbourhood interactions. The fact that an analysis of pattern that ignored other neighbour species found evidence in three of four cases that mortality was spatially non-random is strong evidence for the role of intraspecific interference, be it real or apparent competition (Connell 1990). We will return to the implications of this for tropical forest diversity in Chapter 8.

4.6 SIZE, DENSITY AND FITNESS

The fitness of an individual vlant is strongly related to its size. Small plants not only bear the brunt of mortality in crowded populations but, if they survive, also produce fewer seeds than large plants. If the relationship between plant size and fecundity is linear, the L-shaped size distribution that develops in a dense population will create an L-shaped distribution of fitness. Samson and Werk (1986) examined the relationship between plant size and the biomass of reproductive parts in 30 species, most of them annuals, and found that the relationship varied with species. However, in perennials, fecundity per unit plant weight often increases with plant size, with large plants producing more seeds per unit biomass than small ones (Fig. 4.11). When this is the case, the hierarchy of plant fitness will be even stronger and more unequal than the hierarchy of plant size.

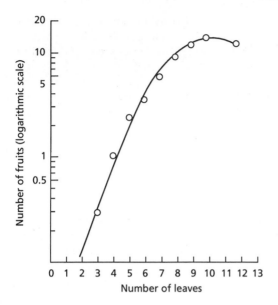

Fig. 4.11 Relationship between plant size and fruit production for *Viola fimbriatula* (from Solbrig *et al.* 1988a).

Hierarchies of plant fitness created by crowding reduce the effective population size N_e because a small fraction of the population makes a disproportionately large contribution to future generations. This has important genetic consequences. Levin and Wilson (1978) and Wilson and Levin (1986) used simulation models to compare the effects of L-shaped and normal fecundity distributions on a number of genetic processes. They found that a fecundity hierarchy resulted in faster response to selection, faster extinction of rare alleles and a greater propensity for genetic drift.

Heywood (1986) calculated N_e for 34 species of annuals, using estimates of the variance in seed crop size to take account of the unequal genetic contributions of different individuals, and compared these N_e values with those to be expected when all individuals contribute equally to the gamete pool. In most cases allowing for the variance in seed crop size reduced N_e by more than half, and substantially increased the potential for genetic drift. However, N_e was probably still overestimated because variance is a poor measure of the inequality in crop size when this is not normally distributed.

A plant's response to density is affected by its shape, size and pattern of growth. If these characters are heritable, density-dependent processes may be selective (Bazzaz *et al.* 1982). Davy and Smith (1985) transplanted *Salicornia europaea* between subpopulations in the upper and lower levels of a saltmarsh in Norfolk, England and found that among several genetic differences between the two levels, plants showed different responses of fecundity to density. Transplants from either level to the other had lower fitness than natives replanted into home ground and the degree of disadvantage varied with density. Families of *Salvia lyrata* from different parents growing in a meadow in North Carolina were planted back into the home turf at a range of densities by Shaw (1987) who found evidence for differential growth in response to density. These experiments suggest that intraspecific competition may be a source of natural selection. However, the evidence is weak that plant size, which is a major determinant of individual success in crowded populations, has high heritability.

Gottlieb (1977) compared the allozyme genotypes of large and small individuals in a dense, natural population of the annual *Stephanomeria exigua* and found no evidence that size correlated with genotype. Antlfinger *et al.* (1985) estimated the heritability of characters associated with plant size and shape in *Viola sororia* and found heritabilities between 0.09–0.39. Since these measures were made in a greenhouse environment they tell us little about the heritability of plant size in the field where it matters. A field study of the annual *Erigeron annus* found that the broad-sense heritability of seedling size was less than 0.1 (Stratton 1992). Low heritabilities for characters closely correlated with fitness are, of course, to be expected (Chapter 3).

Dense plant stands are particularly susceptible to fungal disease. In experimental mixtures of susceptible and resistant genotypes of skeleton weeds *Chondrilla juncea*, Burdon *et al.* (1984) found that the presence or absence of the rust *Puccinia chondrillana* determined which genotype was at the top of the size hierarchy and which at the bottom (Fig. 4.12). In disease-free mixtures the resistant genotype was inferior to the susceptible one, demonstrating that there

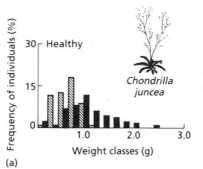
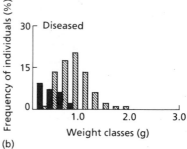

Fig. 4.12 The frequency distribution of plant size for resistant (hatched columns) and susceptible (filled columns) genotypes of *Chondrilla juncea* in: (a) healthy and (b) diseased stands (from Burdon *et al.* 1984).

was a 'cost of resistance' (Harlan 1976). A cost of resistance lowers the relative fitness of resistant genotypes in populations no longer exposed to disease, resulting in selection against them unless disease reappears (see Section 3.3.3).

4.7 POPULATION REGULATION

We have seen that crowding affects individual plants by limiting their growth, that this determines mean plant size and size *variation*, and that in turn plant size affects individual mortality and fecundity. Effects of density on fecundity and mortality affect individual fitness. We will now see that density also has an important influence on population dynamics. Recall from Chapter 1 that we used the following equation to predict population size N_t from its initial size N_{t-1}:

$$N_t = N_{t-1} B - D + I - E$$

When $B + I > D + E$ the population will increase at a rate N_t/N_{t-1} each generation. Now, consider what will happen if N_t/N_{t-1} has a fixed value. First, if $N_t/N_{t-1} > 1$ the population will increase indefinitely at an exponential rate. This is obviously impossible because such a population must eventually run out of space. Second, if $N_t/N_{t-1} < 1$ the population will decline rapidly to extinction and third, if $N_t/N_{t-1} = 1$ exactly the population will, *in theory*, remain stable in size. Actually though, for the last case, random variation in B and D make it impossible for N_t/N_{t-1} to remain forever at unity. To stay at this value $B + I$ must *always exactly* match $D + E$, but this is very unlikely. A severe attack by some herbivore or disease will raise D or decrease B which,

without a compensatory change by the other parameters, will eventually lead to extinction. Therefore, we must conclude that *no fixed* value of N_t/N_{t-1}, be it more than one, less than one *or* exactly one, can adequately account for how real populations behave for long. On the other hand for most populations, the *average* value of R_0 measured over long periods should be unity, unless the population is heading for extinction. Given the hazards of life, how do populations hang on with an average $N_t/N_{t-1} = 1$?

There are two ways in which populations can be cushioned from the consequences of catastrophes that alter B and D. Either (i) by the immigration of plants from areas not affected by the catastrophe; or (ii) by changes in fecundity or mortality that compensate for the losses caused. Compensatory changes operate in a **density-dependent** manner, altering the value of R_0 so that overall density is **regulated** within upper and lower limits.

We have already seen many examples of how B and D change with density. For changes in B and D to *regulate* a population they *must* be density-dependent. Density-dependent mortality increases the per capita death rate as density rises and density-dependent fecundity decreases the per capita birth rate as density rises, with the net result that a wide range of starting densities is reduced to a narrower range of final population size (Fig. 4.13). We will indicate that we are using *rates* of birth, death, immigration and emigration by the italic letters *B*, *D*, *I* and *E*. The calculation of these rates is explained on pp. 73–74.

Notice that the experimental populations of *Bromus tectorum* in Fig. 4.13b obey the law of

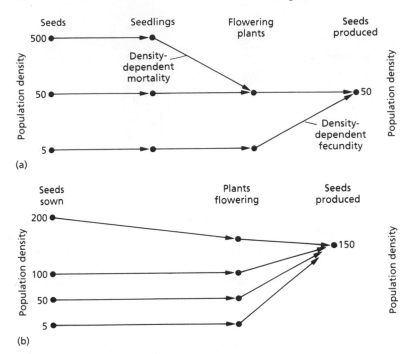

Fig. 4.13 Population regulation by density-dependent mortality and fecundity in: (a) idealized plant populations and (b) experimental populations of an annual grass *Bromus tectorum* (from Silvertown 1987a).

constant final yield: all populations produced the same total number of seeds, regardless of the sowing density. The law of constant final yield and population regulation are simply different facets of a single process: the response of plants to density. There is an important distinction between the two, however, because final yield is expressed in terms of plant size (or weight) whereas the effects of population regulation depend not upon biomass, but upon seed production. So long as seed production remains a constant fraction of total biomass this distinction is unimportant, but in some plants the fraction of biomass allocated to reproduction alters with size (Samson & Werk 1986). Intraspecific competition is not the only source of density-dependence, and *any* factor that operates in a density-dependent manner can regulate a population. Infection by fungal diseases such as damping-off in seedlings is commonly density-dependent (see Fig. 4.16c) (Burdon & Chilvers 1975, 1982).

The effectiveness of an alteration in birth or death rates in regulating a population depends upon how strongly these rates react to density. This can be determined from a graph of *B* or *D* against density. Large ranges of density are normally required to show an effect, so double-log axes are used. If changes in vital rates are density-independent, the graph will be a straight line of zero slope (Fig. 4.14a). The intercept of this line on the y axis tells you what the average value of the density-independent rate is. If changes in *B* or *D* are density-dependent, the graph will have a positive slope for mortality or a negative slope for fecundity. The strength of the regulatory effect is determined by the absolute value of the slope (the steepness) of the line (Fig. 4.14b–d). It is of course possible for birth or death rates to be inversely density-dependent too (Fig. 4.14e).

Only when density-dependence *exactly compensates* for changes in density is a population perfectly regulated and will it exhibit constant population size between one generation and the next. A population where density-dependence is not exactly compensating may fluctuate from one generation to the next, but this does not mean that the population is not regulated.

A population is at **equilibrium** when the death rate and birth rate are equal. The equilibrium density of a population can therefore be determined from the intersection of the death rate and birth rate curves in Fig. 4.15. As this shows, a change in

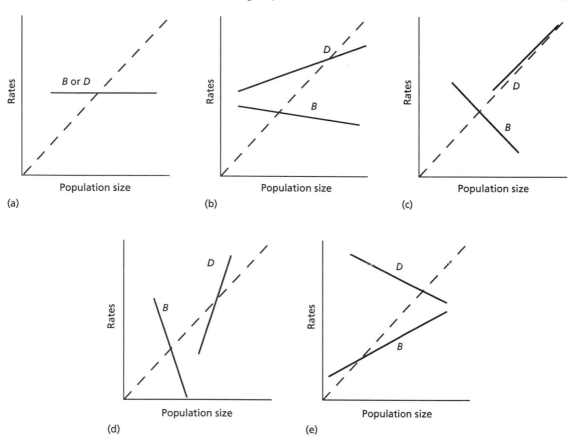

Fig. 4.14 Birth (*B*) and death (*D*) rates that are: (a) density-independent, (b) density-dependent and undercompensate, (c) density-dependent and exactly compensate, (d) density-dependent and overcompensate, and (e) inversely density-dependent.

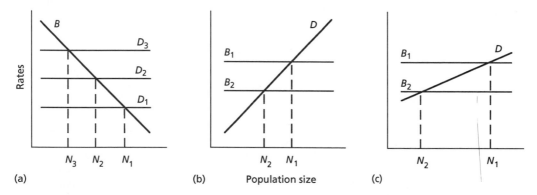

Fig. 4.15 A variety of ways in which birth and death rates interact to determine equilibrium population size *N*. (a) When the birth rate is density-dependent, a density-independent change in death rates (D_1 to D_3) will shift *N*. (b) A density-independent change in birth rates (B_1 to B_2) when the death rate is density-dependent will also change *N*. (c) If the strength of density-dependence (shown by the slope of *D*) is reduced, the same density-independent change B_1 to B_2 will have a bigger effect upon *N* (after Watkinson 1986).

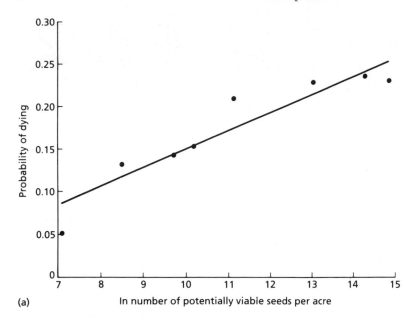

(a)

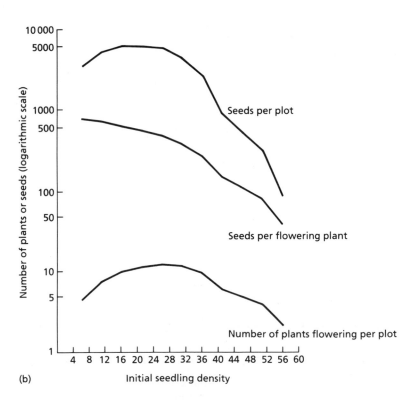

(b)

Fig. 4.16 Density-dependent responses in: (a) seedling mortality in a population of sugar maple *Acer saccharum* (Hett 1971; 1 acre = 0.4 ha), (b) fecundity in the annual *Erophila verna* (data from Symonides 1983a, b), (c) the proportion of *Lepidium sativum* seedlings infected by the damping-off fungus *Pythium irregulare* 7 days after sowing and experimental inoculation with the disease (from Burdon & Chilvers 1975).

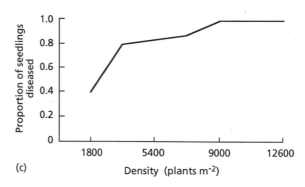

Fig. 4.16 *Contd.* (c)

one parameter by a fixed amount, for example the density-independent birth rate, can lead to widely different population densities depending upon the degree of regulation (the slope of the curve) by the other (e.g. the death rate).

We have argued for the existence of population regulation from first principles, on the grounds that unregulated populations would go extinct. In the field there are essentially two ways that population regulation can be detected. The first is by looking for changes in B and D that correlate with natural fluctuations in population density through time or differences in density between plots (Fig. 4.16a, b). Since regulated populations are cushioned against changes in density, observation alone is not a very effective way of measuring density-dependence because the range of observable densities will be small. It is paradoxical but true that density-dependence is most difficult to detect in those populations that are most strongly regulated.

The second, and better, method of detecting population regulation in the field is to alter density experimentally, and to measure the response of N, B and D to the change. A good example that shows how density-dependent and density-independent factors combine to regulate a population, and how they may create different equilibria at different sites, is provided in the experimental study by Keddy (1981, 1982) of sea rocket *Cakile edentula*, an annual species growing on a sand-dune in Nova Scotia. There were three sites: on the seaward side of the dune at the top of the beach, near the dune crest and on the landward side of the dune (Fig. 4.17). A range of densities was sown at each site and survival and fecundity

were determined for each of them. Using Fig. 4.14 as a guide, you should be able to interpret the results of the experiment shown in Fig. 4.17 for yourself.

Plants were large at the seaward end where there was no competing vegetation and where there was decaying eelgrass that supplied nitrogen. Here, increasing density resulted in a plastic adjustment of plant size and fecundity without causing mortality. Plants were small at the landward end and many of them were too small to reproduce. At this site, increasing density did not alter fecundity but did increase mortality. We shall look at further examples of population regulation in the general context of plant population dynamics in the next chapter.

4.8 SUMMARY

Most population models are based on the assumption that **space** is the primary resource limiting plant growth. At the level of the individual, a plant interacts with others in its **neighbourhood**, but **density** (N, plants per unit area) usually gives a good guide to the average behaviour of plants in a crowded population. At high density the total yield from a crop tends to approach a constant value. This is known as the **law of constant final yield** (Eqn 4.1), and from it we can develop an expression (Eqn 4.4) to describe how mean plant weight (w) varies between plots growing at different densities, called the **competition–density effect** (or C–D effect). Intraspecific competition at very high densities leads to **density-dependent mortality** or **self-thinning**. As plants in very crowded stands grow, mean plant weight rises

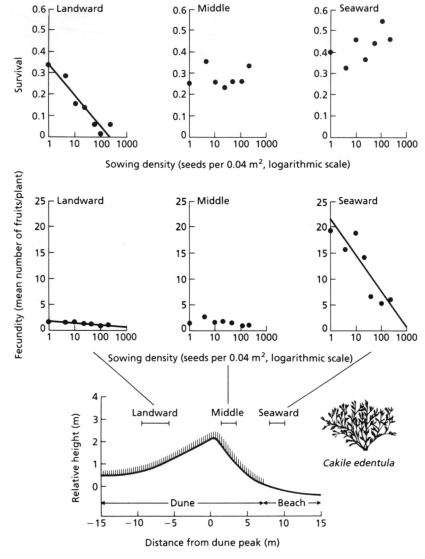

Fig. 4.17 The relationship of fecundity and mortality to density in experimental plots sown with *Cakile edentula* at three sites on a sand-dune. Statistically significant density-dependent relationships are shown by a regression line. There is no significant difference in levels of density-independent mortality at the three sites but there is a significant difference in the level of density-independent fecundity. The size and fecundity of solitary plants are far greater on the beach than elsewhere (from Keddy 1981).

and mean density falls. This relationship takes the form $w = cN^{-k}$ (Eqn 4.5), where c and k are constants. The value of k is often in the region of 3/2, and hence Eqn 4.5 has been called the **−3/2 power law**, though its generality has been questioned.

Competition in crowded stands initially increases the **inequality** of size among members of the population, producing a **size hierarchy**. When self-thinning occurs the smallest plants are usually the first to die, and this may eventually reduce size variation. Size hierarchies tend to develop when the effect of neighbours on each other is disproportionate to their relative size, and competition is **asymmetric**. Competition for light is usually asymmetric, but competition for soil resources is more likely to be symmetric. Asymmetric competition magnifies slight initial differences between crowded seedlings, such as in **priority of emergence**, into much larger differences in plant size. This can weaken the correlation that might otherwise be expected between the size of neighbours, but nevertheless intraspecific competition can produce non-random patterns among neighbours in the field.

The **fitness** of a plant is strongly related to its size, but because seed output may be non-linearly related to total biomass, hierarchies of seed output and fitness can be even more marked than inequalities in size itself. These fitness hierarchies reduce N_e because a few plants make a quite disproportionate contribution to the total seed production of a population. Population size is **regulated** by **density-dependent** processes that affect mortality and fecundity. Density-dependence often arises from intraspecific competition, but other causes such as **disease** may also operate in this manner. The absolute abundance of a population is set by the interaction of **density-independent** factors with density-dependent ones. Population regulation is most readily detected in the field by the experimental manipulation of density.

Chapter 5
Population Dynamics

5.1 INTRODUCTION

Plant populations are dynamic. You may think that this is the least we can tell you in a chapter with this title, but its literal truth has only been grasped fully in the last 25 years. As we have seen in the last chapter, plants occupy space to its limit and so plant communities usually appear full. If you live in the temperate zone or humid tropics, a cursory glance at a lawn or pasture that you last saw a year ago will probably suggest that little has changed. Nothing could be further from the truth. Grass populations have high turnover rates, and it is likely that hardly a single tiller of last year's population is still alive. All will have been replaced. This will also be true of more than half the buttercups (*Ranunculus* spp.) and other broad-leaved herbs. Because these populations are regulated, their total numbers are relatively constant, despite the flux of individuals through them (Fig. 5.1).

Marking or mapping individual tillers or rosettes in permanent plots will tell you what the turnover rates are for ramets, and how net population sizes change. What it will not tell you so easily for most species is how many genets have been lost from the population, because it is usually difficult to tell which ramets belong to which genet. If, as is sometimes the case, seedlings are difficult to distinguish from small ramets, genet birth rates may also be difficult to estimate. The genetic variation in a plant population resides in genets, so we must know genet birth and death rates to discover whether these rates affect genotypes differentially. Of course if they do, we have evidence of natural selection.

Unravelling the intricacies of plant population dynamics will take us three chapters. We will start by constructing a simple model of population dynamics using the demographic parameters you met in Chapter 1 and the yield/density equation introduced in the last chapter. This will be applied to some examples of the simplest kind of population — annuals with no pool of dormant seed. Next, we shall look at the ecology of seeds in the soil and then, in Chapter 6, we shall see how the timing of germination and the pattern of recruitment generate age structure in populations. Chapter 6 also examines populations whose dynamics are influenced by age or size structure, and Chapter 7 looks at spatial structure and metapopulations.

5.2 DEMOGRAPHIC PARAMETERS

As we saw in Chapter 1, the simplest statement one can make about how numbers in a population change between one occasion and the next is that the number at time $t + 1$ is equal to the number at time t plus the number born over the time interval (plus any immigrants), minus the number that died (minus any emigrants). Using B for births, D for deaths, I for immigrants and E for emigrants, this gives us the equation:

$$N_{t+1} = N_t + B - D + I - E \qquad (5.1)$$

Technically this kind of equation is known as a **difference equation**. Population dynamics becomes more complicated than Eqn 5.1 (or we could end this chapter right here!), but this equation is fundamental to the population models used, so you should remind yourself of it if ever things look too complicated. In this section we will see how much can be learned about a population from the basic demographic parameters given in Eqn 5.1.

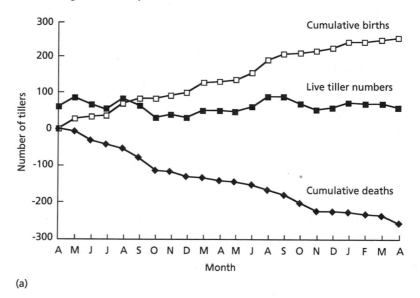

(a)

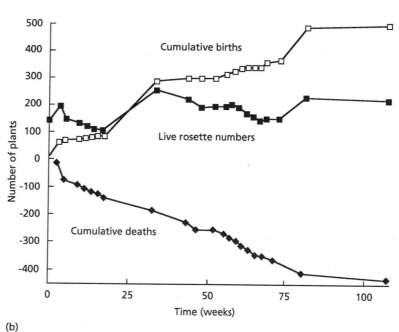

Fig. 5.1 Population flux in two grassland species: (a) *Lolium perenne* (J. Bullock, B. Clear Hill & J. Silvertown, unpublished data), and (b) *Ranunculus repens* (Sarukhán & Harper 1973).

(b)

The ratio N_{t+1}/N_t calculated from the values in Eqn 5.1 is the **annual rate of increase** (λ) (Chapter 1). When the population is stable, neither increasing nor decreasing in size, by definition $N_{t+1} = N_t$ and therefore $\lambda = 1$. When this is true it follows that $B - D + I - E = 0$, and of course $B + I = D + E$. In fact the number of births and deaths in a population tends to be related to N_t, so it is better to express these parameters in terms of *rates*, or in other words numbers per head of population per time interval. We will use the letters B, D, I, E to signify *rates* of birth, death, immigration and emigration respectively. They are defined as follows:

$$B = \frac{B}{N_t} \qquad D = \frac{D}{N_t} \qquad E = \frac{E}{N_t} \qquad I = \frac{I}{N_t}$$

Using these definitions Eqn 5.1 can be rewritten:

$$N_{t+1} = N_t(1 + B - D + I - E) \qquad (5.2)$$

Divide both sides of Eqn 5.2 by N_t and, remembering $\lambda = N_{t+1}/N_t$ you will find:

$$\lambda = 1 + B - D + I - E \qquad (5.3)$$

When $B + I > D + E$ the population will multiply by λ each year. This increase is **exponential**. To calculate N after x years, for a population that starts with N_t individuals, we use the equation that describes exponential increase:

$$N_{t+x} = N_t e^{rx} \qquad (5.4)$$

where e is the base of natural logarithms and r is the **intrinsic rate of natural increase** of the population. In truth there is nothing very 'intrinsic' about r, because it is affected by all the ecological influences that alter B, D, I, E. If we let $x = 1$, divide both sides of Eqn 5.4 by N_t and express the result in terms of natural logarithms we get:

$$\ln \lambda = r \qquad (5.5)$$

Sometimes it is more useful to measure the change in N_t over the length of a generation rather than over one year. A **generation** is defined as the average time τ (pronounced 'tau') between a mother first giving birth and her daughter doing

the same. The rate of population increase over a generation is called the **net reproductive rate** R_0 and is:

$$R_0 = \frac{N_{t+\tau}}{N_t} \qquad (5.6)$$

Since R_0 is measured over τ years, the *annual* rate of increase is given by R_0 root τ:

$$\lambda = R_0^{1/\tau} \qquad (5.7)$$

R_0 may also be calculated from a life table (Chapter 1) using the formula:

$$R_0 = \Sigma l_x m_x \qquad (5.8)$$

No population can increase exponentially for long, for even quite low values of r or values of λ significantly greater than one lead to very rapid population growth which must eventually be checked by lack of suitable habitat, if nothing else. In plant populations we are most likely to see exponential population growth when a species has newly arrived in an uncolonized habitat, such as happened at the end of the Pleistocene when trees were able to reinvade higher latitudes (Fig. 5.2a), or when a weed is allowed to let rip without control in a crop (Fig. 5.2b).

With the definitions of λ and R_0 in our hands, we are now equipped with the basic tools to begin an investigation of the dynamics of some real populations.

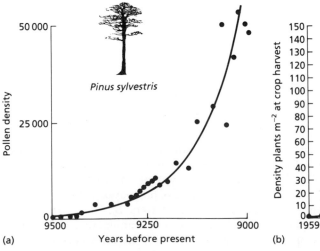

(a)

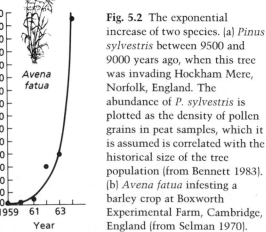

(b)

Fig. 5.2 The exponential increase of two species. (a) *Pinus sylvestris* between 9500 and 9000 years ago, when this tree was invading Hockham Mere, Norfolk, England. The abundance of *P. sylvestris* is plotted as the density of pollen grains in peat samples, which it is assumed is correlated with the historical size of the tree population (from Bennett 1983). (b) *Avena fatua* infesting a barley crop at Boxworth Experimental Farm, Cambridge, England (from Selman 1970).

5.3 ANNUALS WITH
NO PERSISTENT SEED POOL

Annual plants have a generation time of 1 year, so $R_0 = \lambda$. The dynamics of such a simple population can be described with the equation $N_{t+1} = R_0 N_t$, but if R_0 has a fixed value greater than one this relationship only applies at low density when there is sufficient space for the population to increase exponentially. To be realistic, as density rises the value of R_0 must fall. We can quite easily construct an equation to describe this, based upon what we have already learned about yield/density relationships in Chapter 4. There we used this equation to describe the density-dependent adjustment of yield:

$$Y = w_m N (1 + aN)^{-b} \tag{4.3}$$

Seeds are a component of plant yield. The number per plant is a function of plant size and the total number is a function of plant density. Therefore an equation of similar form to Eqn 4.3 can describe the total seed output of a population (Watkinson 1980). First, remember that $Y = wN$, so Eqn 4.3 can be written:

$$wN = w_m N (1 + aN)^{-b} \tag{5.9}$$

Next, we translate the w terms representing plant size in the equation into seed quantities. Let us say that a plant of maximum size w_m produces s_m seeds, and that the seed output of

the average plant w is s. Replacing the weight terms in Eqn 5.9 with their equivalents in seed production we get:

$$sN = s_m N (1 + aN)^{-b} \tag{5.10}$$

The left hand side of this equation is the seed output of a population at density N. If we assume for simplicity that there is no seed mortality and all of them germinate, then this also represents the size of the next year: $sN = N_{t+1}$. Placing a subscript t on N in Eqn 5.10 to make it clear which generation we are referring to, we now have an expression for the density-dependent dynamics of our population:

$$N_{t+1} = s_m N_t (1 + aN_t)^{-b} \tag{5.11}$$

In Chapter 4 we remarked that the yield/density response for whole plants and for their parts may be different. Seeds are only a part of total plant yield, so we should not expect the parameters a and b in Eqn 5.11 to have the same values as they do in Eqn 4.3, even for the same populations. The parameters for yield and for seed production must be separately determined. Parameters for the seed production of four species of annuals, including populations of *Cakile edentula* at the three sites studied by Keddy (Chapter 4), are given in Table 5.1. Figure 4.3 illustrated the relationship between the value of the parameter b in the yield/density equation (Eqn 4.3) and the shape of the curve relating total yield to density. Because Eqn 5.11

Table 5.1 Estimates for the parameters of Eqns 5.8 & 5.9 for four species of annuals. Calculated by Watkinson and Davy (1985) from the sources indicated below

Species	s_m	a	b	m
*Cakile edentula**				
Seaward site	20.8	1.17×10^{-3}	0.43	—
Middle site	1.97	7.86×10^{-3}	0.28	—
Landward site	2.21	1	0.14	—
Rhinanthus serotinus†	19	3.59×10^{-3}	1.11	—
Salicornia europaea‡				
Low marsh	1100	7.57×10^{-4}	1.63	8×10^{-5}
High marsh	33.8	9.23×10^{-4}	0.36	—
Vulpia fasciculata§	3.05	1.93×10^{-4}	0.90	10^{-6}

Sources of data: *Keddy (1981); †ter Borg (1979); ‡Jefferies *et al.* (1981); §Watkinson & Harper (1978)

has the same form as this relationship, Fig. 4.3 also shows how the shape of the curve for the graph of N_{t+1} versus N_t varies with the value of b. Refer to Fig. 4.3 now, and you will see that when $b = 1$, seed production behaves like the law of constant final yield: a constant seed output is achieved, regardless of density. This is **exactly compensating density-dependence**.

Most of the populations in Table 5.1 have values of $b < 1$ and exhibit **undercompensation**, but it is notable that two populations have values of $b > 1$. Referring to Fig. 4.3 you will see that when $b > 1$ the relationship between N_{t+1} and N_t has a hump, and that beyond a certain density total seed yield (N_{t+1}) actually drops. We will see in a moment that this produces some very interesting dynamics. However, we need to make two further modifications to Eqn 5.11. Firstly, we must allow for **density-independent mortality**. This can be included by assuming that only a fraction of the total s_m seeds that an uncrowded plant can produce will survive density-independent mortality, so we will replace s_m by s. Secondly, Eqn 5.11 only accounts for one kind of density-dependent response — the adjustment of fecundity with density — but, as we know, at very high density another response occurs too — **density-dependent mortality**. This can be incorporated by dividing sN_t by an additional term msN_t, where m represents the reciprocal of the maximum value that density can achieve, limited by self-thinning (Watkinson 1980). Adding these extra terms to Eqn 5.11 gives:

$$N_{t+1} = \frac{sN_t}{(1 + aN)^b + msN_t} \qquad (5.12)$$

Only two of the populations in Table 5.1 achieved densities sufficient to cause density-dependent mortality, but the values of m calculated for these populations will be useful as a means of investigating the effect of self-thinning upon population dynamics. We now introduce a simple graphical method that can be used to predict how a population whose dynamics is described by Eqn 5.12, or any function relating N_{t+1} to N_t, will behave. In Fig. 5.3a N_t is represented on the horizontal axis of the graph and N_{t+1} on its vertical axis. The diagonal line shown in Fig. 5.3a passes through all points where $N_{t+1} = N_t$. Any

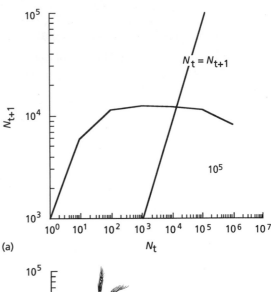

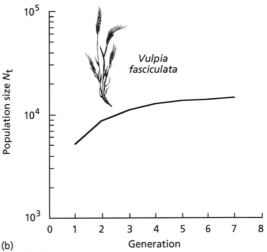

Fig. 5.3 Density-dependent population dynamics of *Vulpia fasciculata*: (a) the recruitment curve obtained by applying the values in Table 5.1 to Eqn 5.12, and (b) dynamics of a population starting with $N = 5000$ and no density-independent mortality, predicted from the recruitment curve.

population falling on this line will by definition be at equilibrium at that point. The curved line in Fig. 5.3a shows the relationship between N_{t+1} and N_t for the grass *Vulpia fasciculata*, calculated by plugging the values for s_m, a, b and m given in Table 5.1 into Eqn 5.12. This curve, which we shall call the **recruitment curve**, can be used to predict how the population will behave through time because the slope of the curve, which is

N_{t+1}/N_t, gives us the value of λ for any value of N_t. Box 5.1 describes how to use a method called **cobwebbing** to predict the behaviour of a population from its recruitment curve. You should read this now, before proceeding.

Analysis of recruitment curves by cobwebbing (Box 5.1) shows that the behaviour of a population is affected radically by the slope of the recruitment curve at the point where it is intersected by the diagonal line $N_{t+1} = N_t$. Unlike the simple

---

Box 5.1 Cobwebbing the recruitment curve

For simplicity in this example we will use a recruitment curve that is a straight line, representing a population with a slope of 0.5. The equation for this recruitment curve is $N_{t+1} = C + 0.5N_t$, where C is the intercept of the curve on the vertical axis and has a value of 2.5. We will start the population off with $N_t = 9$ individuals.

Step 1: *Find N_{t+1}* by drawing a vertical line drawn from the starting value of N_t until it meets the recruitment curve. A horizontal line from this point on the recruitment curve to the vertical axis gives us the value of N_{t+1}, which is 7.

Step 2: *Locate the next value of N_t on the horizontal axis* by retracing a line from $N_{t+1} = 7$ to the diagonal and then vertically from where the line meets the diagonal, down to the horizontal axis. This pinpoints the new value of $N_t = 7$.

Step 3: *Find N_{t+2}* by repeating steps 1 and 2, starting from $N_t = 7$.

Step 4: *Locate the equilibrium point* by continuing to repeat steps 1 and 2 until $N_{t+1} = N_t$, which is at the point where the diagonal and the recruitment curve intersect.

A short cut can be taken by repeating the procedure just described, but by drawing in just the lines that fall between the recruitment curve and the diagonal. These lines summarize the trajectory that will be followed by a population starting at $N_t = 9$ or $N_t = 1$.

Graph the dynamics of the population by plotting the values you have found for N_t against t.

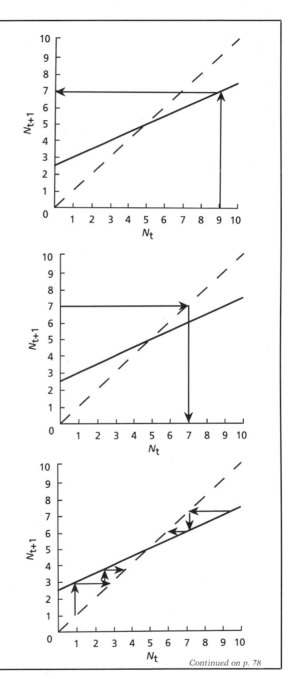

Continued on p. 78

Box 5.1 *Contd*

Other patterns of population behaviour can be seen by cobwebbing recruitment curves where the slope and intercept of the recruitment curve have other values, for example:

Slope = 1.5, $C = -2.5$, the equilibrium point in this population is unstable. For values of $N_t > 5$ the population will increase exponentially, and for values of $N_t < 5$ it will decrease to extinction.

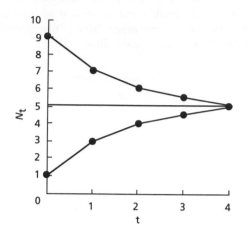

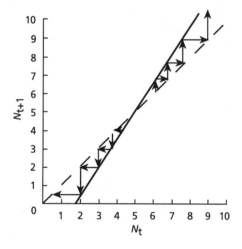

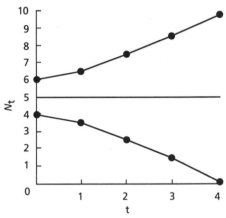

Slope = -1, $C = 10$, there is no stable equilibrium in this population, and the population cycles between $N_t = 9$ and $N_t = 1$.

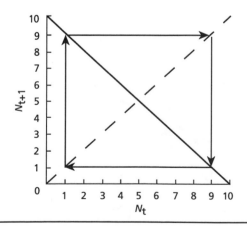

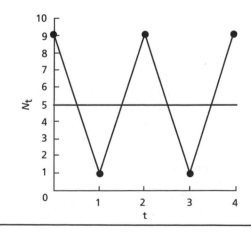

examples given in Box 5.1, real recruitment curves vary their slope with density. If the recruitment curve is intersected where its slope lies between 0 and +1 (i.e. between the horizontal and the diagonal), as is the case for example in *Vulpia fasciculata*, deviations in population size caused by some mortality factor such as herbivory will be followed by a smooth return to equilibrium (Fig. 5.3). An intersection where the slope lies between 0 and −1 leads to **stable oscillations**, as is seen in *Erophila verna* (Fig. 5.4a). When the slope is less than −1, unstable oscillations occur which are described as **chaos**. Chaotic oscillations are very difficult indeed to tell apart from random fluctuations caused by density-independent factors, yet this kind of analysis shows that they are entirely due to deterministic density dependence. The dynamic behaviour of a population can therefore be predicted from the mean value of seed production *s* and the exponent *b* in Eqn 5.12 describing its recruitment curve (Fig. 5.5).

Recruitment curves can be used to investigate how changes in the parameters of Eqn 5.12 will affect a population's dynamics and stability, and to examine the effects of various ecological factors on this. We shall look at two examples: the effect of self-thinning and the effect of density-independent mortality. Because of its peculiar morphology, the saltmarsh annual *Salicornia europaea* can reach extremely high densities without self-thinning (Ellison 1987; Chapter 4), although some density-dependent mortality may have occurred on the low marsh studied by Jefferies *et al.* (1981). Watkinson and Davy (1985) estimated that the reciprocal of maximum plant density, *m*, had a value of 8×10^{-5} (Table 5.1). In Fig. 5.6 we have compared the dynamics of this population with and without self-thinning. With self-thinning the population is stable, but without it the population cycles. This is a particularly interesting finding because it may account for the population cycles that have been reported in a related species *Salicornia patula* in Poland (Symonides 1988).

The level of density-independent mortality affecting a population with a humped recruitment curve may determine whether or not the population cycles. The very regular 2-year population

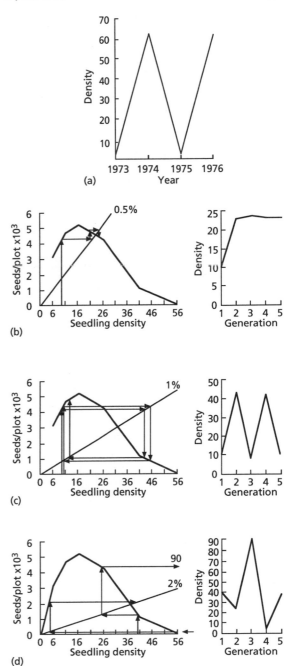

(a)

(b)

(c)

(d)

Fig. 5.4 (a) Regular 2-year cycles of abundance observed in some parts of a population of *Erophila verna*, (b–d) the dynamics of the *Erophila verna* population predicted from its recruitment curve when the germination rate (*p*) is: (b) 0.5%, (c) 1%, and (d) 2% (from Symonides *et al.* 1986).

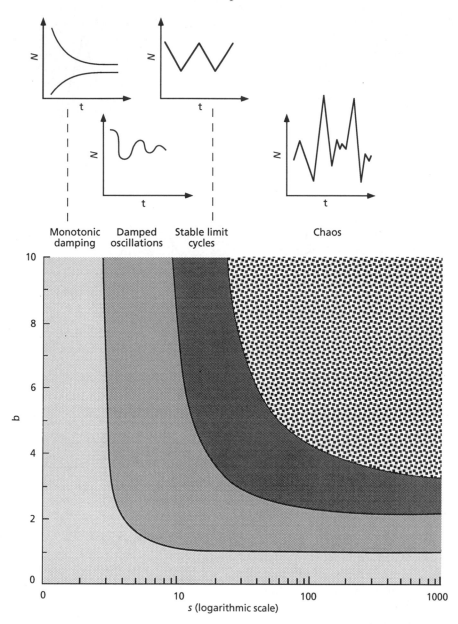

Fig. 5.5 The effect of s and b in Eqn 5.12 on the dynamic behaviour of a population (after Hassell *et al.* 1976).

cycles of *Erophila verna* shown in Fig. 5.4a were found by Symonides (1984) who studied the species on inland sand-dunes in Poland. She found that parts of the population cycled and parts did not. Symonides *et al.* (1986) analysed this population using a recruitment curve where N_t was the initial seedling density and N_{t+1} was the seed

output. The slope of the diagonal that intersects this particular kind of curve represents the proportion of seeds that germinate successfully, or in other words the value s in Eqn 5.12. By varying the slope of the diagonal to represent different rates of germination it was found that when $p = 0.5\%$ the intersection fell on a part of the curve

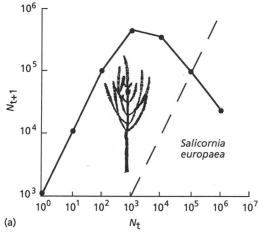

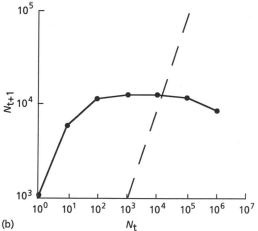

Fig. 5.6 Recruitment curves for *Salicornia europaea*: (a) with self-thinning, $m = 8 \times 10^{-5}$ and (b) without, $m = 0$ (after Silvertown 1991a).

quite a small difference in a density-independent factor such as germination can have a major influence on how a population behaves when it interacts with overcompensating density-dependence.

In the latter part of this section we have concentrated somewhat on populations with humped recruitment curves because these illustrate a number of important general points. It should not be imagined, however, that curves with this shape are very common, or that all plants that have this kind of recruitment curve will cycle. Pacala and Silander (1985) found a humped recruitment curve for the annual old-field weed *Abutilon theophrasti* but discovered that in nature the cycles they expected to find were damped by seed dormancy (Thrall *et al.* 1989). Though an annual plant in its growing phase may lack age structure, the existence of a seed bank complicates matters by introducing one.

Even if it is strongly regulated, a population cannot be totally protected from extinction by density-dependence. Populations of sand-dune annuals such as *Erophila verna* and *Vulpia fasciculata* are ultimately doomed to local extinction as succession proceeds and perennials usurp their space. Nine years from the start of the study, *V. fasciculata* went extinct in the permanent plots where Watkinson originally censused it (Watkinson 1990). Annual species without a seed pool persist in the long term only by colonizing new territory, but this is a very precarious existence. In the particular case of *V. fasciculata* seeds must be buried by wind-blown sand for seedlings to stand a chance of successful establishment. Watkinson (1990) found that *V. fasciculata* had values of $\lambda > 1$ only in plots that were at least 50% covered by sand. For values of $\lambda < 1$ equilibrium population size is, of course, zero. This created a threshold in the relationship between per cent sand cover and equilibrium population size (Fig. 5.7) that would suggest that the density of *V. fasciculata* should be very patchy, and that small changes in the vegetation cover that stabilize sand movement can produce very large changes in the population density of this annual. The behaviour and fate of seeds has important consequences for the dynamics of many plants, so we shall now consider seeds in some detail

where no cycles would occur, but if germination success was 1% or greater, cycles similar to those observed in parts of the field population were predicted (Fig. 5.4). Differences in germination success of this order between shaded and unshaded microsites in the *E. verna* population are easy to imagine and would explain why parts of the population cycled and parts did not. Unfortunately spatial variation in germination success was not measured, but this population analysis illustrates how a population model can help identify the important parameters that should be investigated in the field. The analysis also demonstrates that

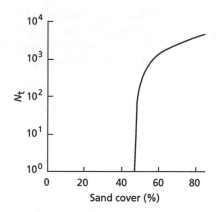

Fig. 5.7 The relationship between percentage area covered by sand and the expected equilibrium population density of *Vulpia fasciculata* (from Watkinson 1990).

before looking at the dynamics of such populations.

5.4 SEEDS IN THE SOIL

While in the soil only two things matter to seeds: the risk of death and the chance to germinate. The irony is that chancing germination also risks death. Dormancy is the safest condition for a seed, but a dormant seed bears no offspring. Germination is always risky, but only a plant that reproduces can transmit its genes and multiply. In Chapter 10 we address some evolutionary strategies that enable plants to resolve this problem. Here we look in some detail at what goes on in seed populations.

Seed populations in soil are often described as a '**seed bank**', a term which gives the erroneous impression that this is a place of safekeeping for plant genotypes. In fact the seed bank is constantly plundered by predators and is highly prone to attack by fungi. Even when they germinate, few seeds make a successful escape. Plants which do invest their progeny in the seed bank retrieve only a tiny fraction of them, and each of these is worth (in terms of fitness) much less than a seed which has successfully produced progeny of its own rather than remain dormant in the soil. In effect the soil is like a bank which offers little protection from thieves or kidnappers and which guarantees that the value of what is

left of a deposit when it is withdrawn will be severely eroded by inflation. A more neutral term for 'seed bank' is a **seed pool**.

5.4.1 Numbers and distribution

Despite its subterranean perils, the soil beneath most kinds of vegetation is replete with seeds. The number of seeds in the soil is determined by the rate of input in the seed rain, the rates of loss due to fungal disease and predation, and of course losses due to germination. The fate through the year of cohorts of seeds belonging to two grassland herbs is shown in Fig. 5.8. The buttercup *Ranunculus repens*, studied by Sarukhán (1974) in a field in North Wales, lost half its seed pool to rodents in the first 6 months. Few seeds had germinated by the end of the 15-month study, though none of the survivors were innately dormant. Pavone and Reader (1982) studied the legume *Medicago lupulina* in a pasture in South Ontario, Canada. By contrast with *R. repens*, seeds of *M. lupulina* were dormant when they entered the seed pool. Most deaths occurred during or after germination, and at the end of 9 months the 20% of the initial seed cohort still remaining in the pool was innately dormant.

With the rare exception of a few apomicts, such as pathenogenetic lineages of the dandelion *Taraxacum officinale*, seeds are always sexual products. Unlike the plants in many populations above ground, every individual in a seed population is unequivocally a genet. Above ground, a single genet may in time cover a large area, but below ground genet density increases with time as seeds accumulate and may reach densities in the tens of thousands per square metre.

Seed densities are highest in frequently disturbed habitats such as arable fields, and lowest in primary forest (Table 5.2). In all habitats the species most heavily represented in the soil tend to be those with the shortest lifespan in the vegetation, because these are the species that produce large numbers of small seeds, which are frequently capable of dormancy. Thus, the seeds in the soil beneath perennial vegetation such as woodland tend to be unrepresentative of the species above ground (Fig. 5.9a), and consist mostly of the seeds of herbs, pioneer trees and

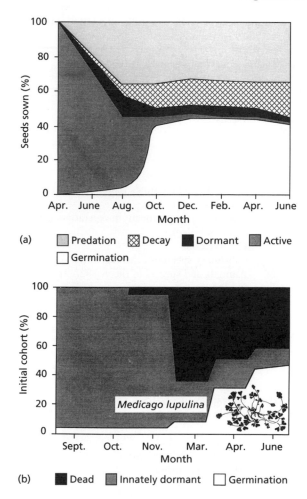

(a) □ Predation ⊠ Decay ■ Dormant ▨ Active
□ Germination

(b) ■ Dead ▨ Innately dormant □ Germination

Fig. 5.8 The dynamics of seeds in the soil for two herb species: (a) *Ranunculus repens* (from Sarukhán 1974) and (b) *Medicago lupulina* (from Pavone & Reader 1982).

the variance in the number of growing plants. Such correlations can be used as a basis for predicting weed infestations, but are only reliable if the cultivation regime is consistent over a period of years as a result of different crops and practices favouring different weeds (Wilson *et al.* 1985).

The spatial distribution of seeds is inherently patchy because most fall near the parent plant (Fig. 5.10) (Chapter 3). Unfortunately, the patchiness of the seed pool is rarely considered in quantitative assessments, and consequently values such as those in Table 5.2 have wide margins of error and may be quite inaccurate. Patchy distributions should be sampled by many small samples rather than by a few large ones (Roberts 1981; Bigwood & Inouye 1988).

Dispersal agents carry seeds away from the parent, but these do not necessarily make distributions more homogeneous because dispersed seeds are often deposited in clumps. In the Sonoran Desert where occasional sheet flows of water and wind as well as animals transport seeds, Reichman (1984) found extreme densities of seeds concentrated in shallow depressions in the soil surface. Local concentrations of seeds are also created by birds and rodents that cache their food supplies, and ants that carry seeds to their nests (Brown *et al.* 1979; Howe & Smallwood 1982). Seeds that survive passage through an animal's gut are deposited in clumps with its faeces. Cow pats from a pasture in a forest clearance in Amazonia contained concentrations of grass and sedge seeds 20 times greater than the density of seeds of the same plants in the soil (Uhl & Clark 1983). Grant (1983) found that 70% of the seedlings appearing in grassland plots at a site in North Wales emerged on the site of worm casts. In a Dutch pasture studied by Reest and Rogaar (1988), germinable seeds of annual meadow grass *Poa annua* and rush *Juncus* spp. were 15 times more concentrated in worm casts sampled from earthworm burrows between 5–15 cm below the surface than in ordinary soil at the same depth.

Animals are important in the burial of seeds and the formation of a dormant seed pool. Reest and Rogaar (1988) calculated that earthworm activity transported about 20% of the grassland seed pool from the soil surface to a depth below

shrubs. On the other hand the seeds beneath arable fields infested with annual weeds may be quite representative (Fig. 5.9b). Debaeke (1988) compared the population sizes of 10 annual broad-leaved weeds with the size of their respective seed pools in fields of winter wheat near Paris, France. Even though no species had a germination rate higher than 14%, of the seeds present, the correlation between numbers above and below ground was statistically significant in all of the 10 species. In half of the species the size of the seed pool accounted for more than 85% of

Chapter 5

Table 5.2 Numbers of seeds and the predominant species present in the seed pools of various vegetation types (from Silvertown 1987a)

Vegetation type	Location	Seeds m^{-2}	Predominant species in the soil
TILLED AGRICULTURAL SOILS			
Arable fields	England	28 700−34 100	Weeds
Arable fields	Canada	5000−23 000	Weeds
Arable fields	Minnesota, USA	1000−40 000	Weeds
Arable fields	Honduras	7620	Weeds
GRASSLAND, HEATH AND MARSH			
Freshwater marsh	New Jersey, USA	6405−32 000	Annuals and perennials representative of the surface vegetation
Saltmarsh	Wales	31−566	Sea rush where abundant in vegetation, grasses
Calluna heath	Wales	17 500	*Calluna vulgaris*
Perennial hay meadow	Wales	38 000	Dicotyledons
Meadow steppe (perennial)	CIS	18 875−19 625	Subsidiary species of the vegetation
Perennial pasture	England	2000−17 000	Annuals and species of the vegetation
Prairie grassland	Kansas, USA	300−800	Subsidiary species of the vegetation, many annuals
Zoysia grassland	Japan	1980	*Zoysia japonica*
Miscanthus grassland	Japan	18 780	*Miscanthus sinensis*
Annual grassland	California, USA	9000−54 000	Annual grasses
Pasture in cleared forest	Venezuela	1250	Grasses and dicot weeds
FORESTS			
Picea abies (100 year old)	CIS	1200−5000	All earlier successional spp.
Secondary forest	North Carolina, USA	1200−13 200	Arable weeds and spp. of early succession
Primary subalpine conifer forest	Colorado, USA	3−53	Herbs
Subarctic pine/birch forest	Canada	0	No viable seeds present
Coniferous forest	Canada	1000	Alder *Alnus rubra*
Primary conifer forest	Canada	206	Shrubs and herbs
Primary tropical forest	Thailand	40−182	Pioneer trees and shrubs
Primary tropical forest	Venezuela	180−200	Pioneer trees and shrubs
Primary tropical forest	Costa Rica	742	Pioneer trees and shrubs

4 cm (Fig. 5.11). Although most seeds occur near the soil surface, redistribution by soil animals can carry some seeds to considerable depths. Only those near the surface, or those brought up by earthworm casts, digging animals or uprooted trees stand any chance of successful emergence (Putz 1983). Shallow burial of seeds often increases germination and aids seedling establishment, but deep burial hampers the emergence of germinating seeds (Fig. 5.12).

5.4.2 Dormancy

There are legendary instances of dormant seeds stored in unusual conditions retaining viability for periods of hundreds, and even thousands of years, but these records have little relevance for the normal ecology of seeds in the soil (Baker 1989). Seed dormancy is of selective value to plants because it allows them to time germination to coincide with conditions favourable to seedling survival, not because it allows the occasional seed to sit out the centuries. Dormancy does have an important impact on the longevity of

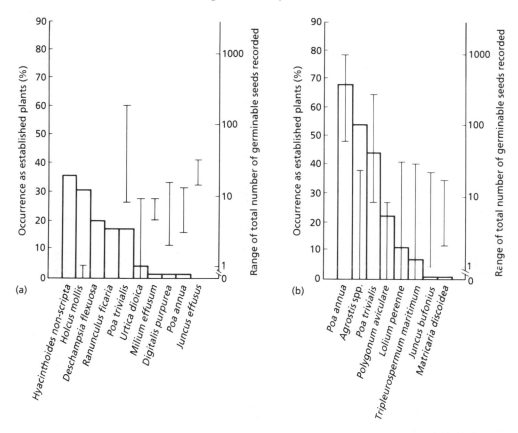

Fig. 5.9 The relative abundance of mature plants (histograms) and seeds in the soil (vertical bars) for the major species of: (a) the herb layer of a deciduous woodland and (b) an arable field in Britain (from Thompson & Grime 1979).

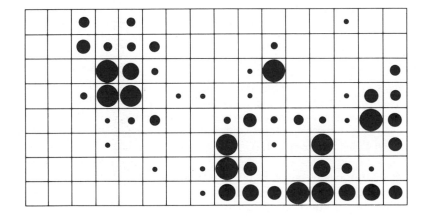

Fig. 5.10 The distribution of seeds of the grass *Danthonia decumbens* in 7 cm × 7 cm, 5 cm-deep blocks of soil taken from a turf of acid grassland on Dartmoor, Devon, England. Smallest dots are one seed, the largest are greater than 20 (from Thompson 1986).

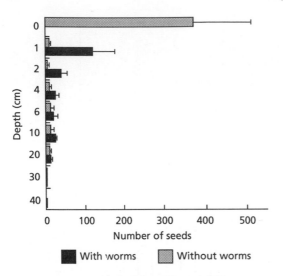

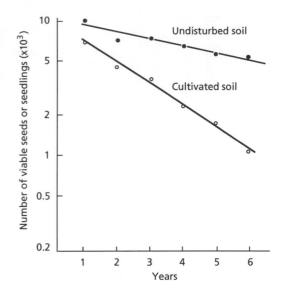

Fig. 5.11 The distribution of viable seeds of *Poa annua* at different depths in the soil at grassland sites with and without earthworm populations (from Reest & Rogaar 1988).

Fig. 5.13 Decay of viable weed seeds in undisturbed soil and cultivated soil (from Roberts & Feast 1973).

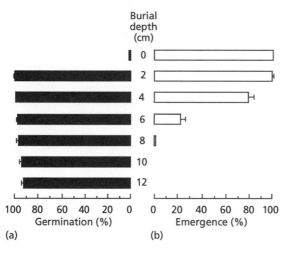

Fig. 5.12 (a) Percentage of seeds of the sand-dune grass *Agropyron psammophilum* buried at different depths that germinated, and (b) percentage of germinated seeds that successfully emerged at the surface (from Zhang & Maun 1990).

seeds in the soil, but rates of decay are generally exponential and populations of even the longest-lived species are severely depleted after a matter of decades (Fig. 5.13).

Species that are long-lived as adults often have short-lived seeds. Seeds of the primary tree species of forests tend to lose viability very rapidly. Ng (1983) estimated that the viability of seeds in the soil of Malaysian rainforest decayed with an average half-life of only 6 weeks. In another tropical forest, at Barro Colorado Island (BCI) in Panama where the rainfall is seasonal, Garwood (1983) found that 80% of species germinated with the arrival of the next rainy season to follow seed dispersal. This involved only a brief period of dormancy for those species which produced seeds outside the rainy season. The rainy season at BCI lasts 8 months, but Garwood (1982) found that pioneer trees, lianas and canopy trees, all requiring gaps for germination, appeared only in the first 2 months of this period, while understorey and shade-tolerant species germinated throughout the rainy season. Among the gap requirers Garwood (1983) found that seedling mortality during the dry season that followed the rains was related to time of emergence. Seedlings that emerged earlier were better able to beat the competition in their gaps and to tolerate the ensuing drought than seedlings that emerged later. No such relationship existed for the shade-tolerant species that germinated throughout the rainy season. The optimal timing of germination for different species at BCI is evidently determined by a combination of seasonally changing climate, biotic interactions and tolerance of shade.

In seeds that have the capacity for prolonged dormancy, dormancy state is often changeable. Seeds from different parts of the same plant, or matured at different times, may exhibit different germination behaviours when shed (Silvertown 1984); seeds may change their behaviour with time and do so in different directions depending upon storage conditions; and all these differences may themselves vary between populations. In many instances the ecological significance of the complex patterns found in laboratory germination tests still requires investigation in the field, but a number of kinds of seed behaviour do have apparent adaptive value for the plant. Fresh seeds of knotgrass *Polygonum aviculare* are dormant and will not germinate before winter, but dormancy is broken by chilling in the imbibed state ('**stratification**') which permits germination as the soil warms in the spring. Stratified seeds in the soil that have not germinated by May become dormant again and require another cold period before they will germinate (Courtney 1968; Roberts 1970). The dormancy of fresh seeds prevents germination too early when seedlings would be exposed to frost, and reacquisition of dormancy in May prevents germination too late when seedlings would probably be at a severe disadvantage to earlier cohorts (e.g. see Fig. 4.10). In fat hen *Chenopodium album* dormancy is seasonally adjusted in a different manner. Non-

dormant seeds are produced early in the season and dormant ones in the main crop.

Seeds of many species with long-lived soil populations undergo annual cycles of dormancy induced by seasonal temperature changes, which permit germination at certain times of year and not at others. Both the winter annual thale cress, *Arabidopsis thaliana*, and the summer annual common ragweed, *Ambrosia artemisifolia*, exhibit dormancy cycles, but these are 6 months out of phase with each other (Fig. 5.14). Roberts (1986) studied the seasonal patterns of germination in 70 species of annual and perennial herbs of the British flora and concluded that about half of them showed cyclic changes in germination requirements through the year.

Baskin and Baskin (1985) suggest a scheme of classification for the different kinds of seed dormancy that explicitly allows for changes in dormancy state. The scheme divides seeds into those with **primary dormancy**, which are unable to germinate when shed from the plant, and those with **secondary dormancy**, which acquire dormancy after leaving the parent. Seeds with primary dormancy may become non-dormant and then acquire (secondary) dormancy again. Both primary and secondary dormancy are divided into two subcategories: **innate dormancy** and **conditional dormancy**. Innately dormant seeds will not germinate under any normal environmental conditions. Seeds with conditional dormancy will germinate in only a narrow range of

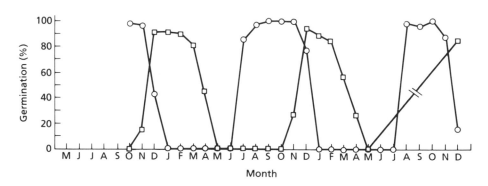

Fig. 5.14 Annual cycles of dormancy in buried seeds of: a strict winter annual *Arabidopsis thaliana* (circles) (from Baskin & Baskin 1983), and the summer annual *Ambrosia artemisifolia* (squares) (from Baskin & Baskin 1980) retrieved from the soil and germinated at their respective temperature optima.

conditions. When innately dormant seeds alter their dormancy state they usually do so by a gradual change which takes them through a phase of conditional dormancy. The transition back again from non-dormancy to secondary dormancy also passes through a conditional phase (Baskin & Baskin 1987).

An important form of innate primary dormancy caused by a hard seed coat is found in many legumes such as *Medicago lupulina* (Fig. 5.8b). With the passage of time in the soil this kind of dormancy may be broken by the decomposition of the testa. Fire or passage through the gut of a vertebrate dispersal agent will frequently break this kind of dormancy too. Conditional secondary dormancy is induced in many species which will normally germinate in the dark but which become light requiring when exposed to light filtered through a leaf canopy (Silvertown 1980a). Half of a sample of 32 common East African weed species tested by Fenner (1980) showed this behaviour. Other environmental cues may also modulate changes in dormancy. Low humidity caused innate secondary dormancy in seeds of the Nigerian tree *Hildegardia barteri*, which became conditional when humidity was raised to 90% (Enti 1968).

Although seed dormancy is so phenotypically plastic, it is a character with strong effects upon plant fitness. One obvious explanation for this apparent paradox is that there is selection for phenotypic variability in germination behaviour when there is some unpredictability about the best time to germinate (Chapter 9). The fitness of seeds with a particular degree or type of dormancy can be estimated from the survival and fecundity of the seedlings that appear when dormancy is broken. Arthur *et al.* (1973) compared the survival and fecundity of autumn and spring cohorts of seedlings of the poppy *Papaver dubium*, an arable weed of British fields. The autumn seedlings emerging from non-dormant seeds were susceptible to heavy winter mortality in some years, but survivors of this cohort produced larger plants and at least 10 times as many seeds as spring cohorts.

There are many examples of annual plants in which seedlings from non-dormant seeds have a lower survival but a higher fecundity than later-appearing seedlings from dormant seeds. This pattern has been observed in prickly lettuce *Lactuca serriola* (Marks & Prince 1981) and in charlock (*Sinapis arvensis*) (Edwards 1980) in Britain, between summer and autumn cohorts of *Leavenworthia stylosa* in Tennessee cedar glades in the USA (Baskin & Baskin 1972) and in two closely related annuals *Erodium botrys* and *E. brachycarpum* that occur in annual grassland in California (Rice 1987). In all instances variability in climate (rainfall, drought or frost) may favour dormancy in some years and non-dormancy in others, causing selection for phenotypic variation. This may arise from phenotypic plasticity or from genetic variation in the population. In *Papaver dubium* seed dormancy has very low heritability, and variation therefore reflects phenotypic plasticity (Arthur *et al.* 1973). The opposite is the case in *Sinapis arvensis* in which dormancy is determined by two alleles at a single locus (Garbutt & Whitcombe 1986). There is also evidence that dormancy has a heritable component in *Erodium* spp. (Rice 1987).

5.4.2.1 *Genetic consequences of the seed pool*

The existence of a seed pool increases effective population size N_e because generation time is increased by the length of time plants spend as dormant seed in the soil, and because the breeding members of the population may be drawn from different cohorts of seeds. In the absence of selection, the effect of dormancy on N_e depends on the average time spent in the seed bank (Templeton & Levin 1979). The importance of a seed pool in increasing N_e will of course depend on which seeds germinate, not simply the existence of a large dormant population. If most of the seeds in the soil are buried too deeply to germinate and recruitment occurs mainly from recent entrants that are near the surface, much of the seed pool may be genetically as well as demographically irrelevant.

Nevertheless, the seed pool can serve as an important reservoir of genetic variation, buffering the effects of local extinction of genotypes in the adult population caused by selection or drift. This should reduce the decay of genetic variance, and retard local subdivision of the population. In a sense, the seed bank can provide a kind of 'genetic inertia' to the population. Templeton and Levin (1979) modelled the role of seed banks in the evolution of adult plant characters and found that seeds produced in 'good' years for the adult plants come to dominate numerically the seed bank. Differential fecundity may bias the contributions of genotypes to the seed pool, reducing N_e and increasing genetic subdivision (Chapter 4).

5.4.3 Recruitment and the 'safe site'

Physiological studies of germination in the laboratory can be some guide to the behaviour of seeds in the field, but this behaviour is sensitive to a range of environmental variables and many of these interact with each other to subtle effect. A demonstration of this was provided by Harper *et al.* (1965) in a simple but compelling experiment. A seed bed was sown with equal quantities of seeds of three plantains *Plantago major*, *P. media* and *P. lanceolata* and divided into plots that were given a replicated series of treatments to the soil surface, as described in the legend to Fig. 5.15. The different treatments had markedly selective effects on the emergence of the three species, depending upon the kind of micro-environment each treatment had created (Fig. 5.15). To explain the results of this experiment Harper *et al.* (1965) introduced the term **'safe site'** to describe the specific conditions that allow the seeds of a particular species to emerge successfully from the soil.

The concept of 'safe site' is similar to the idea of niche and shares with it the difficulty that neither of them can be reliably identified until after they have been successfully occupied. Habitats are obviously not uniformly 'safe' for seed germination, but is it more than chance that seedlings appear in some places and not in others? In some instances it clearly is. Decaying logs ('nurse logs') are safe sites for the establishment of seedlings in a number of tree species. In Hawaii there is a tree fern whose name in the native language means 'Mother of Ohia', because Ohia (*Metrosideros collina*) germinates on its fallen trunks and even on living trees.

In the Cascade Mountains in Oregon almost all the young western hemlock *Tsuga heterophylla* found by Christy and Mack (1984) in their 200 m

square study plot were growing on decaying logs of Douglas fir *Pseudotsuga menziesii*, though such logs occupied only 6% of the area. By sowing hemlock seeds onto soil and different kinds of log Christy and Mack (1984) found that germination was possible on a wide range of surfaces, but that logs with rotten heartwood provided the best substratum for survival. Logs had to be rotten enough to permit hemlock seeds to lodge in them and root, but still sufficiently intact to raise the seedlings above the litter which threatened to bury and kill them.

In a deciduous forest in upstate New York, Handel (1976) found that the sedge *Carex pedunculata* was particularly common on fallen logs, where the species avoided competition from the rest of the woodland ground flora. Like those of many woodland herbs, seeds of *Carex pedunculata* possess an elaiosome (Chapter 3) that attracts ants. *Carex* seeds were carried to rotting logs by ants that nested in them and which discarded the intact seeds there after removing the elaiosome.

In theory, germination safe sites might explain the macro-distribution of species, as well as their micro-distribution. Keddy and Constabel (1986) tested this hypothesis in a group of 10 wetland species that are distributed along a gradient of wave exposure on the shores of the North American Great Lakes. The gradient of wave exposure creates a parallel gradient of soil particle size, from silt in sheltered bays to gravel on the most exposed shorelines. Seeds of the 10 species, which ranged in size from those of *Bidens vulgata* at 14 mm long to those of *Lythrum salicaria* at 0.76 mm long, were sown in trays of graded soil. Soil of the finest texture passed a 0.25 mm sieve and the coarsest an 8 mm sieve. When watered well, most species showed no particular preference for soil type, and when watered sparingly all except one established best on the finest soil. Among these plants then, there was no evidence that species have safe site differences related to substratum texture that could explain their distributions along the wave exposure gradient.

An interaction between climate and the germination ecology of the annual weed corn spurrey *Spergula arvensis* appears to explain an intriguing cline in gene frequency in this species in Britain. Three seed morphs are known, determined at

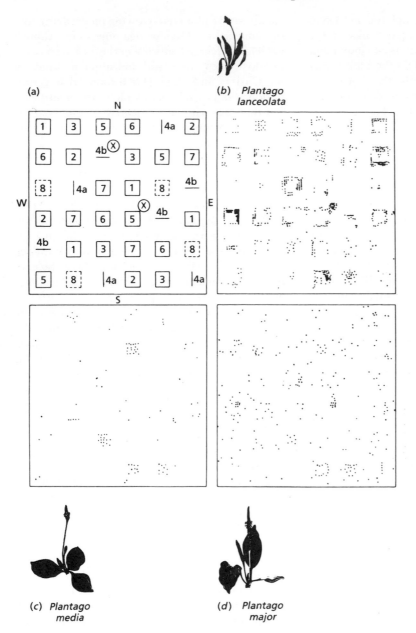

(a)

(b) *Plantago lanceolata*

(c) *Plantago media*

(d) *Plantago major*

Fig. 5.15 (a) The distribution of various treatments to the *Plantago* seed bed: (1, 2) two kinds of depression in the soil surface, (3) a sheet of glass laid on the soil surface, (4a, b) sheets of glass placed vertically in the soil, (5−7) rectangular wooden frames of three different depths pressed into the soil surface, $\otimes$ worm casts. (b−d) The distribution of seedlings of *P. lanceolata*, *P. media* and *P. major* (from Harper *et al.* 1965).

a single locus by two alleles. Seeds covered in papillae are produced by one homozygote, the other homozygote produces seeds with a smooth seed coat, and the seeds of the heterozygote are intermediate with half as many papillae as the papillate morph (Fig. 5.16a) (New 1958). All the seeds produced by a particular plant have the same morph because the seed coat is maternal tissue, which makes it possible to score gene frequencies in the field when the plants are in fruit. In fact the plant is highly selfing so heterozygotes are rare.

New (1958) surveyed populations of *S. arvensis* across Britain in the 1950s and found that the frequency of the papillate morph decreased from the drier regions of the south to the wetter parts

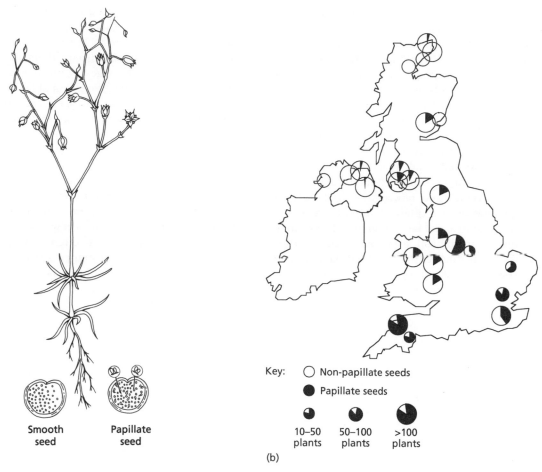

Fig. 5.16 (a) Papillate and smooth seed morphs of *Spergula arvensis*, and (b) the cline in frequency of the papillate form in Britain. Proportions of papillate and non-papillate seeds are shown in black and white respectively (from New 1958).

of the north and west (Fig. 5.16b). When New (1978) surveyed the cline again in the 1970s she found that there had been no significant change after 20 years. New and Herriott (1981) discovered in laboratory experiments that papillate seeds germinate better than smooth ones in dry conditions, suggesting that natural selection in favour of papillate seeds in dry areas could account for their distribution. What advantage smooth seeds have in wetter conditions is not known, so the full explanation for the cline and its stability is not yet clear, but this does appear to be a case where the subtleties of germination ecology affect gene frequencies on a geographical scale.

5.5 SUMMARY

Plant populations are **dynamic** and often have high rates of birth and death. The simple **difference equation** model introduced to describe population dynamics in Chapter 1 can be used to calculate two related measures of population growth: the **annual rate of increase** λ and the **net reproductive rate** R_0. When these measures have a value greater than one the simple difference equation predicts **exponential** increase. In reality, population increase is limited by **density-dependent** decreases in fecundity or density-dependent increases in mortality which, for an annual plant, can be described by a modification of the yield/density equation used in Chapter 4. The modified equation describes the shape of the **recruitment curve** which can be used to predict how a population will behave by a graphical method known as **cobwebbing**. Density-regulated populations may show a variety of dynamic behaviours ranging from **monotonic damping** to **chaos**, depending upon the values of the exponent b and the mean seed output s in the equation for the recruitment curve. **Density-independent** mortality may affect population behaviour through its influence on s.

The behaviour and fate of the seed fraction of a population are often of importance to its overall dynamics, and large numbers of seeds may accumulate in the soil **seed pool**. Seed densities are highest in frequently disturbed habitats and the species most strongly represented in the soil are often those with the shortest lifespan above ground. The **spatial distribution** of seeds is usually clumped because dispersal distances from the parent tend to be limited and **animal dispersal agents** may concentrate them in their faecal deposits. Seed longevity in the soil is prolonged by **dormancy**. Seeds unable to germinate when shed from the plant are said to have **primary dormancy** and those which acquire dormancy after leaving the parent have **secondary dormancy**. The latter may be cyclic. Either of the main kinds of dormancy may be **innate** or **conditional**. The **seed pool** can serve as an important reservoir of **genetic variation** and it may increase N_e if the recruits from it to the active population do not belong to a few numerically dominant genotypes. Seeds often have quite particular germination requirements that restrict the appearance of new plants to **safe sites** that may be limited in number.

Chapter 6
Dynamics of Structured Populations

6.1 INTRODUCTION

Not all individuals in a population make an equal contribution to its finite rate of increase λ because values of B and D vary with the age, size and stage of an individual (Chapter 4). In populations where recruitment is a frequent event compared with the lifespan of adults, such as in many forest trees, there is an **age structure** (Fig. 6.1a). At any one time seeds, seedlings, saplings and adults of various ages may all be found in such populations. Younger ages tend to have higher mortality rates than older ones, so a population of 99 seedlings and one adult has a quite different future to a population of one seedling and 99 adults, even though both have the same total population size.

Plants are highly plastic in their rates of growth and development, so that two individuals of the same age but with different local environments may be at quite different stages of the life cycle. Stunted seedlings of forest trees may persist to 30 years of age on the forest floor without developing to sapling size, while other individuals that germinated at the same time, but happen to have been located in a light gap, may achieve reproductive maturity before age 30 years. Plant populations therefore have a **stage structure** (Fig. 6.1b) as well as an age structure. The size distributions that we have seen develop in crowded stands are an example of the latter (see Figs 4.7 & 4.8). Both age and stage structures may influence plant population dynamics, and we will see in this chapter how these are dealt with. However, not all perennial plant populations have an age structure. For example, even-aged populations are found in semelparous bamboos (Janzen 1976) and pines such as *Pinus banksiana* (Kenkel 1988). Whether a plant population has age structure or not depends upon the pattern and periodicity of recruitment.

6.2 DISTURBANCE AND THE PERIODICITY OF RECRUITMENT

Because large plants suppress the growth of small ones, the entry of new recruits into many populations can only occur when established individuals die or are killed and vacate space. Recruitment may be limited by the events which create this space, which we may loosely gather together and call 'disturbance'. For many, though not all plants (Grubb 1988), recruitment is coupled to disturbance. Populations without age structure inhabit sites where there is periodic, severe and widespread disturbance such as is caused by fire, hurricanes, extreme drought or tillage of the soil for agriculture. These events remove existing vegetation and provide space for the massive, simultaneous recruitment of new populations, usually from seed.

If the disturbed area is large and recruitment into it is rapid, the populations that become established will be approximately even-aged. If the disturbed area is small, the population will consist of a mosaic of patches recruited at different times. How the age structure of such a population is described obviously therefore depends upon the scale of which you are talking. Local stands of *Pinus* spp. in Itasca State Park are even-aged and date from the last fire, but in the Park as a whole the population of this species has an age structure (Fig. 6.2).

Species of pine (e.g. *Pinus banksiana*, Chapter 4), *Banksia* spp. in Australia and Proteaceae of the fynbos heathland of the South African Cape that live in fire-prone habitats store seeds in sealed

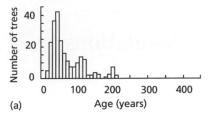

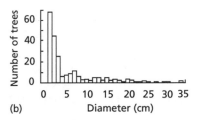

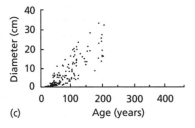

Fig. 6.1 (a) Age structure, (b) size structure, and (c) the relationship between size and age of trees in a population of *Pinus sylvestris* in Sweden (from Ågren & Eriksson 1990).

morphic character, because fire frequencies are variable in space and time and there is strong selection after a fire against trees lacking serotiny, as well as against those possessing the character in an environment where fire is absent and seeds remain trapped.

In populations of *Pinus resinosa* in the New Jersey pine barrens, Givnish (1981) found that serotiny was most frequent in the most fire-prone areas. Similar correlations have been found in *P. coulteri* and *P. torreyana* (McMaster & Zedler 1981; Borchert 1985). In populations of *P. contorta* growing in Montana the frequency of serotiny was related to the kind of disturbance that had preceded recruitment. Serotiny was common in stands established after fire but uncommon in those initiated by windfalls or disease, suggesting that a single generation of natural selection might be sufficient to alter radically the frequency of genes affecting serotiny (Muir & Lotan 1984).

Serotiny, and seed storage in the soil, allow the rapid recruitment after disturbance that creates even-aged populations. If recruitment is slower, because no seeds are available at the site of disturbance itself, we can expect the populations that arise to be uneven-aged, and a succession of species to take place.

The life history characteristics of a species determine whether it may survive over the long term at a site that has a particular frequency and severity of disturbance. Any species with a seed pool can potentially colonize new space when it becomes available, but not all can complete their life cycle before the next episode of disturbance occurs. Where it occurs every year, such as in a ploughed arable field or in a desert, annuals are favoured because they can set seed within the year. In herbaceous communities, slightly longer

(serotinous) cones that accumulate on the tree (Bradstock & Myerscough 1981; Bond 1985). When the heat of a fire unseals them, many years' seed production are released at one go onto the newly cleared ground. Serotiny is usually a poly-

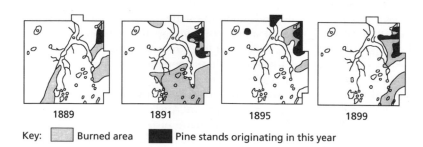

Fig. 6.2 Stands of pine recruited in Itasca State Park in four different years, in relation to the distribution of fires in the same years (from Frissell 1973).

intervals between disturbance events favour semelparous perennials such as evening primroses *Oenothera* spp., mullein *Verbascum thapsus* and teasel *Dipsacus sylvestris*. Populations of these species tend to be evanescent, relying upon a pool of dormant seed to carry them over between one disturbance and the next that occurs at the same spot (Werner & Caswell 1977; Reinartz 1984; Kachi & Hirose 1985).

A useful way to portray the effects of disturbance upon population structure is to graph the average lifespan of adults of a species against the typical periodicity of recruitment (Fig. 6.3). Species falling above the diagonal of this graph have populations with age structure. Those species falling on or below the diagonal tend to form populations that are even aged, at least locally.

On the diagonal of the graph are populations which recruit in a pulse immediately following the death of a cohort. For example, at bottom left on the diagonal are annuals with no persistent seed pool. Further up the diagonal fall the various species of serotinous pines that are, like the mythical phoenix, at one and the same time killed

and recruited by fire. Semelparous bamboos (Chapter 1) also fall on the diagonal because their synchronized flowering and death are followed by massive recruitment from seed produced in the terminal act of reproduction. Species falling below the diagonal must survive between the infrequent events that permit recruitment by means of a dormant seed pool. These populations include desert annuals, semelparous perennials and the early colonizers of large forest gaps such as *Cecropia obtusifolia* and *Piper* spp. in the neotropics. Locally, these species have even-aged populations of adults but if the whole population is considered, including seeds, there is an age structure.

When the period between recruitment events is longer than the lifespan of adults, the finite rate of population increase λ has to be averaged over two phases of the entire life cycle: the phase during which adult plants grow and produce seed, and the dormant phase when the population consists only of seeds in the soil. The best way to compare or combine what happens in these two phases is to measure the size of the population in

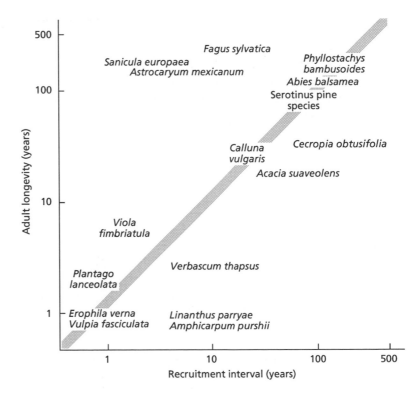

Fig. 6.3 Effect of the relationship between adult lifespan and recruitment periodicity on population age structure, showing species mentioned in chapters 5 and 6.

terms of *seeds* throughout.

Auld (1986a, b; 1987) and Auld and Myerscough (1986) studied the population dynamics of the legume shrub *Acacia suaveolens* that occurs in heathlands and woodlands along the east coast of Australia. The shrub is killed by fire, which is also required to trigger the germination of seeds in the seed pool. As one would expect, populations tend to be even-aged. Adults are short-lived, few reaching 20 years of age. Peak seed production by individuals occurs at age 2 years, and thereafter the proportion of shrubs producing seeds, and the fecundity of those that do, diminish rapidly with time. This decline in seed production combined with the mortality of adults and predation of seeds in the soil leads to the pulse of seed production that occurs early in the life of a cohort being quickly eroded by exponential decay (Fig. 6.4). It can be seen from Fig. 6.4 that if fires recur with a periodicity of 55 years or less, when the seed pool reaches the replacement level of 1000, the population will survive, though of course numbers of both seeds and adults will cycle. Note that these cycles are driven entirely by fire frequency, not by density-dependence as in *Erophila verna*.

6.3 AGE AND STAGE STRUCTURE

The existence of age and stage structure has to be taken into account in population models. We can treat each age or stage class in a structured

population like a population in its own right. If the number of individuals in a class i is n_i, the structure of the population at time t is represented by a list of n_i values called a **column vector**, and given the symbol $\boldsymbol{n_t}$:

$$\boldsymbol{n_t} = \begin{bmatrix} n_1 \\ n_2 \\ n_3 \\ \dots \\ n_i \end{bmatrix}$$

By analogy with the basic equation of population dynamics (Eqn 5.2) we derive $\boldsymbol{n_{t+1}}$ from $\boldsymbol{n_t}$ by multiplying $\boldsymbol{n_t}$ by a coefficient representing the inputs to and outputs from each class. These are gathered together in a matrix called **A**, so:

$$\boldsymbol{n_{t+1}} = \mathbf{A}\boldsymbol{n_t}$$

A is a **projection matrix**, or transition matrix, containing all the stage- or age-specific rates of survival and seed production. The projection matrix is square, with i columns and i rows for a population with i age classes. For a population with three age classes representing, say, seedlings, 1-year-old rosettes and 2-year-old flowering adults the projection matrix representing transitions between the classes is:

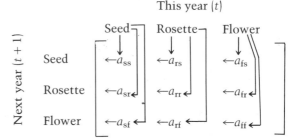

Box 6.1 describes how $\boldsymbol{n_{t+1}}$ is calculated from $\mathbf{A}\boldsymbol{n_t}$.

The precise form of the projection matrix will depend upon the details of the biology of the species and how you choose to represent its population structure. If in our hypothetical plant seedlings, rosettes and flowering plants are true age classes and all 1-year-olds are seedlings, all 2-year-olds are rosettes, and all 3-year-olds flower and then die, then a number of the elements in the matrix will have zero values. You should be able to work out for yourself which these are.

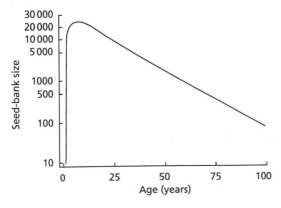

Fig. 6.4 Changes in the size of the soil seed pool of *Acacia suaveolens* for an initial cohort of 1000 seeds (from Auld 1987).

Box 6.1 Matrix analysis

Written algebraically the multiplication of a projection matrix $\mathbf{A}$ by a column vector $\boldsymbol{n}_t$ is carried out as follows:

$$
\begin{array}{ccc}
\mathbf{A} & \times\ \boldsymbol{n}_t & =\ \boldsymbol{n}_{t+1}
\end{array}
$$

$$
\begin{bmatrix} a_{11}\ a_{21}\ a_{31} \\ a_{12}\ a_{22}\ a_{32} \\ a_{13}\ a_{23}\ a_{33} \end{bmatrix} \times \begin{bmatrix} n_1 \\ n_2 \\ n_3 \end{bmatrix} = \begin{bmatrix} (n_1 a_{11}) + (n_2 a_{21}) + (n_3 a_{31}) \\ (n_1 a_{12}) + (n_2 a_{22}) + (n_3 a_{32}) \\ (n_1 a_{13}) + (n_2 a_{23}) + (n_3 a_{33}) \end{bmatrix}
$$

Describing this calculation in words:

the *first* element of $\boldsymbol{n}_{t+1}$ is the sum of each element in $\boldsymbol{n}_t$ times its corresponding element in the *first* row of $\mathbf{A}$;

the *second* element of $\boldsymbol{n}_{t+1}$ is the sum of each element in $\boldsymbol{n}_t$ times its corresponding element in the *second* row of $\mathbf{A}$;

the *third* element of $\boldsymbol{n}_{t+1}$ is the sum of each element in $\boldsymbol{n}_t$ times its corresponding element in the *third* row of $\mathbf{A}$; and so on, if there are more than three classes.

To find the value of $\boldsymbol{n}_{t+i}$, the projection matrix is repeatedly multiplied in a process called **iteration**:

$$\boldsymbol{n}_{t+1} = \mathbf{A}\boldsymbol{n}_t, \quad \boldsymbol{n}_{t+2} = \mathbf{A}\boldsymbol{n}_{t+1}, \quad \boldsymbol{n}_{t+3} = \mathbf{A}\boldsymbol{n}_{t+2}$$

and so on to $\boldsymbol{n}_{t+i}$

Note that $\mathbf{A}$ does not change its value in this process, and that by continually multiplying the population by a constant, iteration produces an exponential increase in population size. The matrix equation $\boldsymbol{n}_{t+1} = \mathbf{A}\boldsymbol{n}_t$ is the equivalent for an age-structured population of $N_{t+1} = R_0 N_t$ for other populations.

Calculation of λ

After a certain number of iterations, the exact number depending upon the projection matrix $\mathbf{A}$, population structure usually attains a stable age distribution with a fixed ratio between $n_1 : n_2 : n_3 : n_i$. Convergence on a stable age distribution is a mathematical, not a biological property of the matrix and is called ergodicity. When the stable age structure has been attained, the finite rate of population increase λ for the population may be found by dividing the size of any age class n_i in one year by its size the previous year. The value of λ depends upon $\mathbf{A}$, but not upon the starting conditions $\boldsymbol{n}_t$. In mathematical terminology λ is the dominant eigenvalue of $\mathbf{A}$.

The stable age/stage distribution

The stable age distribution derived from iteration of the $\mathbf{A}$ matrix can be compared with the actual age distribution of a population to determine whether its present composition is stable. Large discrepancies between the real and projected age distributions imply that the age composition of the population will change, if present trends continue.

All the techniques of matrix analysis described for age-classified populations may also be used for stage-classified ones, although there is an important difference in how age and stage distributions should be interpreted. If an *age* distribution observed in the field has a peak, with many more individuals in one of the older age classes than in younger classes, this suggests that a strong cohort (or 'baby boom') is passing through and that the age distribution is unstable. A similar peak in the *size* or stage distribution of a population does not necessarily imply departure from a stable stage distribution (Caswell 1989). Stable stage distributions can have peaks caused by some stages being of longer duration than others.

Density-dependence

Matrix models incorporating density-dependence can be constructed by replacing the fixed values of a_{ij} in the $\mathbf{A}$ matrix with density-dependent functions. Since the value of the elements in the projection matrix then become specific for the particular value of N at each iteration, the matrix is denoted $\mathbf{A}_n$, and:

$$\boldsymbol{n}_{t+1} = \mathbf{A}_{nt}\boldsymbol{n}_t$$

Continued on p. 98

If the survival and fecundity of all age classes are affected equally by density-dependence, a function such as we used in Eqn 5.11 can be applied by splitting the density-independent matrix **A** into two matrices, one called **P** for the survival probabilities and another called **F** for fecundities, and then

$$A_n = (1 + aN)^{-1}(F + P)$$

Simple density-dependent models such as this are ergodic, but the assumption that all age classes are equally affected by density is highly unrealistic for plant populations (Chapter 4).

The reader is referred to Caswell (1989) for further information.

It may seem that the difficulties involved in incorporating density-dependence and other important aspects of reality into a matrix model limits the usefulness of this technique, but things are not as bad as they might seem! Matrix models can be used for two distinct purposes which should not be confused: (i) for discovering something (e.g. the value of λ) about the population *at the present time* by projecting from its current status represented by **A**, and (ii) for *predicting* the future dynamics of a population.

A projection matrix for an age-structured population has non-zero elements *only* in the subdiagonal, which represents annual survival, and in the top row, which represents births. A matrix of this form is called a **Leslie matrix**. As we have already discussed, plant population structure may often be better described by the distribution of individuals among a series of stages. If seeds, rosettes and flowering plants in the hypothetical population have a range of ages and are identified by their size or behaviour rather than by their exact age, we may use a matrix called the **Lefkovitch matrix**. This has non-zero elements in the main diagonal, representing the proportion of seeds that remain seeds, rosettes that remain vegetative, and flowering plants that flower again the next year. The subdiagonal of the Lefkovitch matrix represents growth or development from one stage to the next, not simply an increase in chronological age. Studies of plant populations generally classify individuals by either age *or* stage, though greater accuracy is achieved by using both classifications simultaneously (Law 1983; van Groenendael & Slim 1988). The kind of matrix which classifies individuals by age and stage simultaneously is called a Goodman matrix, and as you can imagine it is quite complicated, consisting of a matrix whose elements are themselves submatrices.

6.3.1 Annuals with a persistent seed pool

The existence of a seed pool confers an age structure on an annual population, all be it that most of this is buried in the soil. The seed pool is periodically recharged after each successful generation of adults has flowered, and it accumulates a heterogeneous population of seeds that vary in age, depth of burial in the soil and state of dormancy. All of these characteristics affect the probability of a seed germinating, and all change with time.

Annuals with a pool of dormant seed are able to exploit habitats where recruitment opportunities are irregular such as deserts (Kemp 1989; Venable 1989), grassland disturbed by animal activity (Rice 1989) and arable land that is periodically fallowed (Cavers & Benoit 1989). Population densities above ground fluctuate wildly from year to year, depending upon the happenstance of weather and soil disturbance (Fig. 6.5). Because these two factors and the response of dormant seeds to them are so important to recruitment, the role of changes in seed density in generating the patterns of abundance seen among adults can be much more difficult to determine than it is for annuals without a seed pool.

As well as buffering the genetic composition of a population (Chapter 5), a seed pool buffers local populations from extinction and may affect spatial

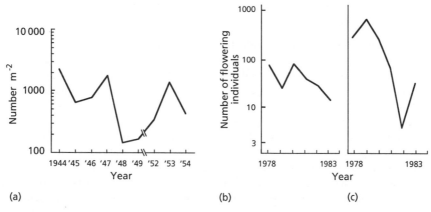

Fig. 6.5 Long-term population dynamics of annuals with a seed pool: (a) *Linanthus parryae* (Epling *et al.* 1960), (b) *Blackstonia perfoliata,* and (c) *Centarium erythraea.* Note the logarithmic scales of abundance (Grubb 1986).

pattern (MacArthur 1972). Annuals with no seed pool can only persist where R_0 is *always* greater than 0. This is only likely in, or adjacent to, places where the average value of R_0 is high. We would therefore expect annuals with no seed pool to be very patchy and where they do occur they should be abundant. This constraint is much less rigid for annuals with a seed pool, which may therefore be less patchy and exhibit a wider range of densities in the years when they are abundant. In summary then, annuals without a seed pool should have high spatial patchiness and low temporal variance, while species with a seed pool should have the reverse characteristics. There is at present no empirical test of these predictions.

We have seen that for a plant with no seed pool λ is influenced by rates of mortality and fecundity. For plants with an age or stage structure, the rate of transitions between classes also influences λ (Box 6.1). In the particular case of an annual with a seed pool, the survival of seeds in the soil, and the proportion of seeds which break dormancy each year, which we shall call G, are both transitions whose value will have an effect on λ. The question is, what contribution do they make to λ, compared to the contributions of fecundity and survival? This is a fundamental question that we shall be asking about quite a number of species in this chapter, and which can be answered using two related techniques called **sensitivity analysis** and **elasticity analysis**. The

bases of these techniques, and how they are applied to projection matrices, are outlined in Box 6.2.

Fluctuations in weed populations from year to year are frequently attributed to the influence of weather, but how intrinsically stable should we expect annuals with a seed pool to be anyway? Rice (1989) analysed an age-structured model of an annual plant with a seed pool to determine the effect of germination delays upon the **transient dynamics** of such populations. This analysis uses a count of the number of iterations it takes for a projection matrix to converge on a stable age/ stage structure to measure the stability of the population. He found that low values of G, which cause large numbers of viable seeds to be carried over from one year to the next in the seed pool, dramatically increase the time it takes for population structure to converge. The results from models of age-structured seed populations point to the importance of understanding events in the soil better and suggest that if there were the empirical data available to use in such models, it might be easier to understand the complicated dynamics of weeds and other annuals with seed pools (Kalisz 1991).

6.3.2 Perennials

The distinction between annual and perennial plants is sometimes an arbitrary one. Many

---

Box 6.2 Sensitivity analysis

Density-independent matrix models can be used to determine the sensitivity of λ to small perturbations in the value of the elements a_{ij} in the projection matrix **A**. Each element a_{ij} of the projection matrix has an associated sensitivity value s_{ij}. These values have two very important applications. First, they tell us the relative importance of different transitions to maintaining population growth rate, so that if we wish to increase or decrease λ, sensitivity analysis indicates the population's most vulnerable points that may either be attacked or protected accordingly. Secondly, because λ may be used to estimate fitness, s_{ij} indicates how changes in a particular transition a_{ij} will affect the fitness of individuals in the population. A high s_{ij} value suggests that natural selection should act on characters affecting the transition a_{ij} because small differences between individuals affect their fitness.

The sensitivity value for a matrix element a_{ij} is the product of the ith element of the vector of reproductive values (**v**) and jth element of the vector (**w**) containing the stable age/stage distribution, divided by the scalar product of the two vectors

$$s_{ij} = \frac{v_i w_j}{<\mathbf{v},\mathbf{w}>}$$

Mathematically, **v** and **w** are respectively the left and right eigenvectors of the matrix **A**. Since $<\mathbf{v},\mathbf{w}>$ is not dependent on i or j, it can be ignored for comparisons between values of s_{ij} within a matrix (Caswell 1989).

Sensitivity values measure the relative effect upon λ of a change in a_{ij} by a small, fixed amount. Values of a_{ij} representing survival probabilities can only range between 0 and 1, whereas an a_{ij} representing fecundity can have any value at all. This introduces a difficulty into the comparison of sensitivity values, because a fixed change of say 0.05 to a probability of 0.5 is of much greater significance than a change of the same amount to a fecundity of 100. To overcome this problem, de Kroon *et al.* (1986) introduced a refinement to sensitivity analysis. Their modified index of sensitivity, called **elasticity** (e_{ij}), is

$$e_{ij} = \frac{a_{ij}}{\lambda} \times s_{ij}$$

Elasticities can be used for the dual purposes already described for s_{ij}, and have the additional useful property that they all sum to one within a matrix. This means that e_{ij} represents the proportion of λ due to the transition a_{ij} though, of course, we should remember that, to have any value at all, λ requires the transitions of the entire matrix just as a plant cannot dispense with a stage of its life cycle.

---

species such as annual meadow grass *Poa annua* (Law *et al.* 1977), wild rice *Oryza perennis* (Sano *et al.* 1980), eelgrass *Zostera marina* (Gagnon *et al.* 1980) and even desert 'annuals' such as *Astragalus lentiginosus* and *Tridens pulchellus* (Beatley 1970) have populations in which some individuals are annuals and others perennate. Perennation itself is an important life history trait that influences λ (Chapter 10).

An important reason why so many herbs sit astride the divide between the annual life history and a longer-lived one is because perennation is inherent in the modular construction of plants, which grow by the addition of new ramets. In *Poa annua*, and indeed all grasses, tillers die after flowering. If all tillers flower simultaneously in the first year, the genet is annual and dies after reproduction. If some tillers remain vegetative in the first year, the plant will perennate. In fact it is generally believed that annual species derive from perennial ancestors and therefore we should really think of annuals as plants that have evolved precocious reproduction, not as plants unable to perennate.

6.3.2.1 Herbs

Most perennial herbs are composed of collections of a few to many ramets. The entire demography of such herbs is dependent upon the birth rates and death rates of their parts, because as long as this remains in positive balance the genet is alive and will grow. The genet dies with the death of its last ramet. Perennial herbs vary greatly in their longevity, but this has little if anything to do with the longevity of parts. Tillers of the grass *Festuca rubra*, for example, are quite short-lived, but Harberd (1962) found that most plants of this species in a 90×90 m area in a Scottish population belonged to one of only a few large clones. One clone at the site had fragments distributed over an area more than 200 m in diameter and must have been quite ancient (Harberd 1961).

Even herbs without the tendency to spread shown by *Festuca rubra* may live to considerable ages. The woodland herbs *Hepatica nobilis* and *Sanicula europaea* were mapped in permanent plots in Sweden by Tamm over a period of nearly 40 years (Inghe & Tamm 1985, 1988). Both species have a rhizome, which slowly advances by tiny increments as the annual shoot dies and a new shoot replaces it. Individuals recorded by Tamm fall into two distinct classes, a group of proven survivors he called the 'old guard' that were present when he first mapped the plots in 1943 and many of which were still alive in 1981, and a group of individuals recruited since 1943 that had a much higher mortality rate. By fitting exponential decay curves to the survivorship curves for the old guard, Inghe and Tamm (1985) calculated half-lives in different plots ranging from $32-360$ years for *Hepatica nobilis* and from $74-221$ years for *Sanicula europaea*. These half-lives may underestimate genet survival because in both species the rhizome branched. Although branching occurred relatively infrequently, it was sufficient to confer virtual immortality on the old guard in some plots.

As Tamm's study illustrates, there may be significant demographic differences between different parts of the same population. Variation in demographic parameters from place to place and year to year occurs in most herbs that have been studied, though few populations have been monitored for as long as Tamm's, so it is difficult to know how important demographic variability is in the long run. The answer should depend upon the longevity of adults and the sensitivity of λ to recruitment, because this tends to be the most vulnerable stage in the life cycle, and consequently also the most variable. Other things being equal, the populations most at risk of extinction from demographic variability should be those with a short lifespan and low population density. How do such populations manage to persist? Rabinowitz *et al.* (1989) investigated this question for four species of prairie grass whose populations are typically sparse (i.e. occur at low density). Over a 9-year period at Tucker Prairie, Missouri, the fecundity of the four sparse species varied far less than that of three common ones at the same site. The reason was that the sparse species flowered early in the year when rainfall was fairly reliable, but the common species flowered later when rainfall was more uncertain. It appears that early flowering buffered the sparse species against reproductive failure due to drought. The common species were probably better able to survive frequent reproductive failure because they are longer-lived and occur at higher population density than the sparse species.

All demographic studies are plagued by the problem that data collected over only a few years may be atypical. Also, variability between years may itself play a role in population dynamics and life history evolution that is obscured by using mean values (Chapter 10). Sensitivity analysis can be used to overcome the first problem to some extent, because it can test the robustness of λ to perturbation caused by changes in the value of particular demographic parameters. Moloney (1988) studied a population of downy oatgrass *Danthonia sericea* growing along a soil fertility gradient in a mown field in North Carolina. Plants were mapped three times at annual intervals, and classified into six size classes for matrix analysis. No plants in the two largest size classes died during the study, even though there was a drought in one of the two years. In the smaller size classes there were significant differences in survival between sites along the gradient, and between years. Elasticity analysis showed that recruitment made little contribution to λ anywhere on the gradient, and that averaging over all locations and both years the leading diagonal of the matrix, representing survival of plants in their existing size class, contributed about 70% and the sub-diagonal (growth to the next size class) most of the remainder. Experimental addition of seeds to cleared and uncleared plots showed that recruitment was largely controlled by the availability of colonization sites (Moloney 1990), competition for which regulated the population. Although Moloney found significant variation from year to year in some parameters of the projection matrix, this may not be important in the long run because the population is regulated and the stages with the largest influence on λ were large plants whose survival was not affected by year-to-year variation. Long-lived adults, like a seed pool, are a cushion against temporal variability.

Natural selection can generate spatial variation in demographic parameters. This kind of evolutionary divergence has been elegantly demonstrated in ribwort plantain, *Plantago lanceolata*, by van Groenendael who compared populations in a dry dune grassland and a wet meadow in the Netherlands. The dune grassland was nutrient poor, subject to drought and had been grazed by cattle and horses for over three centuries. By contrast the wet meadow was on waterlogged peat, was mown once a year and was more fertile. Substantial demographic differences were found between the populations of *P. lanceolata* on the two sites. Juveniles in the dry grassland had much higher survival and adults much lower survival than in the wet meadow; fecundity in the dry site was higher and seed size was half that in the wet meadow; plants of the dry grassland first reproduced at 1 year of age, while in the wet meadow this occurred at 3 years; and dry grassland plants produced vegetative side-rosettes while wet meadow plants did not (van Groenendael & Slim 1988). Elasticity analysis (see Box 6.2) of Goodman matrices for the two populations pointed to some underlying demographic similarities between them, as well as confirming the significance of their differences. Although there was a significant seed pool at the dry site, this was of little importance in either population. Fecundity was subsidiary to establishment and survival in both populations, though it was more

important in the dry grassland (van Groenendael & Slim 1988). Timing of germination was different in the two populations and significantly influenced λ. At the dry site most seeds germinated in spring, while at the wet site most germinated in autumn.

By raising plants from the two populations in a uniform greenhouse environment, van Groenendael (1985) showed that a number of the differences observed in the field had a genetic basis. Seeds, juveniles and adults were also transplanted between habitats to determine the extent to which adaptation to the native habitat created a handicap in a different environment. After 2 years in alien habitats, transplants showed significant decreases in survival compared with natives (van Groenendael 1985). Seedling transplants of *P. lanceolata* were also made by Antonovics and Primack (1982) between different habitats in North Carolina. In this instance the fate of transplants was different between sites, but there was no detectable genetic component to how seedlings of different origin survived in different environments.

Population regulation is an important stabilizing force (Chapter 4) which the majority of studies using matrix analysis omit. A density-dependent matrix model for the woodland perennial *Viola fimbriatula* was compiled by Solbrig *et al.* (1988a) from extensive field studies that included the seed pool as well as the above-ground population (Solbrig *et al.* 1988b). Seeds were classified according to their dormancy status and depth of burial in the soil. With time, dormant seeds worked their way down the soil profile. Near the surface seeds had a high probability of survival and a 10% probability of germination; at lower depths survival increased slightly and germination probabilities dropped rapidly, reaching zero below 5 cm. Growing plants were classified into 10 size classes by the number of leaves, which was closely correlated with fecundity and survival (see Fig. 4.11). Seedling survival was density-dependent, with a cut-off that reduced seedling survival to zero when plants in the model reached a density equivalent to about $50\,\mathrm{m}^{-2}$. The model was too complex for the application of elasticity analysis, so the sensitivity of the population's behaviour to changes in demographic parameters was determined by simulation.

Most versions of the model showed stable oscillations around an equilibrium density, induced by density-dependence and the time lag between the death of adults and their replacement by recruits. Changes to the parameters governing the seed pool altered the amplitude of oscillations in adult numbers, but did not damp them out altogether. This is contrary to the finding of Thrall *et al.* (1989) who found that a seed pool damped oscillations in *Abutilon theophrasti*. Changing the form of density-dependence to one that did not involve a sharp cut-off damped the oscillations. As one might expect then from what we learned about density-dependence in Chapter 5 (see Fig. 5.4), the exact form in which density-dependence operates will alter population dynamics.

6.3.2.2 Forest trees

The recruitment of new individuals into tree populations is controlled by disturbance of the forest canopy and the appearance of gaps. Forest trees fall into two broad categories: pioneers and primary species, both of which depend upon gaps for recruitment, though in different ways (Whitmore 1989).

Pioneers such as *Betula* spp. in northern temperate forests and *Cecropia* spp. in the neotropics appear soon after gaps are created, germinating from seeds already in the soil or from seeds brought in by wind or animals. Pioneers are fast growing, short-lived and produce copious, regular crops of small seeds that recharge a dormant seed population in the soil. These seeds are unearthed by the upheaval of roots when a tree falls (Putz 1983; Whitmore 1983). Seeds of *Cecropia obtusifolia* in Mexican rainforest are triggered into germination by the light conditions in gaps (Vázquez-Yanes & Smith 1982), and in French Guiana *Cecropia obtusa* actually establishes best on the mounds of fallen trees (Riera 1985; Nuñez-Farfán & Dirzo 1988). Germination of another neotropical pioneer *Heliocarpus donnell-smithii* is stimulated by the diurnal temperature fluctuations characteristic of gaps (Vázquez-Yanes & Orozco-Segovia 1982). The pulsed recruitment typical of pioneer species generates local patches of even-aged individuals.

In temperate regions at high elevations, and where disturbance such as fire is recurrent, pioneers may replace themselves (Whitmore 1988), but these trees are relatively intolerant of shade as juveniles and are generally succeeded by primary species that are slow growing, but more shade-tolerant as well as longer-lived. This was the pattern found by Brokaw (1985) in gaps in the seasonal tropical forest at Barro Colorado Island, Panama (BCI).

Seed production by primary species such as the oaks *Quercus* spp. and beeches *Fagus* spp. of the northern temperate zone or the dipterocarps of Southeast Asia is typically irregular (Nilsson & Wästljung 1987; Ashton *et al.* 1988). Every 5–10 years they produce huge crops of seeds, but between times few or none. The seeds are often large and do not remain viable for long. This, and the heavy losses due to animal predation, prevent the formation of a dormant seed pool. Although irregular, fruiting is usually synchronized within a species, and sometimes between species, over large geographical areas. This pattern of seed production is called **masting** and appears to reduce losses to predation by swamping seed consumers with food in mast years and starving them in the intervening years of scarcity (Chapter 10).

With no dormant seed pool from which recruits can germinate when a gap opens, but with the ability to tolerate shade, primary tree species recruit from seeds that germinate under a closed canopy. These produce stunted juveniles that linger around the photosynthetic compensation point, waiting for a gap to throw light down upon them. Like the character Oskar in Gunter Grass' novel *The Tin Drum* these juveniles age but do not grow. Until, that is, a gap appears admitting light. The forest floor in temperate and tropical forests alike may be littered with oskars, many of them decades old. In beech–maple forest in New York State, Canham (1985) found that mature sugar maples had gone through several cycles of suppression and release before reaching the canopy. Hemlock *Tsuga canadensis*, another primary species of forests in Northeast America, typically shows signs of successive bouts of suppression and release in the width of growth rings in the trunks of mature trees (Henry & Swan 1974).

Chapter 6

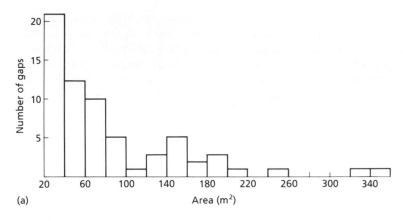

(a)

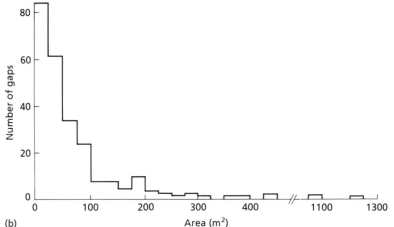

(b)

Fig. 6.6 The size distribution (m² surface area) of forest gaps. (a) Sixty-six gaps occurring over a 5-year period in 13.4 ha of tropical forest on Barro Colorado Island, Panama (from Brokaw 1982). (b) Two-hundred and forty-three gaps occurring over a 2-year period in 34.7 ha of temperate beech forest at Tillaie, Fontainbleau, France (from Faille *et al.* 1984).

Oskars of *Aglaia* spp. growing in Malaysian dipterocarp rainforest also needed several openings of the canopy before reaching maturity (Becker & Wong 1985).

Most forest gaps are small (Fig. 6.6), caused by branch fall, and these are rapidly filled by the growth of the surrounding canopy. Slightly larger gaps are created by trees that die on their feet (called snags) and oskars of primary species may be 'released' by these for a brief period, but pioneers usually require larger gaps. Large gaps occur when trees are blown over, often taking their neighbours with them. At BCI, Brokaw (1985) found that the pioneer *Trema micrantha* was strictly confined to large gaps. In several North American forests studied by Runkle (1982) larger, longer-lasting gaps were colonized by more species than smaller ones that offered a briefer opportunity for recruitment.

Gap-dependent regeneration creates a mosaic of patches of different species and age composition (Fig. 6.7a). At BCI Hubbell and Foster (1986) found that the abundance of shade-intolerant species was correlated with the abundance of canopy gaps. Sarukhán *et al.* (1985) devised an ingenious way to map the mosaic age structure in a 5 ha permanent plot of tropical rainforest in Mexico, using the understorey palm *Astrocaryum mexicanum* which was abundant there. The palm itself can be aged from leaf scars on its trunk. Tree and branch falls in the forest knock over any slender *Astrocaryum* stems beneath them. Palms prostrated in this way are rarely killed in the process, but continue growth by turning the apical shoot to the vertical again. This produces a kink in the stem of the tree that can be used to calculate how long ago the palm was knocked down by counting leaf scars back from the apex. The map produced

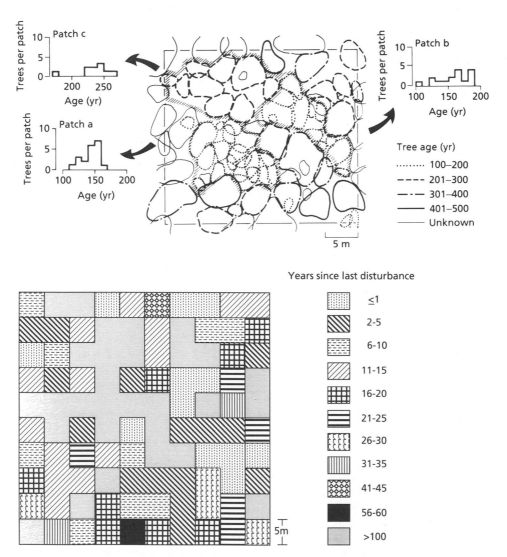

Fig. 6.7 Above: the age-structure mosaic of a 30 × 30 m plot in a subalpine *Tsuga−Abies−Picea* forest in Honshu, Japan. Trees in each patch are similar but not identical in age because they began life as oskars (from Kanzaki 1984). Below: the age-structure mosaic in a 50 × 50 m plot of tropical rainforest at Los Tuxtlas, Mexico (from Sarukhán *et al*. 1985).

by this means showed that a large proportion of the forest had experienced some form of disturbance in the previous 100 years (Fig. 6.7b). Gaps occurred particularly frequently on steep slopes, and their annual rate of formation was correlated with annual rainfall (Martínez-Ramos *et al*. 1988).

The demography of *A. mexicanum* was studied by Piñero *et al*. (1984) who found that its population in the study area as a whole was at equilibrium. However, λ was most sensitive to the growth rate of trees and, although it is a primary species that is tolerant of shade, adult *A. mexicanum* grew faster and produced more fruit in gaps than under a closed canopy. This created local variation in the value of λ, depending upon canopy conditions (Martínez-Ramos *et al*. 1989).

It may seem a paradox that forests where parts of the canopy are continually being torn down and regrown are at equilibrium, but this tension between disturbance and stability is the very essence of how forests work. A study of tropical rainforest at La Selva in Costa Rica, for example, found that over a 13-year period there was an annual loss of 2% of all tree stems greater than or equal to 10 cm diameter at breast height (DBH), nearly a third of which fell and a quarter of which died standing (Lieberman *et al.* 1985a). This remarkably high rate of loss was compensated *exactly* by recruitment. There were 105 deaths ha^{-1} over the period and 104.3 recruits ha^{-1} (Lieberman & Lieberman 1987). This demonstrated numerical equilibrium in the forest as a whole, but there was also evidence that populations of individual species were at equilibrium too. For a sample of 44 tree species at La Selva, Lieberman *et al.* (1985b) found that the rate of recruitment was negatively correlated with the longevity of species (Fig. 6.8), or in other words mortality and recruitment rates were positively correlated for individual species as well as for the forest as a whole.

The mosaic pattern of forest tree populations is a clear example of the importance of spatial structure, which we examine in the next chapter.

6.4 SUMMARY

Populations in which recruitment is a frequent event by comparison with the lifespan of individual plants develop an **age structure**. Even among plants of uniform age, size may vary considerably and create a **stage structure**. A population structured by age or size may be represented by a **column vector** which lists the number of individuals in each age or size class at a particular time. This vector may be used to project the numbers in each class at a later time by multiplying the column vector by a **projection matrix** which contains the rates at which individuals move from one class to another during the appropriate time interval.

Annuals with a pool of dormant seed in the soil have an age structure which provides a demographic buffer against extinction and a genetic buffer against drift and selection. The importance

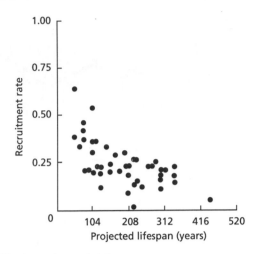

Fig. 6.8 Relationship between recruitment rate and longevity for 44 species of trees at La Selva (from Lieberman *et al.* 1985b).

of the seed pool, or indeed of transitions involving any component of the life cycle, to the value of λ can be calculated by **sensitivity** and **elasticity analysis** of the projection matrix. Because λ can be used as a measure of fitness as well as a measure of population growth, elasticity analysis indicates the contribution of different parts of the life cycle to **fitness**.

The ramets of perennial herbs usually have a relatively brief lifespan. The genets of these herbs may reach large size and live to advanced age, though survival rates tend to vary a great deal in space and through time. Spatial variation in mortality rates can produce **disruptive selection** that leads to **genetic differentiation** between populations. Forest trees fall into two broad categories: **pioneers** and **primary species**. Pioneers are usually the first trees to appear in treefall **gaps**, but are generally replaced by primary species that are able to recruit under a closed canopy. However, primary species do also usually require small canopy gaps to permit recruits to reach the canopy. The effect of canopy gaps on recruitment creates a mosaic pattern of age structure and species composition in forests.

Chapter 7
Metapopulations

7.1 INTRODUCTION

So far in our exploration of population dynamics, we have implicitly assumed that the principal demographic processes that affect numbers are birth and death, and that immigration and emigration are of little numerical importance, even though migration may be significant enough to create gene flow (Chapter 3). In some circumstances this omission is certainly wrong, and all four vital rates significantly affect N.

All populations have a patchy distribution at some scale or other. Patchiness at a local scale often reflects past or present spatial heterogeneity in the habitat. The mosaic structure of forest composition reflects past spatial patterns of disturbance and recruitment (e.g. see Fig. 6.7). In herbaceous vegetation micro-habitats such as gopher mounds in Californian, annual serpentine grassland (Hobbs & Mooney 1985), vernal pools in California (Holland & Jain 1981), badger mounds in tallgrass prairie (Platt 1975), molehills and old dungpats in pastures (Jalloq 1975; Parish & Turkington 1990a, b) and ant hills in British chalk grassland (King 1977) are all colonized by species rare or absent in surrounding vegetation. Should each of these island-like patches be regarded as separate populations, or fragments of a whole? The answer must depend upon how interdependent the population dynamics of different patches are, and how much gene flow there is between them. Populations linked by significant flows of individuals, these are generally seeds in the case of plants, constitute a **metapopulation**, or a population of populations (Levins 1970).

Erickson (1945) compiled what is possibly still the most complete picture of the spatial distri-

bution of any plant. He mapped the perennial herb *Clematis fremontii* var. *Riehlii* at a range of scales from the location of individuals within populations to the entire geographical range of this variety, which was limited to an area of $1129 \, \text{km}^2$ in Missouri. Within this area the plant was confined to rocky outcrops called 'barrens' where the broad-leaved forest cover was broken, forming open glades. Thus, only about 1.5% of its geographical range was colonizable to *C. fremontii* var. *Riehlii* and its distribution was highly fragmented. The plant occurred in four major regions, each region contained on average 50 glade clusters, each cluster contained a mean of 30 glades, and each glade contained a population averaging 10 distinct patches (Fig. 7.1).

At the smallest scale patches were probably close enough together for pollen flow to occur between them, and maybe for seed dispersal to carry seed from one to another. Dispersal between glades must be very limited and some glades within the plant's range were found where *C. fremontii* var. *Riehlii* was absent (Erickson 1943). The complex structure of this plant's spatial distribution is probably typical of rare taxa.

Just as the size of a population is determined by the birth rate and death rate of individuals, the size of a metapopulation is determined by the rates of establishment and extinction of component populations. These rates in turn depend, among other things, upon rates of migration by individuals between patches.

7.2 DISPERSAL BETWEEN PATCHES

In open habitats high density patches may act as a source of seeds that disperse into nearby marginal areas where plants can establish but not maintain

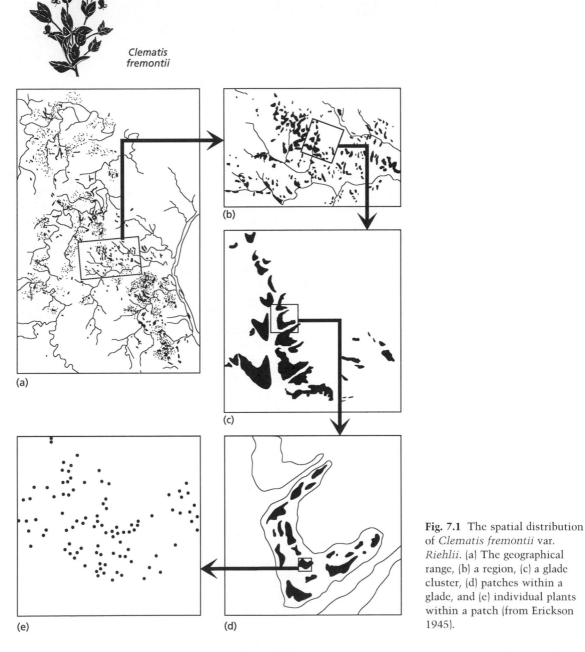

Clematis fremontii

Fig. 7.1 The spatial distribution of *Clematis fremontii* var. *Riehlii*. (a) The geographical range, (b) a region, (c) a glade cluster, (d) patches within a glade, and (e) individual plants within a patch (from Erickson 1945).

$R_0 > 1$. High density patches are **source populations** and marginal patches are **sink populations** (Pulliam 1989). The landward population of *Cakile edentula* on the transect studied by Keddy (1981)

(see Fig. 4.17) was a sink maintained by seed migration from a source at the seaward end of the transect (Watkinson 1985). Kadmon and Shmida (1990) assessed the relative contributions of seed dormancy, seed dispersal and *in situ* seed production to λ in two populations of the annual desert grass *Stipa capensis* in Israel. The seed pool made only a minor contribution to either population, but in one it was estimated that 90% of plants arose from seed dispersed from elsewhere. In the savanna woodlands of Northwest Australia seeds migrating from dense patches of the grass *Sorghum intrans* sustain sink populations of lower density in poor quality patches (Watkinson *et al.* 1989).

Conceptually, the spatial structure of a metapopulation can be represented by a series of life-cycle graphs, one for each population, linked together by paths that represent migration between their appropriate stages (Fig. 7.2). Such metapopulations can be analysed using an extension of the Lefkovitch matrix methods already applied to age- and stage-structured populations (Caswell 1989). The metapopulation dynamics of the pioneer tree *Cecropia obtusifolia* colonizing treefall gaps in rainforest at Los Tuxtlas, Mexico was analysed in this way by Alvarez-Buylla and García-Barrios (1991) (Fig. 7.3). Although this species does have a significant pool of dormant seed in the soil these are relatively short-lived, and seeds brought into new gaps by birds were the

source of about half of all recruits (Alvarez-Buylla & Martinez-Ramos 1990).

Horvitz and Schemske (1986) studied the demography of the clonal herb *Calathea ovandensis* in the ground layer of another evergreen tropical forest in Mexico. Plants grow best in new treefall gaps, and the forest mosaic creates patchy conditions on the ground which gradually deteriorate for *Calathea* as canopy gaps close. Seeds of the plant are dispersed by ants which moved a small proportion over 10 m at their study site — far enough for dispersal between patches to take place. Horvitz and Schemske (1986) constructed a transition matrix model incorporating 10 submatrices to represent 10 different kinds of patch and examined the effect on λ for the whole metapopulation matrix of varying migration between patches (Fig. 7.4). They found that dispersal between patches lowered λ and should therefore be selected against in this particular species, despite the patchiness of its habitat.

7.3 METAPOPULATION DYNAMICS

Metapopulation dynamics — or changes in the number of *populations* with time — can be dealt

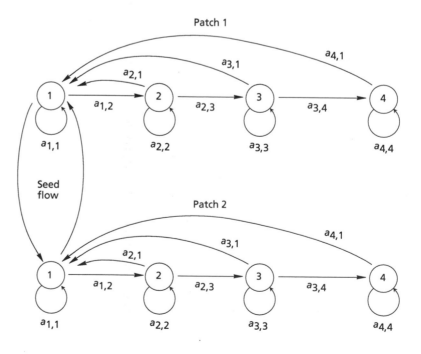

Fig. 7.2 Life-cycle graphs for a metapopulation of two patches (populations) linked by seed flow. The populations each have four stage classes. Within each population the values $a_{x,1}$ *are fecundities, values* $a_{x,x}$ *are the* rates of survival for plants that do not change stage class and values $a_{x,x+1}$ *are the rates at* which plants move up a stage class (compare with Fig. 1.5b).

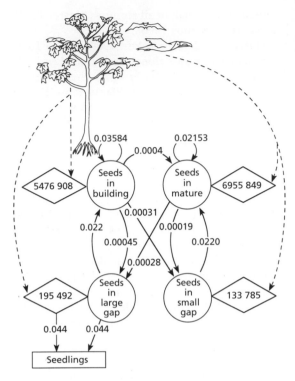

Fig. 7.3 A metapopulation model for *Cecropia obtusifolia* based upon rates of seed production and transport in four types of forest patch (circles); *C. obtusifolia* only recruits successfully in large gaps. Diamonds enclose numbers of seeds falling into each patch type, arrows show probabilities of seeds in one patch type being transferred to another by dispersal, and by a patch changing from one type to another due to treefall or canopy development (from Alvarez-Buylla & García-Barrios 1991).

with in an analogous manner to population dynamics (Chapter 5, Eqn 5.1). Imagine a species that requires a particular type of site for successful establishment and reproduction and that these sites have a scattered distribution. The plants of badger or gopher mounds, forest gaps and vernal pools are all examples. We will call the number of populated sites at time t P_t and the number of vacant sites V_t. Populated sites become extinct at a rate x per population per time interval and vacant sites are colonized at a rate c per population per time interval. The number of populated sites at time $t + 1$ is then:

$$P_{t+1} = P_t + cP_tV_t - xP_t \qquad (7.1)$$

where cP_tV_t is the number of new sites that are colonized and xP_t the number of sites where existing populations become extinct. This simple equation makes some interesting predictions about how metapopulations will behave. At equilibrium $P_{t+1} = P_t$ so from Eqn 7.1:

$$cP_tV_t = xP_t \text{ so, } V_t = \frac{x}{c} \qquad (7.2)$$

which means that if a species is to spread ($P_{t+1} > P_t$), the density of vacant sites V_t must exceed the **relative extinction rate** x/c. Only very slight changes in extinction or colonization rates are needed to cross the **threshold** from a situation where $V_t > x/c$ and a species is spreading, to a violation of these conditions that will cause extinction of the metapopulation when $V_t < x/c$.

Carter and Prince (1981, 1988) have pointed out that the existence of a threshold makes metapopulations sensitive to small changes in relative extinction rate and may explain why some short-lived species, such as the wild lettuce *Lactuca serriola* that they studied in England, have very abrupt distribution limits. When experimentally introduced, such plants typically grow and set seed quite successfully beyond the boundaries of their natural distribution limits, but the species still fail to establish self-sustaining populations. Seeds of the annual *Phlox drummondii* which is confined to sandy, nutrient-poor sites were sown by Levin and Clay (1984) at stations along a transect crossing an abrupt part of the species' boundary that coincided with a change in soil type near San Antonio, Texas. Sites just beyond the boundary were able to support self-sustaining populations ($R_0 > 1$) where the soil was locally poor, but further away such sites were rare and competition from other plants made establishment impossible (Fig. 7.5). Pockets of sandy soil habitable by *P. drummondii* did lie well beyond the species' boundary, but Levin and Clay concluded that colonization rates there were too low and extinction rates too high for them to remain occupied.

The majority of plant species' distribution limits appear to correlate with geographical or topographical trends in climate, but trends in

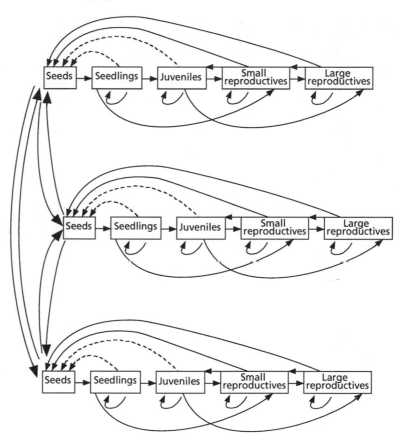

Fig. 7.4 Life-cycle diagram for metapopulation of *Calathea ovandensis* in three patches (after Horvitz & Schemske 1986).

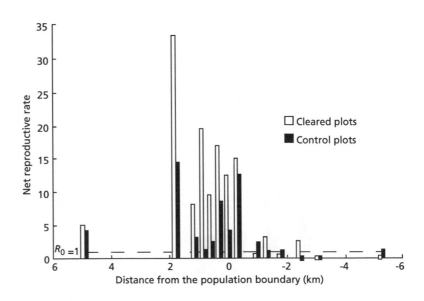

Fig. 7.5 Net reproductive rate R_0 of experimental populations of *Phlox drummondii* sown in plots cleared of vegetation and in control plots along a transect crossing the species' boundary in Texas. The boundary of the natural population is at zero (data from Levin & Clay 1984).

temperature and rainfall tend to be less abrupt than plant distribution limits. If climatic conditions near an abrupt distribution limit alter the relative extinction rate x/c very slightly, or if the density of vacant sites V_t is lower as in *P. drummondii*, neither need cause a measurable deterioration in the growth of individual plants at the boundary but a metapopulation may be unsustainable. The cause of boundary limits may lie in metapopulation dynamics, not the performance of individual plants or populations. This hypothesis is difficult to test experimentally, but two general observations support it. Populations of many species become less frequent near their distribution limits (Hengveld & Haeck 1982) (Fig. 7.6), suggesting that V_t may be falling or x/c increasing near their boundaries. Secondly, the habitat requirements of plant species often become more demanding at the edge of the range. For example, the box tree *Buxus sempervirens* reaches the northern limit of its European distributions in Britain where it is confined to calcareous soils, while in continental Europe it occurs across a broader pH range. In Hudson Bay, Canada where the saltmarsh annual *Salicornia europaea* is at the edge of its range, populations are confined to south-facing shores (Jefferies *et al.* 1983). Such narrowing of habitat distribution must decrease V_t.

Narrower physiological tolerances at distribution limits may be forced upon plants because climate interacts with other environmental variables such as soil type in determining plant performance, or because peripheral populations have lower genetic variation than those in the centre of a species' range. For example, Yeh and Layton (1979) found that allozyme polymorphism in lodgepole pine *Pinus contorta* was lower in marginal populations in the Yukon, northern Canada than in more central populations in British Columbia. Such a pattern is likely to be a consequence of smaller effective population size, founder effects, drift and perhaps directional selection in extreme environments. By contrast Levin (1977) found that allozyme variation in *Phlox drummondii* was not reduced at its boundary in Texas.

Since population biology is concerned with abundance, it should have something to say about why some particular species are typically rare

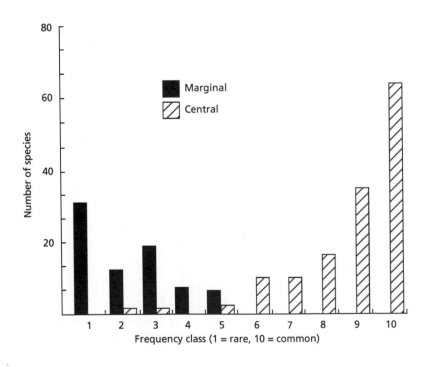

Fig. 7.6 Relationship between the frequency of localities for 818 plant species in Warwickshire, England and the position of Warwickshire in relation to the centre of each species' range (from Hengveld & Haeck 1982).

and others common, yet demographic studies of rare plants generally find satisfactory explanations very elusive. Many rarities that are of conservation interest are actually on the edge of their geographical range. Populations are few but individuals in them are abundant. Demographic studies that focus on these individual populations may be misdirected if rarity arises at the level of metapopulation dynamics. A little manipulation of Eqn 7.1 can show how sensitive the abundance of populations is to the relative extinction rate. Let the total number of sites $V_t + P_t = T$, and the *proportion* of sites that are populated $P_t/T = p$. Then $V_t/T = 1 - p$, and from Eqn 7.2 we can replace V_t with the relative extinction rate x/c to give:

$$p = 1 - \frac{x}{cT} \qquad (7.3)$$

Rare species occupy only a few sites and will therefore have low values of p. For a rare species with $p = 0.01$ only a 10% fall in the value of x/cT, from 0.99 to 0.90, is required to increase the proportion of occupied sites tenfold to $p = 0.1$ (Levins & Culver 1971). It is easy to imagine how slight ecological differences between species in the traits that influence x, c and T could be difficult to detect and yet radically affect their commonness or rarity.

In a study of a group of five perennials colonizing soil disturbances made by badgers in tallgrass prairie, Platt (1975) describes a community in which metapopulation dynamics appears to be responsible for differences between the species in their spatial distribution and abundance (Platt & Weiss 1977, 1985). The study site at Cayler Prairie, Iowa was on sloping ground where the density of ground squirrels was highest at the top and decreased progressively downslope. The density of badger disturbances, made by these animals digging for ground squirrels, followed the same pattern and decreased downslope, while soil moisture increased in the same direction. None of the five plant species studied had a persistent seed pool so all relied on seed dispersal to reach new disturbances. The species were segregated along the density gradient of disturbances in a remarkable way (Fig. 7.7). Only *Mirabilis hirsuta* occurred at one extreme of the density gradient

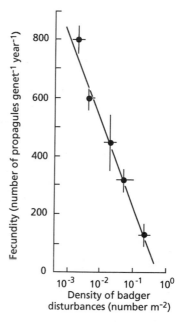

Fig. 7.7 Relationship of mean plant fecundity to mean density of sites occupied by five species of plant colonizing badger disturbances (from Platt & Weiss 1985).

because other species could not colonize the very dry sites that occurred there. Only *Apocynum sibiricum* occurred at the other extreme of the density gradient because no other species had sufficiently good dispersal to reach remotely spaced sites (Platt & Weiss 1977). Why though, did not all species occur at intermediate disturbance densities?

The answer seems to be that rates of colonization and extinction for each species were negatively correlated. The well-dispersed species such as *Asclepias syriaca* had high rates of colonization, but low rates of establishment and reproduction when they arrived at sites where other species were present (Platt & Weiss 1985). A high rate of extinction thus kept them out of areas where disturbance density was great enough for more competitive species to colonize. The more poorly dispersed species were confined to areas of high disturbance density because their colonization rates were too small in areas of low disturbance density. Their competitive superiority over other species ensured low rates of extinction where colonization was possible.

7.4 INVASIONS

Like the end of the rainbow, the tail of the seed dispersal curve is impossible to reach. The occasional seed is carried by chance events quite extraordinary distances, but these seeds are so few that we can only ever know where they end up when they attract attention by starting a new population in an alien site. Invasions are a lesson in the sometimes overwhelming significance of very rare events. Alien plant species are constant hitch-hikers on international trade, and botanists with an eye for the exotic have monitored these 'adventive' floras for nearly a century (Crawley 1987). We know from such records that, in Britain at any rate, about 20% of vascular plant introductions have become established and widespread, and of these about 25% are considered pests (Williamson & Brown 1986).

There is often a significant time lag between a species' first arrival and its subsequent spread. For example, the Brazilian tree *Schinus terebinthifolius* was first introduced into Florida in the 19th century, but it was not until the 1950s that it began an explosive spread across South Florida (Ewell 1986). Such a time lag could be more apparent than real if it represents the early phase of exponential population increase (see Fig. 5.2), or it might be an actual delay reflecting the existence of a threshold constraining the establishment of a metapopulation. Most invasive plant species cannot invade closed vegetation and human disturbance generally provides them with the opportunity to spread. The annual cheatgrass *Bromus tectorum* invaded Northwest USA with great rapidity at the beginning of the 20th century, permanently replacing native perennial bunchgrasses when they were overgrazed (Mack 1981). A similar story must have occurred in the history of the California annual grasslands that are composed almost entirely of aliens (Jackson 1985), though this replacement happened earlier than in the northwest and the event was not chronicled.

Once plants do begin to spread they can do so with incredible rapidity. The initial population of an invading weed typically spawns new colonies at a distance (e.g. Forcella & Harvey 1988), and it is these small outliers that push the range of an invading species forward so rapidly. Such invasions are a metapopulation phenomenon in which the rate of colonization consistently exceeds the rate of extinction. Moody and Mack (1988) have pointed out that control measures against invading alien plants usually target the largest, oldest foci of infection and ignore small populations that they term 'nascent foci'. However, numbers of scattered nascent foci make a bigger contribution to spread than a single large population of the same total area so control programmes that ignore nascent foci because they are small and inconspicuous are likely to fail.

Most of the plants of the northern temperate zone have spread to the northernmost parts of their ranges in geologically very recent time following the retreat of the ice at the end of the last glaciation (e.g. *Picea abies*, see Fig. 1.1). The rates of this spread have been calculated from pollen maps (e.g. see Fig. 1.1a) for the major tree genera and turn out to have been remarkably fast — of the order of 7 or 8 km per generation for oak *Quercus* and beech *Fagus* in North America for example (Johnson & Webb 1989). In a pioneering paper modelling dispersal Skellam (1951) found that after the retreat of the ice, oak spread into Britain at a rate far too fast for the process to have happened by diffusion along a single advancing front. A metapopulation model is required to account for such rapid spread. Today, the large seeds of species such as beech and oak in both North America and Europe are dispersed by jays and pigeons that may well have been responsible in the past for creating small nascent foci well beyond the trees' main front of advance, thus accelerating their spread and allowing them to move as fast as species with much smaller, wind-dispersed seeds (Bossema 1979; Johnson & Webb 1989). The passenger pigeon, that existed in North America in such huge flocks that they are said to have darkened the sky as they flew overhead, may have been important long-distance dispersers of seeds until the sudden extinction of the species under pressure from hunting at the beginning of the 20th century (Webb 1986).

Pollen maps do not yet have sufficient spatial or temporal resolution to reveal the structure of postglacial metapopulations, but there are gaps in some historical or present day distributions that suggest long-distance dispersal did take place. Webb (1987) suggests that an isolated population of *Fagus grandifolia* in eastern Wisconsin that lies 70 km beyond the current main distribution limit of the species in Michigan may have arisen by a very rare event of long-distance dispersal over Lake Michigan or the Illinois Prairie Peninsula. Birks (1989) found that Scots pine *Pinus sylvestris* arrived in Scotland and alder *Alnus glutinosa* arrived in Wales 8000 years BP, well ahead of other areas nearer the source of colonists for these species, and he suggests that these populations may also have been founded by long-distance dispersal. Lodgepole pine *Pinus contorta* may also have spread up the Northwest coast of Canada by this means (MacDonald & Cwynar 1991).

The manner in which new populations are founded has important genetic consequences (McCauley 1991). Founder effects may occur when new populations arise from small numbers of individuals (Chapter 2), and the dispersal process itself may select for characters that favour dispersal. For example, Cwynar and MacDonald (1987) found that the winged seeds of *P. contorta* became progressively smaller and more dispersable along a transect from southern British Columbia to the edge of its range near Alaska, where they estimated the species had only been resident 100 years. Of course there was a parallel change in climate as well as time in residence along this transect, so dispersal is not the only selective force that may have caused this pattern.

Because most seeds that are dispersed to a distance from an isolated population die, once colonization has taken place the direction of selection on dispersal-related characters reverses. Olivieri and Gouyon (1985) studied populations of the thistle *Carduus pyncnocephalus* that

produces seeds with a plume and seeds without in a ratio that is partly genetically determined. A higher proportion of plumed seeds was produced in recently founded populations than in old ones. It is intriguing that *Clematis fremontii* var. *Riehlii*, which lives in such isolated populations, has fruit that lack the pappus that aids dispersal in other *Clematis*. The hazards of long-distance dispersal for plants living in small, isolated patches would also seem to explain the poor dispersal of some desert plants (Ellner & Shmida 1981), and the paradox that plants on remote oceanic islands such as Hawaii have lost the dispersal powers of their mainland ancestors that got them there in the first place (Carlquist 1974). Metapopulation dynamics is a promising area for the integration of genetics with population ecology (Olivieri *et al.* 1990), but at the moment the genetic theory for spatially structured populations is significantly ahead of our knowledge of actual metapopulation dynamics in plants (Silvertown 1991b).

7.5 SUMMARY

The spatial structure of a group of populations at a large scale can be thought of as a **metapopulation**, which can be represented by a group of linked life-cycle graphs. Marginal **sink populations** may be sustained by seed flow from high density **source populations**. The dynamics of a metapopulation is determined by the rates of **colonization** and **extinction** of its component populations. The balance between colonization and extinction can be a fine one that generates a **threshold** which determines where the boundary of a species' **distribution** lies. Colonization depends upon the rate at which individuals move between populations, and rare events of long-distance dispersal may be of great demographic and genetic significance.

Chapter 8
Competition and Coexistence

8.1 THE VARIETY OF INTERACTIONS BETWEEN PLANTS

Plants usually live in mixtures of species. Populations are components of communities. Try as some might to maintain tidy monocultures, farmers generally fail to eradicate weeds entirely, and their successes are only as long-lived as the crop. Natural monocultures are often equally ephemeral, for example when stands of a single species spring up at the site of a disturbance where propagules happen to be plentiful. Clearly, the activities of farmers and the multifarious sources of disturbance in natural vegetation alter the balance between species, but what happens in mixtures when no external forces intervene? This is a deceptively simple question, which it has proved surprisingly difficult to answer. At least one reason for this is that *real* populations that are free from any 'external forces' are rarer than monocultures of hen's teeth!

There are multifarious definitions of competition, and the different approaches that may be taken to the subject continue to cause controversy (Keddy 1989; Grace & Tilman 1990), but to the simple mind of a population biologist what chiefly matters is the *outcome* of any interaction. What we want to know is how the interactions between the plants in a mixture affect the net reproductive rate R_0 of the component species. Straight away we can envisage a range of possibilities for the individual species, each of which may gain (+), lose (−) or be unaffected (0) by the presence of the other. In a two-species mixture both populations may be depressed (in shorthand this is a − − interaction); one may increase and the other lose (+−); both may benefit (++); or the interaction may be neutral for one species

and beneficial (+0) or detrimental (−0) to the other. These combinations define five kinds of interaction:

competition	− −
parasitism	+−
mutualism	++
commensalism	+0
amensalism	−0

Goldberg (1990) has argued that most interactions between plants occur through some intermediary such as light, nutrients, pollinators, herbivores or microorganisms. She suggests that we should dissect interactions between plants into the **effect** of each species on the abundance of the intermediary, and its **response** to an increase in the abundance of the intermediary. So, for example, a grass population limited by nitrogen availability will deplete that resource (− effect) and will increase if the nitrogen supply increases (+ response). The various permutations of positive and negative effect and response define interactions between plants based upon the net result for the intermediary. In this example the net result for the plant is negative (− effect × + response = net −), and two species interacting via nitrogen experience resource competition (Table 8.1).

Not every type of interspecific interaction between plants involves an intermediary, for example the parasitic, rootless dodders *Cuscuta* spp. rob their hosts directly from the phloem. However, for the majority of interactions to which the analysis in Table 8.1 does apply, the dissection of the interaction into effect, response and net result is useful because it highlights the *mechanism* underlying the interaction. This reveals that not all − − interactions need be due

Table 8.1 Types of interaction between plants, that involve an intermediary. The net result is the product: effect × response (after Goldberg 1990)

Type of interaction between species	Intermediary	Effect on intermediary	Response to intermediary	Net result (effect × response)
Resource competition	Resources	−	+	−
Apparent competition	Natural enemies	+	−	−
Allelopathy	Toxins	+	−	−
Positive facilitation	Resources	+	+	+
Negative facilitation	Resources	−	−	+
Apparent facilitation	Natural enemies	−	−	+

to straightforward competition and not all ++ interactions need be due to mutualism. Natural enemies may generate **apparent competition** and **apparent mutualism**. Proven examples of this are hard to find, because investigators of competition in the field scarcely ever investigate natural enemies as well. Connell (1990) suggests that the segregation into different habitats of scarlet oak *Quercus coccinea* and white oak *Q. alba* found by Futuyma and Wasserman (1980) in a forest in New York State could be a case of apparent competition. Both species are defoliated by larvae of the moth *Alsophila pometaria*, but in any particular habitat attacks are heaviest on the rarer of the two species present. Whether this pattern of attack by a shared natural enemy is the cause or the consequence of habitat segregation between *Q. alba* and *Q. coccinea* is not known. Other examples where competition between species may be mediated by natural enemies are discussed by Price *et al.* (1986, 1988).

As in the instance with two species of oak, interactions between plants often first come to the attention of ecologists when positive or negative patterns of association are noticed in their spatial distribution. However, patterns can be quite misleading about the processes that generate them. Some species, such as beech drops *Epifagus virginiana* that grows on the roots of beech trees, are associated because one is a parasite of the other. Others, such as the annuals *Minuarta uniflora* and *Sedum smallii* that are found together in shallow depressions in granite outcrops in North Carolina and Georgia, are positively associated because of similar habitat requirements. At a finer spatial scale these annuals

are segregated along a gradient of soil depth due to competition between them (Sharitz & McCormick 1973). In the Chihuahua Desert the pencil choya *Opuntia leptocaulis* is strongly clumped under the canopies of the creosote bush *Larrea tridentata*. This pattern is probably caused by birds which perch in the bushes and deposit choya seeds in their droppings. The nature of any direct interaction between the two plant species is unclear (Yeaton 1978).

When individuals of two plant species meet, it is generally assumed that the interaction between them will be antagonistic, and that to some degree each suffers from the presence of the other. This is indeed probable, because we know that plant species have very similar resource requirements to one another, but it should not blind us to the possibility that there is a richer variety of plant–plant interactions out there, waiting to be discovered (Hunter & Aarssen 1988). In the meanwhile the reader will find that we have a lot more to say about competition than about other types of interaction between plants. Such is the state of the science.

8.2 COMPETITION

The composition of mixtures is dynamic because populations are dynamic. Competition changes the composition of mixtures, or sometimes stabilizes it. The start-point and the end point of an experimentally contrived interaction can therefore be marked by the starting densities and the finishing densities of the species. A simple way to describe the composition of any mixture of two species is to plot the density of one against the density of the other on a two-dimensional graph called a **joint-abundance diagram** (Fig. 8.1a). Any point on this graph can be described by a pair of coordinates N_i, N_j denoting the density of species i and the density of species j in the mixture. The total density is, of course, $N_i + N_j$, and this total is constrained by the same limitations that constrain the density of a monoculture (Chapter 4). With increasing density, mixtures reach an upper limit just as monocultures do. If

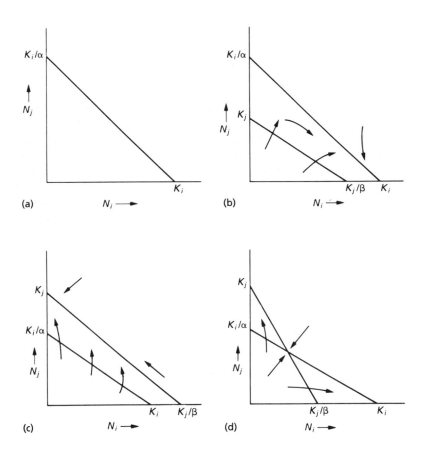

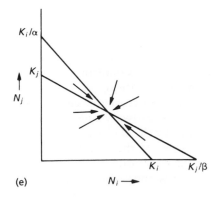

Fig. 8.1 Joint-abundance diagrams, illustrating various theoretical outcomes of competition between two species: (a) an isocline for one species, described by Eqn 8.1, (b) competitive exclusion of species j by species i, (c) competitive exclusion of species i by species j, (d) an unstable equilibrium, and (e) a stable equilibrium.

we assume, for the moment, that the precise ratio of the two species does not affect the upper limit on total density, then $N_i + N_j$ is a constant, which we will call K. This is the maximum sustainable density, or the **carrying capacity**, for any kind of mixture. All possible values of $N_i + N_j = K$ are represented on the graph by a straight line of slope -1 (Fig. 8.1a). This line intersects the horizontal axis at $N_i = K$ and intersects the vertical axis at $N_j = K$. Remembering that any straight line of negative slope can be defined by an equation of the form $y = a - bx$, we can write the equation for the line $N_i + N_j = K$ as:

$$N_i = K - \alpha N_j \qquad (8.1)$$

where K is the intercept and α is the slope of the line. Ecologically, the coefficient α represents the number of species j that must be lost to increase species i by a single individual. When we made the assumption that the ratio $N_i : N_j$ does not affect K we implicitly decided that the two species were equivalent, so $\alpha = 1$ in this example. The line $N_i + N_j = K$ can also be expressed in terms of N_j:

$$N_j = K - \beta N_i \qquad (8.2)$$

Here, the coefficient β represents the number of species i that must be lost to increase species j by a single individual. Again, because we have assumed that the competitors are equivalent to one another, its value is also one. The coefficients α and β are called **competition coefficients** and are fundamental to most theoretical formulations of competition between species in population biology.

Although in the simplest of examples $\alpha = \beta = 1$ and the intercepts on both axes are the same (K), in most mixtures the competing species will not be completely equivalent. In such cases α and β will have different values from each other, and the carrying capacities (K_i, K_j) that determine the intercepts will also have different values. When α and β, and K_i and K_j are different, Eqns 8.1 and 8.2 describe different lines (e.g. Fig. 8.1b). The lines, which are called **zero isoclines**, pass through the points on the graph that represent mixtures at carrying capacity. The lines are crucial boundaries between mixtures below the line where populations have net reproductive rates greater

than 1 (and will therefore increase), and points above where net reproductive rates are less than 1 (where populations will decline).

Using the isoclines as a tool, we can now explore how mixtures will behave, according to the relationships between the competitors that are described by the competition coefficients and carrying capacities of the species. This exploration is best tackled using the graphs in Figs 8.1b–e. First, note that now we have abandoned the assumption that the competitors are equivalent, Eqns 8.1 and 8.2 must be rewritten:

$$N_i = K_i - \alpha N_j \qquad (8.3)$$

$$N_j = K_j - \beta N_i \qquad (8.4)$$

You should be able to see how the intercepts of the lines, shown in Figs 8.1b–e, can be calculated from these two equations. The arrows drawn on the graphs indicate how the joint abundances of the competing species change, depending upon their starting densities. Figure 8.1b illustrates the conditions for the competitive exclusion of species j by species i. This will occur when $K_i > K_j/\beta$ and $K_i/\alpha > K_j$. When these two inequalities are reversed, species j will exclude species i (Fig. 8.1c). In the relationship shown in Fig. 8.1d either species can exclude the other, depending upon starting conditions, or upon chance perturbation (inevitable, sooner or later) of the 50/50 mixture where there is an unstable equilibrium. The conditions for this situation are $K_j > K_i/\alpha$ and $K_i > K_j/\beta$. Reversing these two inequalities creates a stable equilibrium mixture at the intersection of the two isoclines, as shown in Fig. 8.1e.

In what ecological circumstances are the conditions for stable coexistence of competitors fulfilled? For stable coexistence to occur, the inequalities $K_j < K_i/\alpha$ and $K_i < K_j/\beta$ must be satisfied. If the species are ecologically equivalent, then $K_i = K_j$ and $\alpha = \beta = 1$. By inserting these values into the inequalities you will see that the conditions for stable coexistence are violated. From this we draw the important conclusion that ecologically equivalent species cannot coexist. This is sometimes called the **competitive exclusion principle** or **Gause's principle**.

The simplest way both inequalities may be

met is if the maximum monoculture densities K_i and K_j are similar, but $\alpha < 1$ and $\beta < 1$. This implies that each species is more sensitive to its own density than to that of its competitor. The most obvious way that this can happen is if there is some kind of **niche separation** caused, for example, by differences in resource use between the species. Niche separation is not easily proved in plants, but one of the best examples has been demonstrated by Berendse (1981, 1982) who found that competition between two grassland species that commonly grow together, the grass *Anthoxanthum odoratum* and *Plantago lanceolata*, was ameliorated by *P. lanceolata* having deeper roots than the grass. When separation of the rooting zones was experimentally prevented in pot or field experiments, the grass had a significantly greater negative effect upon its competitor than in controls where deep rooting by *P. lanceolata* was possible. There was some evidence too that the presence of *A. odoratum* caused *P. lanceolata* to develop deeper roots

than it did when growing by itself. Competitive relationships have to be elucidated through experiments of this kind that compare plants growing in mixtures with plants in monocultures.

8.2.1 Competition experiments

There are four basic designs of competition experiment used to explore the behaviour of mixtures. The essential features of each can be described by plotting the mixtures they use on a joint-abundance diagram. The simplest experiments use the **partial additive** design, in which a constant density of one species is combined with a range of densities of another (Fig. 8.2a). The total density of the mixture and the relative proportions of the species vary in parallel in this design, which makes it impossible to disentangle their separate effects. This problem is partly alleviated by the **replacement series** design which maintains the *total* density of mixtures constant, and varies the ratio of species to each other

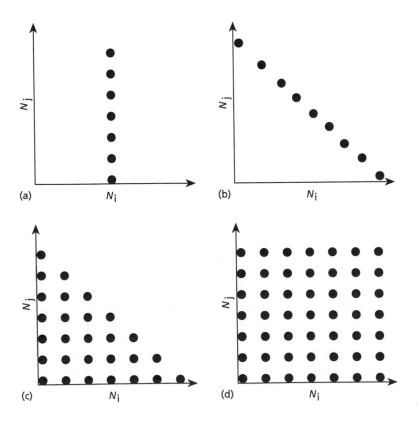

Fig. 8.2 The four basic designs of competition experiment plotted on joint-abundance diagrams for two species: (a) partial additive, (b) replacement series, (c) additive series, and (d) complete additive.

(Fig. 8.2b). This design has the drawback that it cannot be assumed that the proportions in a mixture have the same influence at different densities. Only the **additive series** and **complete additive** (Fig. 8.2c, d) designs explore the effects of varying density and proportions of species independently of each other.

Competitive interactions first influence the performance of plants, that is their size or their yield, and as a consequence of this may later affect density by altering survival or fecundity. Depending upon the time over which an inter-action is observed, changes in either performance or density may be chosen to measure the effects of competition. In agricultural mixtures of crops and weeds, performance is the usual choice.

8.2.1.1 Partial additive designs

This is the design often chosen to determine the effect of weeds on a crop, for example the effect of sicklepod *Cassia obtusifolia* or redroot pigweed *Amaranthus retroflexus* on cotton yield (Fig. 8.3a). The partial additive design has also been used to look at the effect of different numbers of neigh-bours of a competitor on a single individual of a 'target' species planted in their midst (Goldberg 1987; Goldberg & Fleetwood 1987; Miller & Werner 1987; Pacala & Silander 1990). Goldberg and Fleetwood (1987) used the design to compare the effect of different neighbour species upon five annuals. The results of these and similar experiments typically show a hyperbolic relation-ship between the weight of the target plant and the density of its neighbours (Fig. 8.3b). This indicates that the first few neighbours have a large effect upon the target, but additional ones have proportionately much less effect. The hyper-bolic response curve of a target plant to the density of competitors around it has essentially the same shape as the response of yield per plant w_i to density N seen in monocultures (see Fig. 4.1c). The equation we used to describe this in Chapter 4 was:

$$w_i = w_m (1 + aN)^{-1} \qquad (4.2)$$

where w_m is the maximum weight of an isolated plant and a is the area needed for a plant to achieve w_m. Equation 4.2 can describe the effect

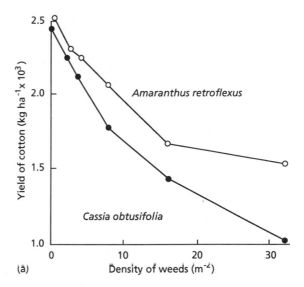

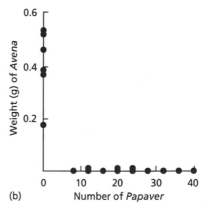

Fig. 8.3 Results of partial additive competition experiments between: (a) cotton and *Amaranthus retroflexus*, and cotton and *Cassia obtusifolia* (from Buchanan *et al.* 1980), and (b) *Papaver rhoeas* and *Avena sativa* (from Goldberg & Fleetwood 1987).

of competitors on a target of another species if we substitute N by an equivalent density of competitors αN_j. Recall that α is the competition coefficient that measures the equivalence of the two species in their effects on the target species:

$$w_i = w_{mi} (1 + a_i \alpha N_j)^{-1} \qquad (8.5)$$

The extra subscripts in the terms w_i, w_{mi} and a_i indicate that these measures apply to species i.

They would, of course, have different values for species *j*.

As can be seen from Fig. 8.2a, partial additive competition experiments are very limited in the range of mixtures they explore, and are really only useful if there is a particular reason to be interested in the effect of competitors on a fixed density of a particular species. One particular situation in which the partial additive design is useful is in separating the effects of root and shoot competition.

8.2.1.2 *Root and shoot competition*

There are a variety of ways of testing for the effects of competition for resources above and below ground. Soil resources in the form of water or nutrients may be added and the effects observed, and estimates of rates of utilization can be made (Eissenstat & Caldwell 1988). Alternatively, partitions can be used to separate competitors above or below ground, and to compare their performance with controls that are interacting fully, or not at all (Snaydon & Howe 1986). Groves and Williams (1975) used an additive design to look at the effect of root and shoot competition between subterranean clover *Trifolium subterraneum* and skeleton weed *Chondrilla juncea* which is a weed of cereal crops in Southeast Australia. *Chondrilla juncea* is able to persist through the fallow period between cereal crops when fields are used as pasture, so suppressing it with subterranean clover during this period could aid in its control as a cereal crop weed. Strains of the rust fungus *Puccinia chondrilliana* have been used as biological control agents of *C. juncea*, so this was included as a treatment in the competition experiment (Fig. 8.4a).

Trifolium subterraneum did not suffer significantly from competition with *C. juncea*, but the latter was suppressed by both root and shoot competition. Rust-free plants were reduced to 65% of control weight by root competition, to 47% of control weight by shoot competition, and to 31% (= 65% × 47%) by the combination of the two. An identical synergism between the effects of competition above and below ground occurred in the infected plants (Fig. 8.4a). An experiment using the replacement series design was used by Martin and Field (1984) to look at root and shoot competition between white clover *T. repens* and perennial ryegrass *Lolium perenne* (Fig. 8.4b). *Lolium perenne* greatly benefited in this interaction, and its suppressive effects on *T. repens* showed synergism between root and shoot competition (Fig. 8.4b). The generality of such interactions between root and shoot competition is still an open question.

Wilson (1988) surveyed the results of 23 studies in which investigators had used partitions to compare root and shoot competition. In the majority of cases (68%) root competition had a greater adverse effect on growth than shoot competition. Some of the competition experiments were repeated at a range of densities using the partial additive design, and in over 70% of these the effects of root competition intensified more rapidly as density increased than the effects of shoot competition. All of the species studied were agricultural crops (including pasture grasses) or weeds, so competitors would have been approximately equal in stature, limiting the potential for one species to shade the other and making it more likely that root competition would be detected. Wilson and Tilman (1991) grew three species of grass in field plots along a nitrogen gradient and compared their growth when competing only below ground with growth when competing both above and below ground. In all three species root competition from neighbours was relatively more important at lower levels of soil nitrogen than at higher levels. Root competition may therefore be more important in infertile natural communities than experiments with agricultural species suggest.

Just as different rooting depths can promote niche separation and coexistence between species, Ennos (1985) showed that the same might occur under certain conditions between genotypes of white clover with different rooting depths. He found that length of root had a high heritability in this species, and in an experiment he evaluated the significance of this trait for plant fitness. He selected plants having either relatively long (L) or short (S) roots and planted these in a replacement series competition experiment. Although the *total* number of stolons produced in monocultures of L or S was the same

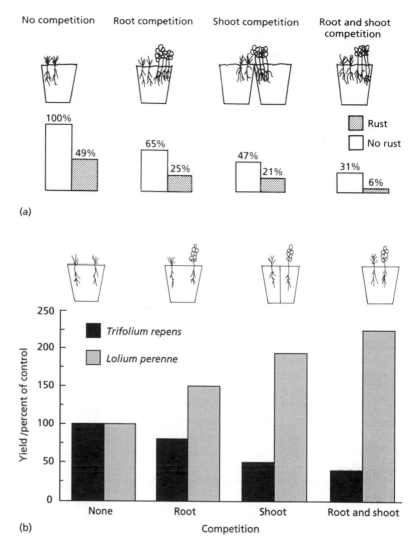

Fig. 8.4 Competition experiments separating the effects of root and shoot using two different designs. (a) Effects on skeleton weed of root and shoot competition with subterranean clover (after Groves & Williams 1975). (b) Effects on white clover (*Trifolium repens*) and perennial ryegrass (*Lolium perenne*) of root and shoot competition between them (Martin & Field 1984).

as in mixtures, L in mixtures produced 22% more stolons than S. Under conditions of drought though, the dry matter yield of a 50:50 mixture was significantly greater than that of either monoculture. Depending on the occurrence of drought then, the exploitation of different soil layers by the two genotypes could promote their coexistence and help maintain genetic diversity.

8.2.1.3 Replacement series

This design, which was first advanced by de Wit (1960), has been used very widely indeed to

compare the performance of species in mixtures with their performance in monoculture using their relative yields:

$$\text{Relative yield of species } i = \frac{\text{yield per unit area of sp. } i \text{ in mixture}}{\text{yield per unit area of sp. } i \text{ in monoculture}}$$

$$\text{Relative yield of species } j = \frac{\text{yield per unit area of sp. } j \text{ in mixture}}{\text{yield per unit area of sp. } j \text{ in monoculture}}$$

The sum of the relative yields for the two species in a particular mixture (usually 50:50) is called the **relative yield total** (RYT). Unfortunately, its value has often been misinterpreted. In fact RYT can only be interpreted under some rather restrictive conditions. The calculation of relative yields is based on the assumption that yields per unit area in monocultures do not change across the range of single-species densities used in the experiment, or in other words that these densities lie within the range that generates constant final yield (Chapter 4). *On condition* that this is true, an RYT > 1 indicates that there is a yield advantage in the mixture, *at that particular density and proportion* of the species. This may or may not apply to other densities and proportions, even if they meet the condition of constant final yield (Taylor & Aarssen 1989).

Validly interpreted, an RYT > 1 may indicate that there is some niche separation between the species (e.g. *Plantago lanceolata* versus *Anthoxanthum odoratum*), but this may not occur at other densities (Willey 1979; Connolly 1986), so one cannot conclude from RYT that mixtures of the two species meet the general conditions for coexistence that we derived earlier (Inouye & Schaffer 1981). Despite these handicaps, the replacement design can be useful for examining the effects of herbivory, disease or some other agent on competition.

For many purposes it is desirable to be able to determine when mixtures of crops produce a greater yield than would an equivalent area planted with separate monocultures. Because of the limitations of RYT as a measure of the performance of mixtures in relation to monocultures, Connolly (1987) proposed an alternative index called the **relative resource total** (RRT) to answer this kind of question. This requires yield/density curves for the two monocultures to be available. RRT is derived as follows. The yields per individual of two species in a mixture are w_i and w_j, when they are growing at densities N_i and N_j. For species i there will be some monoculture density N_{i0} that produces the same yield w_i as i in the mixture. The area occupied by a plant in this monoculture is a measure of the resources needed to produce a plant with yield w_i, and is given by the reciprocal of the monoculture

density, $1/N_{i0}$. To produce the equivalent yield of N_i individuals with yield w_i requires an area N_i/N_{i0}. The corresponding quantity for the other species is N_j/N_{j0}, and:

$$\text{RRT} = \frac{N_i}{N_{i0}} + \frac{N_j}{N_{j0}}$$

If the RRT > 1, the mixture is more efficient at turning resources into yield than are the monocultures. In effect, this measures the extent to which the species utilize different resources, and is a measure of niche separation. If RRT > 2, the species are yielding as if the other species in the mixture did not exist! Note that RRT > 1 does *not mean* that the mixture necessarily yields better than both of the monocultures. One of the monocultures may still perform better than the mixture (Connolly 1987).

The **land equivalent ratio** (LER) is another index, based upon a similar idea to RRT, that has been used a great deal in evaluating mixtures of agricultural crops. For species i and j producing yields per unit area in a mixture that are w_iN_i and w_jN_j:

$$\text{LER} = \frac{w_iN_i}{w_{i0}N_{i0}} + \frac{w_jN_j}{w_{j0}N_{j0}}$$

where $w_{i0}N_{i0}$ and $w_{j0}N_{j0}$ are the yields of i and j in the highest yielding monocultures, regardless of density. As with RRT, an LER > 1 implies that there is a benefit in combining the species, though this might not produce a yield greater than one of the monocultures. Like RYT, the LER of a mixture also has the disadvantage that its value is dependent upon the particular monoculture chosen as a standard of comparison. Riley (1984) describes a method for comparing LERs that have been calculated using different monoculture yields in the denominators.

8.2.1.4 Additive series and complete additive designs

Additive series and complete additive competition experiments are similar in design and effectiveness, so we will refer to both by the shorthand 'additive'. Additive experiments vary N_i and N_j independently of each other, and can therefore be used to map the isoclines in a joint-abundance diagram. However, whereas in Fig. 8.1 we first established the locations of the isoclines, and then used these to derive trajectories (the arrows in Fig. 8.1) for different mixtures, additive experiments allow us to work in the other direction. This provides a useful conjunction of theory and experiment. Each mixture in the experiment has a starting location on the diagram, determined by the densities N_i, N_j when the plants are first sown, and a finishing location determined by the densities the mixture ends up with. In practical terms this might actually be the number of seeds produced by each species in the mixture, if the plants are annuals. Drawing an arrow between the starting and finishing densities for each mixture indicates where the isoclines must lie. This is how the position of the isoclines shown in Fig. 8.5a, for a competition experiment between the two annuals *Salvia splendens* and *Linum grandiflorum*, were derived. Notice that the *Salvia* isocline is curved. Straight lines were used in Fig. 8.1 only for mathematical simplicity.

The results of this complete additive experiment showed that there was an unstable equilibrium between the species at densities of about 10 *Linum* : 8 *Salvia*. A replacement series experiment that happened to utilize a total plant density near 20 would pass through the equilibrium point and would give the misleading impression that the two species could coexist indefinitely, because the instability of the equilibrium cannot be detected using the replacement series. It is even more likely that a replacement series exper-

iment at an arbitrarily chosen density would fail to detect the existence of any kind of equilibrium at all, which would lead to an equally erroneous conclusion. The results of an additive series experiment with two annual grasses (Fig. 8.5b) illustrate the competitive superiority of *Vulpia fasciculata* over *Phleum arenarium* in all mixtures, although these grasses do occur together in sand-dunes.

The performance of a plant in a mixture is affected by both intraspecific competition and interspecific competition. In partial additive experiments only the interspecific component can be estimated. In the discussion of this experimental design above we found that the yield of a target species could be described by modifying a model used to describe density effects in monoculture (Eqn 8.5). Since additive experiments vary densities of both competing species simultaneously, intraspecific as well as interspecific effects on the yield of each species can be estimated (Watkinson 1981). Intraspecific effects may be incorporated in Eqn 8.5 simply by inserting a_iN_i in the right hand side:

$$w_i = w_{mi} [1 + a_i(N_i + \alpha N_j)]^{-1} \qquad (8.6)$$

For species j, the equivalent expression is:

$$w_j = w_{mj} [1 + a_j(N_j + \beta N_i)]^{-1} \qquad (8.7)$$

The yield/density response (Eqn 4.2) from which we have derived this equation assumes that yield approaches an asymptote but, particularly if the yield we are interested in is some plant part such as the seed, the yield/density response may have

Fig. 8.5 (a) A joint-abundance diagram with trajectories showing the outcome of competition between *Salvia splendens* and *Linum grandiflorum* in different starting mixtures (after Antonovics & Fowler 1985). (b) A joint-abundance diagram showing the isoclines derived for an additive series experiment with *Phleum arenarium* and *Vulpia fasciculata* (Law & Watkinson 1987).

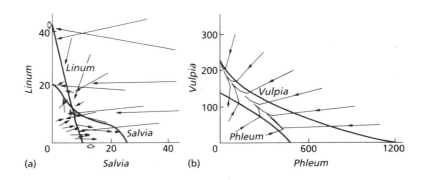

a different shape. As we saw in Chapter 4, this may easily be accommodated by replacing the exponent -1 with a variable $-b$. This exponent may assume a different value for each species, so in Eqn 8.6 it would be denoted b_i, and in Eqn 8.7 it would be b_j. Firbank and Watkinson (1985) fitted Eqns 8.6 & 8.7 to data from a competition experiment between wheat *Triticum aestivum* and the broad-leaved weed corncockle *Agrostemma githago*. The exponents for both species were less than unity, producing the yields per plant shown in the response surfaces in Fig. 8.6.

If the exponents of the yield/density responses are unity, then a simpler model than those in Eqns 8.6 & 8.7 can be used to calculate competition coefficients. This is the **reciprocal yield model** of Wright (1981) and Spitters (1983). Using the reciprocal of yield per plant $(1/w)$ instead of yield per plant (w) as the dependent variable produces a more friendly-looking formulation of yield/density relationship that has found favour in agricultural research (Radosevich & Rousch 1990). In a monoculture:

$$\frac{1}{w} = B_{i,0} + B_{i,i}N_i \qquad (8.8)$$

where $B_{i,0}$ is $1/w_m$ and $B_{i,i}$ measures intraspecific competition (the effect of species i on itself). Comparison of this equation with Eqn 4.2 should convince you that they are equivalent. Equation 8.8 can readily be expanded to two or more species. For species i in a mixture with species j:

$$\frac{1}{w_i} = B_{i,0} + B_{i,i}N_i + B_{ij}N_j \qquad (8.9)$$

where B_{ij} measures the effect of j on i and N_j is the density of j. The corresponding equation for species j is:

$$\frac{1}{w_j} = B_{j,0} + B_{j,j}N_j + B_{ji}N_i \qquad (8.10)$$

These two equations are the equivalents of Eqns 8.6 & 8.7, but separate the effects of intraspecific and interspecific competition more clearly. The ratio of interspecific effects to intraspecific effects in Eqns 8.9 & 8.10 gives us the competition coefficients of Eqns 8.6 & 8.7:

$$\frac{B_{ij}}{B_{ii}} = \alpha \text{ and } \frac{B_{ji}}{B_{jj}} = \beta$$

The reciprocal yield model has been fitted to a number of weed−crop mixtures. For example,

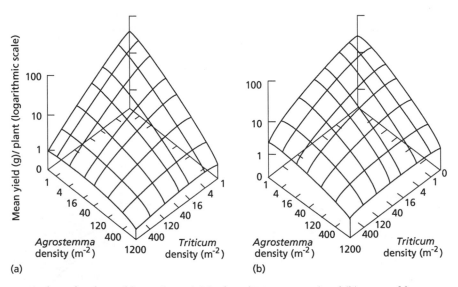

Fig. 8.6 Response surfaces for the yield per plant of: (a) wheat (*Triticum* spp.) and (b) corncockle (*Agrostemma githago*) in competition experiments (from Firbank & Watkinson 1985).

Rejmánek *et al.* (1989) used the model to describe the results of an additive competition experiment between tomato *Lycopersicon esculentum* cultivar 'Ace VF55' and Japanese millet *Echinochloa crus-galli* var. *frumentacea* which is a close relative of barnyard grass, one of the world's worst weeds. The model fitted the yields of both components of the mixtures well, and showed that tomato was more sensitive to interspecific competition than to intraspecific competition, while the reverse was the case for Japanese millet (Table 8.2). The competition coefficients show that each millet plant had an effect on tomato equivalent to 3.7 other tomatoes. Tomato was equivalent to only 0.14 millet plants in its effect on millet. In another experiment, with wheat and annual ryegrass, wheat had a greater effect on the weed (ryegrass) than vice versa (Table 8.2).

Analysing the results of competition experiments is, at least at the moment, very much an empirical exercise in finding which models best fit the data (Cousens 1985; Connolly 1987). This can make it difficult to generalize about plant competition, though there is also a lesson to be learned from the very difficulty itself. The outcome of interspecific competition can be highly contingent upon the conditions of the experiment, including the density and proportions of the species in a mixture. In their experiment with *Vulpia fasciculata* and *Phleum arenarium*, Law and Watkinson (1987) found that the model that best fitted their data allowed competition coefficients to vary with frequency and density. Equations such as the reciprocal yield model were inadequate to describe the interaction between the two species across the full

range of values of $N_i N_j$ that their experiment explored. In a similar competition experiment between two annuals, Connolly *et al.* (1990) found that competition coefficients varied with total density and time.

It is also important to caution that competition coefficients by themselves do not predict the outcome of competition or the relative fitness of the competitors. This depends upon the configuration of the isoclines in the joint-abundance diagram (Fig. 8.1), which is not determined only by competition coefficients.

Firbank and Watkinson (1985) adapted their yield model in Eqn 8.6 to describe the effects of interspecific competition on plant densities and seed output, and applied it to the results of an experiment by Marshall and Jain (1969) who grew two species of wild oat, *Avena fatua* and *A. barbata*, in an additive series competition experiment. Competition between the species affected their mortality and their seed yield per plant, which were treated separately. To model mortality, mean plant weights in Eqn 8.6 become plant densities, and the parameter *a* becomes the reciprocal of maximum plant density *m* after self-thinning (as introduced in Eqn 5.12). The equations relate the harvest densities of the species N_F (*A. fatua*) and N_B (*A. barbata*) to their initial densities N_{iF} and N_{iB}:

$$N_F = N_{iF} [1 + m_F(N_{iF} + \alpha N_{iB})]^{-1} \qquad (8.11)$$

$$N_B = N_{iB} [1 + m_B(N_{iB} + \beta N_{iF})]^{-1} \qquad (8.12)$$

In these equations $1/m$ is the carrying capacity K, and α and β are competition coefficients, so we can use them to see if the *Avena* species in this experiment meet the criteria for stable coexist-

Table 8.2 Parameter values for the reciprocal yield model fitted to two additive competition experiments, between tomato and Japanese millet, and between wheat and annual ryegrass

Species	Intraspecific effect B_{ii}	Interspecific effect B_{ij}	Competition coefficient B_{ij}/B_{ii}	Fit of the model R^2	Source
Tomato	0.003	0.011	3.67	0.94	Rejmánek *et al.* (1989)
Japanese millet	0.001	0.00014	0.14	0.95	
Wheat	1.18	0.17	0.14	0.90	Concannon in Radosevich
Annual ryegrass	3.21	4.51	1.41	0.43	& Rousch (1990)

ence or not. The parameter values calculated by fitting Eqns 8.11 & 8.12 to the experimental data were $m_F = 2.28 \times 10^{-3}$, $m_B = 1.83 \times 10^{-3}$, $\alpha = 0.61$, $\beta = 0.78$. The carrying capacities calculated from $1/m_F$ and $1/m_B$ are $K_F = 439$ and $K_B = 547$. The reader may verify that these values satisfy the conditions given by the inequalities on p. 119 for stable coexistence between species. This is not the whole story though, because Eqns 8.11 & 8.12 refer only to changes in density between seeds germinating and plants reaching maturity. As already mentioned, the two species also influenced each other's seed yield. This effect can be modelled by converting the competition equations that predict yield (Eqns 8.6 & 8.7) to a form that predicts seed output, using the same argument we applied to the yield equation for a single species in Chapter 5 (Eqns 5.9–5.11). Analysis of such equations showed that *A. fatua* had a greater depressive effect on the seed yield of *A. barbata* than vice versa, but that *A. barbata* made up for this disadvantage by producing more seeds per plant over a wider range of densities (Marshall & Jain 1969; Firbank & Watkinson 1985). All in all then, the competition experiments between the two species of wild oats give a satisfactory explanation of their coexistence in the field in South California (Marshall & Jain 1969).

8.2.1.5 *Field experiments and diffuse competition*

Two-species interference experiments or pairwise combinations among a set of species are the simplest design for cultivation experiments, but such experiments are often totally impractical in the field. Field experiments unavoidably involve many species. Competitive effects on a species which derive indiscriminately from all or many of the other species in a community have been described as **diffuse competition** by MacArthur (1972). More or less weak diffuse competition between all the plant species in a community is probably a common situation, with more severe competition occurring between particular species for particular limiting factors (Mitchley 1987).

The simplest field experiments on interspecific competition involve removing individual species from a community and monitoring the response of the remaining species. A number of experiments of this kind have been performed and the majority have demonstrated the existence of some degree of competition (Goldberg & Barton 1992). Such experiments also demonstrate that the behaviour of a plant in a mixture of two species may be quite different from its behaviour in more diverse mixtures. For instance, when plantain (*Plantago lanceolata*) was removed from field plots in a grassland community in North Carolina, USA, the abundance of winter annuals increased. However, this *only* happened if sheep's sorrel (*Rumex acetosella*) was absent from the experimental plot. Where *Rumex* was present and *Plantago* was removed, *Rumex* and not winter annuals benefited from the removal (Fowler 1981). Hence in the field situation the relationship between specific pairs of species is contingent upon the presence or absence of other species. Furthermore, computer simulations of model communities containing five competing grass species have shown that the spatial arrangement of species with respect to each other (who sits next to whom) may greatly influence population dynamics and community composition over the medium term (Silvertown *et al.* 1992).

Competition between several species at once can lead to the apparently paradoxical result that a species may decrease in abundance when another species is removed because of the effect this removal has on a third species. Such an effect was found in the study of a community of desert annuals by Davidson *et al.* (1985) and was also observed by del Moral (1983) when he removed *Carex spectabilis* from plots where it was dominant in a subalpine meadow community in Washington State. The grass *Festuca idahoensis* increased and produced a significant decrease in four other species. This does imply that the *Carex–Festuca* interaction was a specific and not a diffuse one. Effects of this kind plainly depend upon which species occupy a particular experimental plot. The pattern of plant distribution is important. Fowler (1981) found that up to 67% of the variance in the response of a species to the removal of another from her plots was due to differences in plant distribution between plots. On the whole, field experiments involving species removal demonstrate very few specific interactions between species that cannot be accounted for by the disposition of individual plants before removal was carried out. In the short term, the species to respond most strongly to the local removal of another may be whichever one happens to be in the gap which has been created. In the longer term, gaps may be colonized by plants regenerating from seed, and on this time-scale the size of gap may well determine which species appears in it. Removal experiments in old-field communities conducted by Pinder (1975), Allen and Forman (1976), Abul-Fatih and Bazzaz (1979), Hils and Vankat (1982) and experiments in grassland by Fowler all suggest that specific competitive relationships between particular species are few. For instance, Pinder found that all remaining species increased their net production by about three times when clumps of the dominant grasses in the community were removed.

In removal experiments at five different sites on a marsh on the coast of North Carolina, Silander and Antonovics (1982) found some specific responses to the removal of particular species, but these responses varied between sites. Differences in the spatial pattern and abundance of species between sites before removal probably determined the outcome. In this and the other studies, the few specific interactions between species are often not symmetrical. Goldberg and Werner (1983) argue that interspecific competitive interactions in the field are usually size-specific rather than species-specific and that the relative size of a plant, and whether it or its competitors are seedlings, juveniles or larger, determines the outcome of competition. This is borne out by del Moral (1983) who found that adult transplants survived better than seedling transplants and that survival of both depended upon the productivity of different sites, and hence the size of competing plants, within his subalpine meadow community. Grace (1985) found that the outcome of competition between two cattail *Typha* species was different in mixtures of juveniles raised from seed and in mixtures of adults. Even where relative size does determine competitive rankings in short-term experiments, this does not necessarily mean that relative size is the ultimate arbiter of community composition because many other ecological variables are involved (Silvertown & Dale 1991).

All of the removal studies have been short-term ones and we do not know what effects the selective removal of individual species over many years would reveal (Bender *et al.* 1984). It is possible that they might be different. However, forest communities in the East USA where chestnut was a dominant tree changed little when this species was eliminated by chestnut blight in the first half of this century. Gaps left by chestnut were mostly occupied by bordering trees (Shugart & West 1977).

8.2.2 Competition and natural selection

By reducing each other's fitness, competitors exert selection pressures on each other. If genotypes differ in competitive ability and competitors have grown together over many generations this should lead to the evolution of ecological differences that reduce niche overlap and the magnitude of competition coefficients (Roughgarden 1979). The best evidence that this does occur comes from a multi-generation experiment with barley (Allard & Adams 1969). The experiment was initiated in the 1940s by creating a genetically heterogeneous population from crosses among 31 different barley varieties. Seeds produced from this experiment, called a **composite cross**, were harvested and resown each year without any conscious selection. Although the plants are highly selfing and the genotypes in the composite cross compete with one another, substantial genetic variation still existed in this population, and in other composite crosses begun earlier, after 40–50 generations (Allard 1990).

Allard and Adams (1969) took eight genotypes from the 18th generation of the barley composite cross and grew them in mixtures and monocultures to compare their seed yields. None of the genotypes consistently suppressed the yields of others when compared with monoculture yields, indicating that competition coefficients were rarely greater than unity. In fact, in 40% of the genotype combinations yields were significantly larger in mixtures than monocultures, suggesting that competition coefficients were often less than unity. Four barley varieties with no common history were grown together for comparison with the results of the composite cross, but none showed significant yield increases in mixtures. This experiment suggests that during 18 generations of sympatry in the barley composite cross, natural selection had reduced competition between genotypes. This occurred without any reductions in yield, so it seems likely that niche differences were involved. Niche differentiation among genotypes would have contributed to the maintenance of genetic diversity in the population as a whole, though evolutionary interactions with a fungal pathogen certainly made a major contribution to this also (McDonald *et al.* 1989a, b) (Chapter 3).

Kelley and Clay (1987) and Taylor and Aarssen (1990) have shown that there is genetic variation for competitive ability in semi-natural grassland communities, so the scope for coevolution between long-time competitors exists; however the evidence that it has actually occurred in natural populations is weak. Martin and Harding (1981) compared the seed production of the two annuals *Erodium cicutarium* and *E. obtusiplicatum* in experimental mixtures of plants from sites in California where the species were sympatric and sites where they were allopatric. There was some evidence that *E. cicutarium* performed better when matched against sympatric genotypes of *E. obtusiplicatum* than allopatric ones, but no effect of this kind was shown by *E. obtusiplicatum*.

An interesting pattern of genetic variation in two competing species of wild oat was found by Jain (1976) in central California. Pure populations of *Avena fatua* and of *A. barbata* were more polymorphic than the same species in sympatric populations (Fig. 8.7). This pattern could arise if sympatric populations had higher selfing rates than allopatric ones (Chapter 2). On the other hand if the polymorphic loci are not neutral, the pattern could also reflect differences in habitat between allopatric and sympatric populations, or an evolutionary response in each species to competition from the other. In the last case lower genetic variation would be correlated with a narrower niche for each species caused by the selective effect of its competitor. Competition experiments between plants from sympatric populations and from allopatric populations

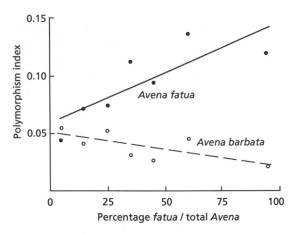

Fig. 8.7 Polymorphism in populations of *Avena fatua* and *A. barbata* in relation to the relative frequency of the two species at seven sites in California (from Jain 1976).

failed to find significant evidence in favour of this hypothesis (Yazdi-Samadi *et al.* 1978). There have been insufficient tests of the coadaptedness of competing plants to draw any final conclusion on this question. In view of the many other ecological processes that affect the fitness of competitors, and indeed that affect which species (if any) wins, it would not be too surprising if competitor coevolution is a rare phenomenon in plants.

8.2.3 Effects of symbionts on competition between plants

Most experiments on plant competition are done in pots where the outcome is constrained. Just as when two boxers enter the ring there has to be a contest, and there has to be a victor. In the field plants may pass each other by, like boxers passing on opposite sides of the street. The outcome of competition between species is probably also much less deterministic than pot experiments imply. One reason for this is that populations confronting each other in the field do not do so alone. There are other plants present that may interfere in the fight, there are herbivores that can turn a heavyweight contestant into a flyweight, and there are symbiotic microorganisms that can subvert the whole match by channelling nutrients from one species to another (Chapter 1).

Grazing can influence competitive outcome, but the reverse may also occur and competition may influence grazing damage. This was demonstrated by Cottam *et al.* (1986) who found that the beetle *Gastrophysa viridula* significantly reduced leaf area and weight of *Rumex obtusifolius* compared with ungrazed controls, when the plant was growing in a grass sward, but not when it was growing free from competition.

Herbivory exercises the most obvious influence over the outcome of competition between plants, but the influence of pathogens, mycorrhizal symbionts and nitrogen-fixing bacteria can also be strong. Their effects are less obvious than those of insect and vertebrate herbivores that chomp holes in leaves or defoliate plants, so their importance can easily be underrated.

The balance between competing plant species may be determined by their relative palatability to herbivores. In many grasses, such as perennial ryegrass *Lolium perenne* and the fescues *Festuca* spp. that are normally palatable to herbivores, infection by endophytic fungi reduces their palatability to vertebrates and invertebrates because the fungi produce toxic alkaloids (Clay 1990). The relationship between endophytic fungi and the plants they infect ranges from the parasitic to the mutualistic, depending upon the fungi involved and the prevalence of herbivory. Heavy grazing favours infected plants which increase at the expense of uninfected individuals.

Rice and Westoby (1982) suggest that another group of pathogens, the heteroecious rust fungi, may become mutualists to their hosts in certain competitive circumstances. Heteroecious rust fungi infect two different plant host species during their life cycle. Although rusts are highly specific in which host species they attack, the two hosts are invariably quite unrelated, as for example in the white pine blister rust *Cronartium ribicola* whose second host is the currant *Ribes triste*. *Ribes triste* survives infection, but releases rust spores that infect nearby pines that are often killed by the disease. Since pines are able to shade currant bushes, Rice and Westoby (1982) argue that currants use rust infection as an agent of biological warfare against their larger competitors: 'my enemy's enemy is my friend'. Many

heteroecious rusts that infect herbs and shrubs attack trees that may potentially shade them. If heteroecious rusts operate as Rice and Westoby suggest, then the interaction between their plant hosts is an example of apparent competition.

Quite the reverse effect on competition has been suggested for another group of microbial symbionts that infect plants. Endomycorrhizal fungi (VAM), which are able to colonize many different plant species, could potentially form functional connections between competitors that might channel resources from one to another. Grime *et al.* (1987) grew experimental mixtures of 18 herb species in treatments with and without VAM infection and with and without clipping to simulate herbivory. Both VAM infection and clipping reduced the dominance of the most abundant species in the mixture (the grass *Festuca ovina*), causing other species to increase in biomass as a consequence of competitive release. Transfers of radio-labelled carbon occurred between plants with mycorrhizal infection and not between those without, but a *net* transfer was not established. The effects of VAM on this experimental community could be accounted for by the direct nutritional benefit conferred on some species, and the release from competition of others caused by reduced growth in the dominant grass. The role of mycorrhizal *links* between species remained an open question (Bergelson & Crawley 1988).

Hetrick *et al.* (1989) grew two grasses of tallgrass prairie together in pots, with and without an inoculum of the VAM fungus *Glomus etunicatum*. With the fungus present, the warm-season grass *Andropogon gerardii* suppressed the cool-season grass *Koeleria pyranidata*, but in the absence of mycorrhizal infection *A. gerardii* grew so poorly that *K. pyranidata* was unaffected by competition. Although both grass species were infected by *Glomus*, and its presence was essential to *A. gerardii*, the mycorrhizal association was of no nutritional importance to *K. pyranidata*. In the field, the two grass species appear to coexist because their main growth periods are phenologically separated. In a review of interspecific competition experiments that contained treatments with and without mycorrhizal infection, Allen and Allen (1990) found that the result was altered by infection in most cases, although in only one case out of eight was the outcome actually reversed.

Although mutualists tend to be much less host-specific than pathogens, some microbial genotypes are more infective or more beneficial to some hosts than others. Chanway *et al.* (1989) studied the relationship between three genotypes of white clover *Trifolium repens* sampled from different parts of a Canadian pasture, and three genotypes of *Rhizobium* and perennial ryegrass *Lolium perenne* growing with them. When all combinations of the *Trifolium*, *Rhizobium* and *Lolium* genotypes were grown in a glasshouse, genotype combinations from the same source outyielded others by 27%. Strangely, this effect was almost entirely due to a specificity between *Rhizobium* and *Lolium* genotypes, and it made little difference with which *Trifolium* genotype they were combined. These results throw some interesting light upon field studies by Turkington and Harper (1979), Aarssen and Turkington (1985) and Evans *et al.* (1985), all of whom found that yields of *T. repens* in experimental mixtures were highest when clover genotypes were grown with the grass species or *L. perenne* genotype occurring with them in the source field (Fig. 8.8). It now seems likely that these specificities are due

to two- or three-way interactions involving *Rhizobium* genotypes.

8.3 COEXISTENCE

Competitive exclusion and coexistence are merely opposite sides of the same coin, or at least alternative states derived from the same model of competition (Fig. 8.1). This model predicts that species should only coexist in a community when each inhibits its own population growth more than that of its competitors. In this section we shall look at some ways in which this condition can be met, and see whether a variety of theories match up to what we know about coexistence in real plant communities.

The challenge to theory is immense. The richest plant communities contain very large numbers of species: 1316 have been recorded in the 15 km² of tropical forest at Barro Colorado Island (BCI, Chapter 6) for example. This is roughly the number of vascular plant species in the whole of the British Isles, which has an area of 314 375 km², but the problem of explaining coexistence is not confined to tropical communities. In British calcareous grasslands up to 30 species can be found coexisting in 0.125 m², and in pine-wiregrass savanna in North Carolina 42 species may be found in plots of the same area (Walker & Peet 1983).

Explaining the maintenance of species diversity and explaining the maintenance of genetic diversity are allied problems. The chief difference between them is that species are usually genetically isolated from one another, while genotypes of course may cross. Some proportion of genetic diversity may be maintained by heterozygote advantage, by selective neutrality or by the sheltering of recessive alleles (Chapter 3), which are mechanisms that work only at the level of the genotype. However, all of the ecological mechanisms that, at least in theory, maintain diversity at the level of the species could also operate at the level of the genotype.

8.3.1 Resource partitioning

The theoretical conditions for coexistence derived from our model of competition may be met if the

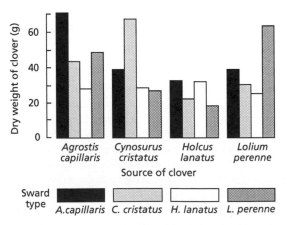

Fig. 8.8 The dry weight of plants of *Trifolium repens* from a permanent grassland sward, sampled from patches dominated by four different perennial grasses and grown in all combinations of mixture with the four grass species (Turkington & Harper 1979).

population growth rate of each species is limited by a different resource. The coexistence of n species requires n different limiting resources. Each available resource can be represented by an axis on a graph or, if it is a nutrient, a gradient of concentration. These **resource axes** define a **niche space**. Two axes, for example one for nitrogen and one for phosphorus, define a two-dimensional niche space (Fig. 8.9). Resource axes are often referred to as **niche dimensions** and there is no theoretical limit to their number. The resource use of competing species can be mapped onto a graph of niche space if we know which resources limit population growth rate, and where the limits of each species consumption of each limiting resource lie. For the coexistence conditions to be satisfied, competing species must be sufficiently different in their use of resources to avoid overlap in niche space. How much overlap is allowed has been a subject of controversy (Abrams 1983).

Resources can be partitioned between the species in a community in a variety of ways. First, a community can usually be divided up into **guilds**, or groups of species that exploit resources in a particular way. For example, at BCI (Chapter 6) plants fall into groups that live in different horizontal strata of the forest, and into a group that regenerates in gaps and a group that regenerates in a closed-canopy environment (Table 8.3). This kind of partitioning of physical space is relatively easy to see, but cannot explain how 93

shrubs or 171 lianas coexist within their respective guilds.

Plants of species-rich grasslands can generally also be divided into a guild that colonizes disturbances and a guild of 'matrix-forming' species (Grubb 1986). Grasses and legumes can be considered separate guilds because the former tend to be limited by nitrogen, while legumes, which have nitrogen-fixing symbionts, tend to be limited by phosphorus (Fig. 8.9). At first sight the scope for further partitioning of mineral elements is greatly limited because all plants share a requirement for a small number of nutrients. However, Tilman (1982) has suggested that small-scale spatial patchiness in the concentration of essential elements could permit many more than n species to coexist on n essential resources.

Tilman's **resource ratio hypothesis** is based on the idea that which nutrient limits the growth of particular species depends upon the rates of supply. If a nitrogen-limited species is dosed with plentiful N, it will grow until it is limited by some other resource, say P. Which nutrient limits the plant in any particular place depends on two ratios: (i) the ratio of N/P concentrations at which the switch in limiting factor occurs in that species; and (ii) the ratio of N/P available in the soil at that spot. Two species requiring the same two nutrients, but switching between them at different ratios of supply, will be able to coexist wherever the ratio of nutrients available in the soil lies between the critical values for the two species, because at this point they will be limited by different nutrients. When more than one species in a patch is limited by the same resource, the one with the lowest requirement (called R^*) for that resource will win (Tilman 1982; Tilman & Wedin 1991). The species with the lowest R^* wins because it can deplete the supply of the resource in the soil until the species with a higher requirement can no longer survive. According to the resource ratio hypothesis, more than two species can coexist on two essential resources if the species all have different critical ratios and if the habitat is heterogeneous, supplying different ratios of the resources in different patches.

Measurements of resource partitioning in plant communities have been relatively few (e.g. Werner & Platt 1976; Cody 1978, 1986, 1989;

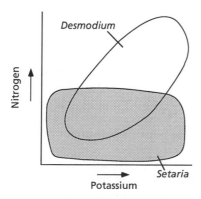

Fig. 8.9 A two-dimensional niche space showing the hypothetical use of two mineral resources by a legume (*Desmodium*) and a grass (*Setaria*).

Table 8.3 Life forms of plant species at Barro Colorado Island, Panama (Croat 1978)

	Number of species	Percentage
Cryptogams		
Epiphytes	41	3.1
Hemi-epiphytes	1	0.1
Aquatics	6	0.5
Vines	4	0.3
Other terrestrials	47	3.6
Tree ferns	5	0.4
Total cryptogams	104	7.9
Phanerogams		
Trees >10 m tall	211	16.0
Trees <10 m tall	154	11.7
Shrubs 2(−3) m tall	93	7.1
Epiphytic or hemi-epiphytic trees and shrubs	16	1.2
Parasitic shrubs	7	0.5
Total arborescent spp.	481	36.6
Lianas or woody climbers (including 10 climbing trees)	171	13.0
Vines	83	6.3
Epiphytic or hemi-epiphytic vines	11	0.8
Total scandent spp.	265	20.1
Epiphytic herbs	135	10.3
Aquatic herbs	54	4.1
Herbs of clearings	197	15.0
Forest herbs	75	5.7
Parasitic herbs	1	0.1
Saprophytic herbs	4	0.3
Total herbs	466	35.4
Total native plant species	1316	100

Rogers & Westman 1979; Silvertown 1983a; Russell *et al.* 1985). There are often distinct ecological differences between competing species when they are compared at the geographical scale (Fig. 8.10), and these can explain coexistence if the habitat is patchy. A major difficulty is to identify the important resource axes *within* a community that permit coexistence at a local scale. In a study of species growing in British calcareous grassland, Mahdi *et al.* (1989) sought differences between eight species on six niche dimensions by measuring their phenology, soil depth, pH and levels of available N, P and K where each plant grew. An ordination technique was used to find the greatest possible separation of the eight species using the variables measured (Austin 1985). The first two axes of this ordination accounted for nearly 80% of the variance among species, but there was substantial niche overlap (Fig. 8.11). As a test of the resource ratio hypothesis, ratios of N : P were compared for the eight species, but there were no significant differences among species.

There are always at least two interpretations

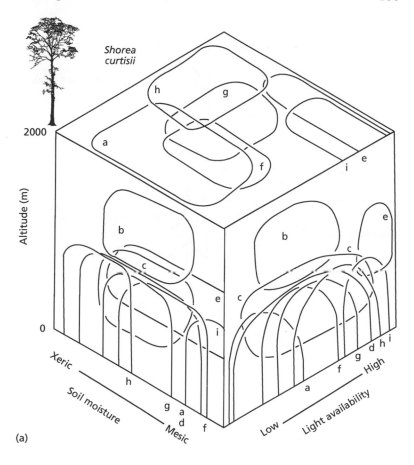

(a)

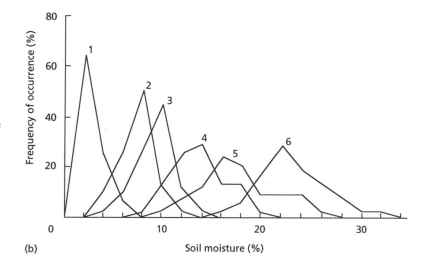

(b)

Fig. 8.10 Separation of related species along environmental gradients: (a) occurrence of nine (a–i) dipterocarp tree species along axes of altitude, soil moisture (xeric–mesic) and light availability (low–high) in wet zone forests of Sri Lanka (Ashton 1988), and (b) occurrence of six goldenrods *Solidago* spp. in tallgrass prairie (Werner & Platt 1976).

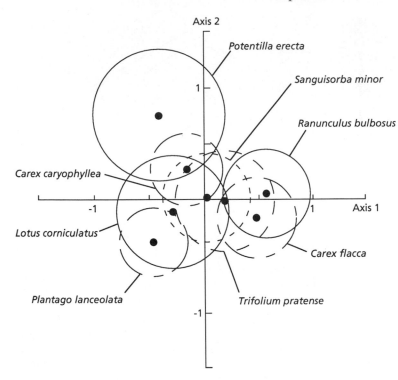

Axis 2

Potentilla erecta

Sanguisorba minor

Ranunculus bulbosus

Carex caryophyllea

Axis 1

Lotus corniculatus

Carex flacca

Plantago lanceolata

Trifolium pratense

Fig. 8.11 Position of eight calcareous grassland herbs on two niche axes obtained by canonical variate analysis; 95% confidence limits are shown by circles. Axis 1 is mainly influenced by phenology, axis 2 by pH and phenology (from Mahdi *et al.* 1989).

that can be put upon a failure to find niche separation among coexisting species: (i) the wrong niche axes were examined; or (ii) there is no niche separation. Because there is no *a priori* basis on which to choose correct niche axes, it is impossible to decide between (i) and (ii) and the more general hypothesis that niche separation is required for coexistence cannot be falsified. The resource ratio hypothesis makes a more specific statement about what kind of niche differences should exist between species and consequently it is falsifiable, assuming that it is possible to determine which mineral resources limit each species and what all the relevant ratios are.

8.3.2 Density- and frequency-dependent recruitment

Resource-based theories of competition and coexistence assume that populations increase until they are limited by resource availability, but what if populations are limited below this carrying capacity by some other agency — for example predation? If predation operates in a

density- or frequency-dependent manner, culling a disproportionately large share from the competitive dominants in a community, it will delay or even prevent competitive exclusion of more minor species. Grazing is usually essential to the maintenance of species diversity in species-rich grasslands, almost certainly due to the control that grazing animals exert over grasses which dominate and exclude other species when grazers are removed (e.g. Watt 1974).

In tropical rainforests seed predators and other herbivores may have a similar effect. Many of these herbivores are specialists, feeding on only one or a few species of plants. The headquarters of such animals are in the adults of their foodplants, and most of the progeny of these adults appear near them where they are most vulnerable to the herbivores feeding on their parents (Chapter 2). Janzen (1970) and Connell (1971) have suggested that heavy predation on juveniles that are in the proximity of adults results in density-dependent mortality among rainforest trees and keeps species at densities so low that they cannot oust each other (Fig. 8.12a). Wright (1983) found at BCI

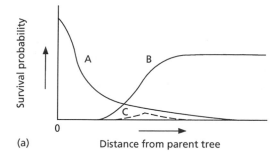

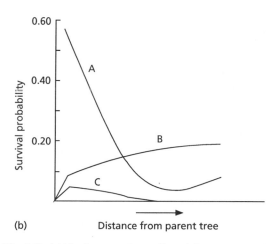

Fig. 8.12 (a) The Janzen–Connell model. Curve A is the distribution of seeds, curve B is the probability of seeds and seedlings escaping predation as a function of the distance from the parent, and curve C, which is the product of curves A and B, is the distribution of surviving plants (after Janzen 1970). (b) Actual curves of seed distribution A, the probability of seedling establishment B, and the distribution of surviving juveniles C, for the tropical dry forest tree *Cecropia peltata* (from Fleming & Williams 1990).

that seeds of the *Scheelea* palm were attacked by bruchid beetles most severely near parent trees and had to be 100 m away to escape. There are many examples of such mortality patterns in tropical trees caused by pathogenic fungi (Augspurger 1983, 1984) as well as herbivores (Clark & Clark 1984), though density-dependence is rarely strong enough to remove aggregation entirely, and it may thin out juveniles, but still leave them clustered around adults (Fig. 8.12b) (Hubbell 1980). Because predation reduces but does not eliminate aggregation, and because of

the weaknesses of correlative studies of density-dependence mentioned in Chapter 4, density-dependence cannot always be detected from the spatial relationships of conspecific juveniles and adults. Studies in Australian tropical rainforest (Connell *et al.* 1984) and at BCI (Hubbell 1979) that have been based on these spatial relationships alone have found little evidence of density-dependence.

Hubbell and Foster (1986) established a 50 ha study plot at BCI to address the question of co-existence in the forest there. All trees and shrubs greater than 1 cm DBH were censused and mapped in 1982 and again in 1985. The first census found 235 895 individuals belonging to 306 species. Lianas were not counted. Nearly 17% of all stems belonged to a single shrub species *Hybanthus prunifolius*, but 21 species were represented by single individuals (Hubbell & Foster 1990a). During the 3-year interval between censuses, saplings of some but not all of the commoner species survived and/or grew significantly better if they were beneath a tree of a different species than if they were beneath a conspecific adult (Hubbell *et al.* 1990; Condit *et al.* 1992). The same pattern significantly affected the mortality of rare species as a group, though not their growth (Hubbell & Foster 1990b). Population regulation in the very commonest tree species did seem to be strong enough to account for their existing densities in the BCI plot, but the evidence that density-dependence was present or powerful enough in other species, including the very abundant shrub *H. prunifolius*, was still lacking (Hubbell *et al.* 1990). The results of future censuses should clarify the situation.

8.3.3 Density-independent mortality

Most types of vegetation, and particularly forests, are prone to periodic disturbance that kills adults and creates opportunities for recruitment (Pickett & White 1985). This mortality is density-independent, but if the frequency and intensity of these episodes of mortality are right and rates of population increase are low, this kind of community disturbance can delay competitive exclusion among similar species almost indefinitely (Fig. 8.13) (Huston 1979).

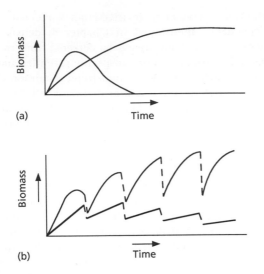

Fig. 8.13 The outcome of competition between two species based upon a model similar to the one described in Section 8.2: (a) without disturbance, and (b) when there is a disturbance producing a periodic, density-independent population reduction (after Huston 1979).

High rates of disturbance will eliminate those species with populations unable to recover quickly, and low rates of disturbance will allow inter-specific competition to take its toll. This idea is known as the **intermediate disturbance hypoth-esis** (Connell 1978), and is in accord with patterns of diversity in several plant communities. The highest diversities of plants in pine-wiregrass savanna occur in annually burned sites (Walker & Peet 1983) and local disturbances by frost-heave increase diversity in subalpine and alpine communities (Fox 1981; del Moral 1983). In herbaceous plant communities studied by Grime (1979) in England, a 'humped-back' relationship between species richness and standing crop biomass was found, which also supports the idea that coexistence is favoured between ex-tremes of disturbance (which removes biomass) and competition.

8.3.4 Spatial refuges from competitors

Our simple model of competition contains no explicit assumptions about the spatial structure of competing populations, but in reality spatial structure can determine how often species en-

counter each other and therefore affect the im-portance of interspecific interactions. This is particularly important because most plant popu-lations are clumped at some spatial scale. Density-dependent processes that thin out clumps of common species promote coexistence, but there are also circumstances in which spatial aggre-gation can promote coexistence too. This can occur in two ways.

A **spatial refuge** from competitors can be pro-vided by the fact that only the individuals on the edge of a monospecific clump compete with other species. Those inside a clump compete with each other. Consequently, the more clumped a species' distribution grows, the less intense *inter*specific competition will become and the more intense *intra*specific competition will become. If this process proceeds to the point where intraspecific competition is stronger than interspecific competition for all competitors, the conditions for coexistence could be met (Shmida & Ellner 1984) without niche separation.

Spatial structure can promote coexistence in quite another way that also involves reducing the frequency of encounters between competing species. As we have seen in chapters 5 and 6, many plant species rely upon vegetation gaps and sites of disturbance for recruitment, and this leads to intraspecific aggregation. Theoretical models have shown that competing species that form this kind of aggregation can coexist, even if there are no niche differences between them, because the meeting of competitors is a probabilistic process. Under the right conditions, there will always be some populations of each species protected from competition because competitors are missing, or they arrive at the site too late to matter (Atkinson & Shorrocks 1981; Rosewell *et al.* 1990). Neither of these ideas about spatial refuges has yet been tested in plant communities.

8.3.5 Temporal refuges from competitors

Fluctuations in recruitment can promote co-existence if the good years for one species are the bad years for others. Such a difference is one aspect of what Grubb (1977) has called the **regen-eration niche**. Because plants become less vul-nerable to competition once they are established, in a good recruitment year a species can establish a cohort of juveniles that will be much less sensitive to its competitors when these are recruited later. This can be thought of as a **temporal refuge** from competitors. In effect the recruits of good years are stored over bad years species. For the **storage effect** to work and to promote coexistence, competitors must have overlapping generations and competition between adults must be weak (Warner & Chesson 1985). The first condition applies to many perennials and to annuals with a seed pool. The second condition may be more difficult to satisfy, unless spatial refuges from competition also exist.

Many of the broad-leaf forests of North America are dominated by oaks. Whittaker (1969) noticed that, though different oak species are dominant in different forests, the two most abundant oak species usually belong to two different subgenera: one to the white oaks and one to the black. Mohler (1990) put this observation to the test using data from 14 forest stands from all over the coterminous USA and found it to be true in 12 of them. Whittaker had suggested that the pattern could reflect niche differentiation between the two subgenera and Mohler (1990) also examined this idea. Oaks, like many trees, vary a good deal in the size of seed crop from year to year and Mohler found that seed crops of species in the same subgenus were consistently better correlated with each other than were the crops of species in different subgenera. This is probably because black oaks usually require 2 years to mature their acorns whereas white oaks need only one. The asynchrony in recruitment that this difference generates may create a storage effect that helps to explain the coexistence of white oaks with black.

8.3.6 Conclusion

It is no doubt obvious by now that the simple model of competition and coexistence with which we began this chapter ignores a lot of situations that can affect the outcome of interspecific competition. The model has still proved valuable because it is based upon explicit assumptions that produce a clear statement of the conditions required for coexistence to occur.

Many of the theoretical mechanisms of coexistence not included in the model can nevertheless be understood by looking at how they would affect the model's parameters. Coexistence in natural plant communities is still very poorly understood, but it seems unlikely that the maintenance of diversity at BCI, in calcareous grassland or in other species-rich communities will be accounted for by any of the mechanisms we have mentioned operating on its own. The challenge is to understand these communities in the round, and to test this understanding experimentally.

8.4 SUMMARY

The types of interaction between plants may be classified according to whether R_0 in each interacting population is increased (+), decreased (−) or unaffected (0) by the interaction. Five types of pairwise interaction are thus defined: **competition** (− −); **parasitism** (+ −); **mutualism** (+ +); **commensalism** (+ 0); **amensalism** (− 0). The classification may be further refined to include the effect of populations on an **intermediary** in the interaction, such as a nutrient, which may reveal the mechanism of the interaction. This approach shows how natural enemies may generate **apparent competition** and **apparent mutualism**, for example.

The dynamics of two competing populations may be represented in a **joint-abundance diagram**. Using simple equations which quantify the numerical effect of competitors on each other in a mixture in terms of **competition coefficients** and the **carrying capacities** of the environment for each population when they are growing alone, the conditions which permit stable coexistence of competing species may be defined. This leads us to the **competitive exclusion principle** which states that ecologically equivalent competing species cannot coexist. Significant ecological differences between species, or **niche separation**, permit stable coexistence.

Two-species **competition experiments** of various designs may be used to explore the parameter space of a joint-abundance diagram. The **partial additive** design consists of mixtures in which the density of one species is fixed and the other varies. In the **replacement design** the total density of the mixture is fixed and the proportions of the two species vary. In **additive series** and **complete additive** designs densities and proportions of both species are systematically varied. The additive design has been used to determine the effect of the number of neighbours on a target plant, and shows that the effect of each additional neighbour tends to diminish rapidly as their number rises. Additive designs are also often used to compare the relative effects of root and shoot competition. The replacement design may be used to calculate the **relative yield total** (RYT) of a mixture, which under restricted circumstances may indicate the existence of niche separation between competitors. **Yield advantage** in mixtures can be assessed by the **relative resource total** (RRT) or by the **land equivalent ratio** (LER).

Additive series and complete additive designs can be used to determine the location of the **zero isoclines** in a joint-abundance diagram. They may also be used to calculate the parameters of the **reciprocal yield model** which allows competition coefficients to be calculated and the effects of intraspecific and interspecific competition to be clearly separated. Competition experiments generally demonstrate that competition coefficients are sensitive to growing conditions and density and that the outcome is contingent upon these. Competition in the field usually involves simultaneous interactions between many more than two species. **Removal experiments** suggest that these interactions are not species-specific and that **diffuse competition** occurs. The outcome of competition between two species in a community may often depend upon their spatial arrangement and upon the presence and response to competition of other species.

Plant populations contain **genetic variation** for competitive ability, and at least one experiment has shown that natural selection can reduce competition coefficients when competing genotypes are grown together over many generations. However, the evidence that selection operates in this way in the wild is weak. A great variety of interactions between plants and **symbionts**, such as herbivores, endophytic fungi, pathogens, mycorrhizas and nitrogen-fixing bacteria, affect the outcome of competitive interactions between plants.

Coexistence among competitors seems very common in the wild, and a wide assortment of theories exists to reconcile this fact with the competitive exclusion principle. Coexistence is consistent with competition if the competitors **partition resources** between them or occupy different **niches**. Alternatively, **density-** or **frequency-dependence** can prevent competitive exclusion, **density-independent mortality** affecting a whole community can delay the extinction of inferior competitors, and **spatial refuges** provided by aggregation can reduce the rate of encounter between competitors. **Temporal refuges** from competition may be provided by differences in **regeneration niche** and by the **storage effect**. It seems unlikely that a single mechanism can account for the coexistence of competitors in all the different kinds of plant community where high species diversity occurs.

Chapter 9
Life History Evolution: Sex and Mating

9.1 INTRODUCTION

Life histories are particularly suited to evolutionary analysis because the two major traits — survival and reproduction — are components of fitness, and other life history characteristics are more or less closely correlated with them. Since net reproductive rate R_0 can be used as a measure of fitness and, as we saw in Chapter 5, $R_0 = \Sigma l_x m_x$, an optimal life history under particular ecological conditions can be defined as one which maximizes $\Sigma l_x m_x$. If we take a simple-minded view of the relationship between life history traits and fitness it is easy to describe what a plant with an 'unbeatable' life history should be like: it should reproduce asexually, but if sexual it should be cosexual and selfing; it should produce seeds early and often, and these should be copious in number, large in size and germinate without delay; the plant should possess clonal growth, and live for ever. In reality, of course, plants display an immense variety of life histories, and their modes of reproduction, birth, growth and death each span a wide range of alternative strategies (Table 9.1). In all this rich diversity, not one species has *all* the characteristics of the unbeatable phenotype. Why does no unbeatable phenotype exist? The answer to this deceptively naive question provides the key to explaining the rich variety of life history strategies found among plants.

9.2 THE PRINCIPLE OF ALLOCATION

There is no unbeatable phenotype because it is physiologically impossible for any organism (even an autotroph!) to do everything at once.

There is always only a limited amount of time (especially important in annual plants), energy and other resources to spend on growth, maintenance and reproduction, so a plant must allocate its resources among these alternative demands. This is known as the **principle of allocation**. Conflicting demands lead to **trade-offs** between different activities: a plant can use a meristem or a gram of carbon to produce a flower or a leaf, but not both simultaneously. This kind of allocation problem has many different solutions, hence the variety of life histories. Two classes of trade-off are virtually universal: (i) a trade-off between reproduction and other activities, which manifests itself as a **cost of reproduction**, and (ii) a trade-off between the **size** and **number** of offspring. The first of these will be dealt with here, and the second in Chapter 10.

Trade-offs can be measured in three primary ways: by **phenotypic correlation** between traits such as reproduction and growth, by **experimental manipulation** of a plant's activities to observe the effect on other traits, and by **genetic correlation** which measures the covariance of traits due to the pleiotropic effects and linkage disequilibrium of the underlying genes. Each method has its advantages and drawbacks (Pease & Bull 1988; Stearns 1989; Reznick 1992). In practice, phenotypic correlation is the method that has provided us with the most evidence for life history trade-offs, but we must recognize that trade-offs should have a genetic basis if they are to shape evolutionary change. Negative phenotypic correlations between reproduction and growth are often quite marked. In a study of Douglas fir (*Pseudotsuga menziesii*), grand fir (*Abies grandis*) and western white pine (*Pinus monticola*), large seed crops reduced the growth of wood in the year of seed

Table 9.1 Life history strategies discussed in chapters 9 and 10. The strategies in italics are those we might expect an imaginary, unbeatable life history to possess, other things being equal (see text)

Life history stage	Trait	Range of alternative strategies		Section
Reproduction	Sex	*Asexual*	Sexual	9.3
	Mating	*Selfing*	Outcrossing	9.4
	Gender	*Cosexual*	Dioecious	9.4
	Maturity	*Early*	Delayed	10.1
Birth	Seed crop frequency	*Annual*	Masting	10.2
	Seed size	*Large*	Small	10.3
	Germination	*Immediate*	Delayed	10.3
Growth		*Clonal*	Aclonal	10.4
Death		*Iteroparous*	Semelparous	10.5

production (see Fig. 9.1) (Eis *et al.* 1965), and in another study of beech (*Fagus sylvatica*) there was a decrease in wood growth for more than 2 years following a large crop (Holmsgaard 1956). In the palm *Astrocaryum mexicanum*, Piñero *et al.* (1982) found a clear inverse correlation between fecundity in a given year and the probability of being alive 15 years later. Root growth is also affected by reproduction, and even light crops of fruit have been observed to reduce root growth in apple trees.

A danger of relying entirely upon phenotypic correlations to measure trade-offs is that negative relationships may be hidden if the characters in question are both positively correlated with a third trait such as size. To take an economic analogy, a person on a small budget may have a small house and a small car, but a person with a large house can generally afford a large car. In the population as a whole this produces a positive correlation between size of house and size of car, even though a person of limited means certainly has to trade-off car against house in allocating their budget. In plants, such trade-offs can be revealed by experimental means. Horvitz and Schemske (1988) created low- and high-reproductive effort individuals in the neotropical perennial herb *Calathea ovandensis*, and then

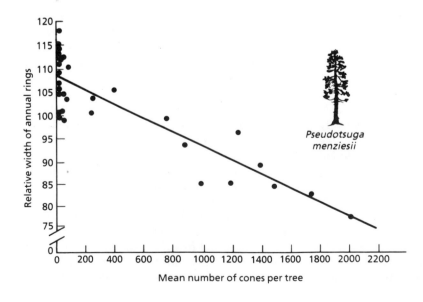

Fig. 9.1 The relationship between cone crop size and annual growth increment in a population of Douglas fir (from Eis *et al.* 1965).

measured subsequent growth, survival and reproduction. They found no significant differences in the following season in any of these demographic parameters. The problem with short-term studies of reproductive cost in perennials such as *Calathea* is that storage in the rhizome is an alternative sink for resources that do not go into reproduction or growth, and this is difficult to measure. In *Tipularia discolor*, a woodland orchid of the Southeast USA, Snow and Whigham (1989) found that experimental manipulation of fruit set had significant effects on the size of the underground storage organ which is formed after flowering each year. In a similar but longer study of the slipper orchid *Cypripedium acaule*, Primack and Hall (1990) also found a clear cost of reproduction, but not until the third successive year of experimental treatment.

All a plant's resource-consuming activities, even the accessories of sexual reproduction, have potential fitness costs as well as benefits. Pyke (1991) measured the cost (in terms of seed number) of nectar production in an Australian perennial herb *Blandfordia nobilis* whose flowers are visited by birds, honey-bees and ants. When nectar is removed from a flower it produces more, so Pyke compared the seed set of plants that he pollinated but protected from nectar removal, with the seed set of plants that were pollinated and had their nectar removed daily. These plants produced nearly three times the volume of nectar produced by protected plants and set significantly fewer seeds, so there was a trade-off between the amount of nectar produced and seed set.

Trade-offs form a major **constraint** on the evolution of life histories. How a plant resolves the conflicting demands upon it shapes its life history, and affects its fitness. Natural selection favours the genotypes with a phenotype that optimizes the balance between costs and benefits to achieve the highest fitness. An important reason why there is such a range of life histories is because the optimum phenotype depends upon ecological circumstances and these differ so much (Partridge & Harvey 1988). If we want to test the hypothesis that a particular life history is adaptive in a particular environment, we must measure the costs and benefits of the life history traits in question in terms of their negative and

positive effects on fitness. The traits in question must be heritable for natural selection to act on them, but there are two distinct ways in which inheritance may be dealt with in evolutionary models that describe how evolutionary change occurs. In **population genetic models**, of which the Hardy–Weinberg equation is a very simple example, allele frequencies are explicit variables, while in **phenotypic selection models** the frequencies of different phenotypes are the variables of interest, and relevant traits are simply assumed to have a genetic basis. The approach we now take to the evolution of the life history traits listed in Table 9.1 is largely based on phenotypic selection models.

9.3 EVOLUTION OF SEX

There is a deep evolutionary paradox about sex: according to theory, it simply does not pay. Despite its ubiquity, the evolutionary *disadvantages* of sexual reproduction are much easier to see than are the advantages. The mystery is deepened by the fact that there are many plants, such as some microspecies of the common dandelion *Taraxacum officinale* or the blackberry *Rubus fruticosus*, which have lost the ability to reproduce sexually, but still retain its trappings in the form of showy flowers and pollen-producing anthers.

Molecular mechanisms which facilitate crossing-over and the production of novel genotypes probably evolved about two billion years ago in bacteria. There is good evidence that the enzymes involved in crossing-over are also responsible for the repair of damage to DNA, and sex may originally have evolved as an incidental by-product of a mechanism for the repair and maintenance of genes (Bernstein *et al.* 1988).

To explain why most plants reproduce sexually we need to resolve the apparent contradiction between two simple facts:
1 Natural selection favours those genotypes that transmit the most copies of their genes to future generations.
2 Sexual reproduction produces offspring that are *different* from their parents.

If we look at things from the maternal perspective, because every sexual offspring produced by outcrossing carries equal genetic contributions of nuclear genes from its mother and its father, the mother transmits only *half* her genes to a sexually produced offspring. By contrast, a mother who reproduces asexually will transmit *all* of her genes. There is therefore a **twofold disadvantage** intrinsic to sex when its consequences for fitness are compared with asexual reproduction; the mother is diluting her genes by half. In addition, sexually reproducing individuals bear the physiological cost of producing male organs and pollen, or male individuals if there are separate sexes. A plant in an asexually reproducing population can, in theory at least, dispense with these disadvantages.

If a sexually reproducing plant self-fertilizes, the twofold disadvantage of sex decreases in proportion to the degree of inbreeding (Charlesworth 1980). Plants with high selfing rates generally have reduced male organs (e.g. Lovett Doust & Cavers 1982a; Charlesworth & Morgan 1991), thus saving on this cost too. In species such as *Impatiens capensis* where chasmogamous and cleistogamous flowers occur on the same plant, chasmogamous flowers are generally the larger of the two. However, it is interesting that some **apomictic** plants (which produce seeds asexually), such as some dandelions, still produce copious amounts of pollen. This may simply be because with the loss of sex they have also lost the main source of the genetic variation that is necessary if a reduction in pollen production is to evolve. Apomicts are in an evolutionary dead end, although selection may occur between clonal lineages.

There are many different theories which seek to resolve the paradox of sex, and nearly as many books have been devoted to the subject (e.g. Ghiselin 1974; Williams 1975; Maynard Smith 1978; Bell 1982; Shields 1982; Stearns 1987; Michod & Levin 1988). Faced with so many plausible explanations for the advantages of sex, some investigators have concluded that there may be no single right answer to the paradox and that which applies to one species may not apply to another (Bierzychudek 1987a; Gouyon *et al.* 1988). The theoretical arguments divide roughly into those which posit a long-term advantage to sex that only shows up when fitness is measured over many generations; and those based upon a short-term advantage that shows up within a generation and benefits the individual. We will deal here with only a few of the ideas, concentrating on those aspects which are most easily tested experimentally. These necessarily tend to be short term. Experimental comparison of the advantages of different sexual systems, and of sex itself, is made easier in plants by the considerable variation that occurs within as well as between species.

9.3.1 Long-term advantage

In the long term, the fitness of sexual populations is greater than that of related asexual populations because a sexual lineage is able to purge accumulated deleterious mutations. Similarly, sex speeds up the rate at which a population can evolve in response to selection (Maynard Smith 1978). Rare, beneficial mutations that arise in different individuals remain genetically isolated from each other in an asexual population, but can be brought together in new genotypes in a sexual population (Fig. 9.2). These are both group- or species-level advantages to sex. All group selection mechanisms share an important weakness because the population is vulnerable to invasion by a mutant favoured through short-term advantage — in this case, asexual genotypes that are capable of rapid clonal spread. However, there are theoretical conditions in which, once some sexual lines have become established, the short-term disadvantages of sexual reproduction are outweighed by long-term benefits because sexual lineages with a high tendency to switch to asexuality are doomed to a high rate of extinction, whereas sexual lineages with a low tendency to switch to asexuality will have a higher speciation rate and lower rate of extinction (Nunney 1989).

9.3.2 Short-term advantage

If sexual reproduction is maintained by short-term advantage we should be able to detect significant fitness benefits for sexually-, as opposed to asexually-produced offspring. This prediction has been tested experimentally in the outbreeding grass *Anthoxanthum odoratum*, by Antonovics and his associates. Their work supports the hypothesis that the rare or unique genotypes

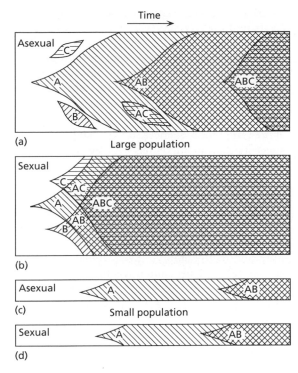

Time →

Asexual

C

A

B

AB

AC

ABC

(a)

Large population

Sexual

C

AC

A

ABC

AB

B

(b)

Asexual

A

AB

(c)

Small population

Sexual

A

AB

(d)

Fig. 9.2 Beneficial mutations that arise in an asexual lineage (a) cannot be combined unless they occur sequentially, but (b) in a sexual population sex can rapidly create genotypes with new combinations of beneficial mutations. This benefit of sex would be more effective in a large population (c) than in a small one (d) because mutations are so rare. From Crow (1986).

produced by sexual reproduction confer an evolutionary advantage on plants having this mode of reproduction because the fitness of progeny with different genotypes is **frequency-dependent** (Chapter 3). For example, Ellstrand and Antonovics (1985) planted separate plots of sexually-derived and asexually-derived tillers of *A. odoratum*, in density gradients at two natural sites from which the source material had been taken. The sexual progeny showed a significant fitness advantage compared to the asexual progeny, particularly at the lower densities. Kelley *et al.* (1988) extended this test of the short-term advantage of sexual reproduction. They simulated the natural dispersal pattern of progeny of *A. odoratum* around parents, when these progeny were either genetically uniform and identical to the parent, or

sexually-produced by those parents and thus genetically variable. The sexually generated progeny had reproductive rates, summed over 2 years, that were nearly one and a half times those of their asexually generated siblings, thus approaching repayment of the twofold cost, but not enough, nonetheless.

Competition between sibs can favour sex in at least two ways. The first mechanism, the **elbowroom model**, proposes that when groups of siblings compete, local genetic diversity (and hence sexual as opposed to asexual females) should be favoured because resource partitioning between sibs, as well as some advantage from avoidance of pathogens and predators, may lead to higher fitness for the sexual groups. Rare genotypes, particularly those unlike their parents or their sibs, may be protected from infection by pathogens that are genetically adapted to the commoner host genotypes (Levin 1975; Rice 1983). This kind of advantage is again frequency-dependent (McCall *et al.* 1989).

The second mechanism, the **lottery model**, proposes that when groups of both sexual and asexual offspring co-occur under conditions of intense local competition, selection favours the genotype which most closely matches local environmental conditions, and sexual females are more likely than asexual females to produce such a genotype. According to this model an asexual parent is analogous to a person who buys several tickets to a lottery, all having the same number; a sexual parent buys tickets that all have different numbers.

In an experimental test of the sib-competition hypothesis, Kelley (1989) compared the performance of genetically variable and genetically uniform tillers of *A. odoratum* (i.e. sexually derived and asexually derived tillers) by planting them *together*, randomly, at 12 sites in a regularly mown field. The planting design was such as to maximize potential competition between the two tiller types. The results showed a substantial fitness advantage for sexual tillers. Other experiments with this species by Schmitt and Antonovics (1986) have shown that seedlings have lower survival rates following aphid infestation when surrounded by siblings than they do when surrounded by non-siblings.

Tonsor (1989) grew seeds of the perennial herb *Plantago lanceolata* in pots containing three competing individuals and only limited amounts of water and nutrients. The experimental design compared pots with full-sib, half-sib and unrelated plants. By the end of the growing season, relatedness among competitors did not affect above-ground dry weight. However, there was a significant decrease in the within-pot variance of vegetative and total dry weights, and an increase in the number of plants flowering per pot with an increase in relatedness. Fruit production also depended upon local relatedness. In pots containing plants that were more closely related, more individual fruits were produced, with no significant differences in either the weight of fruits or the number of seeds per fruit, as a function of the relatedness treatments. This study does not support the sib-competition model, and at present the other experimental evidence available is conflicting, so the matter is still unresolved (Bierzychudek 1987a; Schmitt & Ehrhardt 1987; Willson *et al*. 1987; McCall *et al*. 1989).

9.3.3 Apomixis and the balance between long- and short-term advantage

Apomixis can be fun, can be fun
You don't bother any one, any one
You just take things as they come, as they come
Apomixis, Apomixis —
Scrap that pollen — be just like your mum

Apomicts are plants that can produce seeds without the necessity of sex, generating them by mitosis or by a modified meiosis lacking any reduction division. Some apomicts are facultative and retain the sexual option (e.g. *Oxalis dillenii* ssp. *filipes*), many are polyploid, and nearly all are perennials (Bierzychudek 1987b). Apomicts are especially common in the Asteraceae, Poaceae and Rosaceae (Nygren 1967), but also occur in other angiosperm families. It is significant that all apomicts have close relatives that are sexual, often in the same genus or species, and apomicts can outnumber the related sexual taxa by an order of magnitude or more. For example, the blackberry *Rubus fruticosus* agg. is represented

in Britain by one sexual form and 368 apomicts (Watson 1958). Ninety per cent of known dandelion 'species' are obligate apomicts (Richards 1973), although the species concept is unhelpful in this context as the microspecies of apomicts are really clonal lineages. The existence of such closely related sexual and asexual taxa offers ideal opportunities to compare the long- and short-term consequences of sex and asex and suggests that apomixis has evolved recently and separately in a number of taxa.

Although there is no fusion of gametes, most apomicts, such as *Rubus fruticosus*, are 'pseudogamous' and do require pollination for successful seed production because a pollen nucleus is needed to fertilize the endosperm that supplies nutrition to the embryo. Even though no longer sexual, these plants cannot recoup the cost of the sexual trappings needed to attract pollinators, and in any case they may lack the evolutionary means to drop this atavistic baggage.

The seed production of sexual females and of apomicts in a population of pussytoes *Antennaria parlinii* was compared by Michaels and Bazzaz (1986), and a similar comparison was made by Bierzychudek (1987c) for *Antennaria parvifolia*. In theory apomicts need only produce half the seeds of outcrossing sexual females to match their fitness, but in both studies of *Antennaria* apomicts produced more seeds than females altogether. Bierzychudek (1987c) suggested that female seed set was pollinator limited in both of these *Antennaria* species, in which case the uncertainties of pollination may be an additional factor favouring apomixis over sex. Indeed this could apply to all apomicts because they occur only in self-incompatible taxa that are susceptible to pollinator limitation of seed set. In their study, Michaels and Bazzaz (1986) also compared seedling survival and ramet dynamics of sexual and asexual plants. The seedlings and the ramets of sexual plants both survived significantly better than apomictic ones, redressing some but not all of the advantage of apomicts.

The evidence of these experimental studies makes it all the harder to understand the short-term benefits of sex, but another kind of evidence does suggest that giving up sex altogether must have long-term disadvantages. Quite simply, all apomicts seem to be recently evolved. For example, Bayer (1987) studied the taxonomic, genetic and phylogenetic relationships of North American *Antennaria* species and concluded that the five major apomictic species of the genus (including *A. parvifolia* and *A. parlinii*) were derived by hybridization among living sexual species. One of the asexual species, *A. rosea*, had no fewer than six sexual species in its ancestry, with possible contributions from two more as well (Bayer 1990). If asexuality gave apomicts a long-term advantage over sexual taxa we would expect many apomicts to have no living relatives that resemble them. The relatives should have been displaced by competition from apomicts, or should have diverged during evolution. Apomicts and their related sexual taxa often have distinct geographical distributions, with apomicts predominant in geologically more recent habitats in regions that were glaciated (Fig. 9.3) (Bayer & Stebbins 1983; 1987; Bierzychudek 1987b). This also suggests that apomicts may be better at colonization than at long-term persistence.

The source of the long-term disadvantage of apomixis is unknown, but it may simply be that short-term advantages are temporary for taxa that are unable to evolve resistance to pathogens and other natural enemies. However, work by King and Schaal (1990) suggests that somatic mutation can generate genotypic variation in asexual lineages of *Taraxacum officinale*. Furthermore, Bierzychudek (1987b) has pointed out that polyploid apomicts may not be as genetically depauperate as one might expect (Roose & Gottlieb 1976; Lyman & Ellstrand 1984). Since apomicts are polyploid derivatives of two or more diploid sexual ancestors they may have high levels of heterozygosity as well as the wide ecological tolerances typical of polyploids in general (Levin 1983; Bayer 1989; Bierzychudek 1989; Chapter 2). Although, because of their polyploidy, some apomicts may have a broader ecological tolerance to spatial variation than their progenitors, their capacity to evolve in response to temporally changing conditions must be more limited than that of sexual taxa (Fig. 9.2), particularly if these are also polyploid but have not given up sex as is the case in *Antennaria parlinii* and *A. parvifolia* for instance. The need to keep genetic track of a changing biotic environment could be decisive in the contest between sex and asex.

9.4 EVOLUTION OF MATING SYSTEMS

Variation in sex traits often has a heritable component and sex habits are subject to evolutionary change in response to mutation and selection.

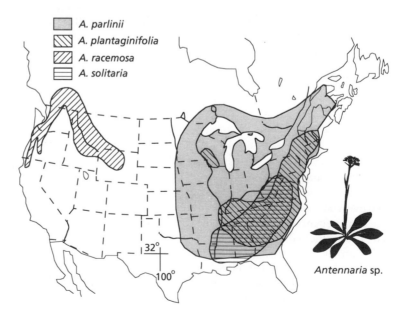

Fig. 9.3 The distribution of *Antennaria parlinii* and its three sexual progenitors *A. plantaginifolia*, *A. racemosa* and *A. solitaria* (after Bayer & Stebbins 1987).

A. parlinii
A. plantaginifolia
A. racemosa
A. solitaria

Antennaria sp.

148

Chapter 9

Like other characters then, we should expect the sex habit and the mating system in any particular population to converge on an optimum combination of traits that maximizes individual fitness. Fitness is determined by the *quantity* and the *quality* of offspring.

Offspring quantity is a function of two processes of allocation:

1 The allocation a plant makes between reproduction and other activities.

2 The allocation a plant makes between male and female function.

We shall look at these two allocation questions later in this section, and then see how the payoffs of different optimal allocation strategies may dictate whether mating systems with one gender class or two yield the higher fitness.

Offspring quality on the other hand is strongly influenced by *which* plants become mates, and in particular by the degree of inbreeding.

9.4.1 Inbreeding depression and selfing

The genes of an outcrossing, cosexual plant are transmitted via pollen and ovules. Because all offspring have one mother and one father each cosexual plant will, *on average*, transmit two sets of genes to the next generation in a population at demographic equilibrium. If a mutant genotype with a gene for self-fertility appeared, this plant would be able to transmit *one* gene for self-fertility by fertilizing another plant with its pollen, and it would also pass *two* genes for self-fertility to the next generation in its selfed seed. In other words the selfer would transmit three genes to the next generation for every two transmitted by an outcrosser. At least initially, while the gene for selfing is rare, the frequency of selfers should increase by 50% each generation. This suggests that natural selection should favour selfing over outcrossing.

In fact, as we have seen in Table 2.1, plants have evolved an impressive range of mechanisms that can prevent selfing. It is generally believed that the principal selective force driving the evolution of outcrossing mechanisms is the detrimental physiological effect of inbreeding, known as **inbreeding depression**, which reduce the fitness of inbred progeny. If there is hetero-

zygote advantage, inbreeding depression can be caused by the loss of heterozygotes that accompanies selfing (Chapter 3), but it is more likely to be due to the expression of deleterious recessive genes that become homozygous in inbred progeny. Although he did not understand its genetic cause, Darwin (1876) observed inbreeding depression in many plant species and first suggested that it was the selective force behind the evolution of outcrossing in plant sexual systems.

Inbreeding depression is observed in all outcrossing species when they are selfed. For example, selfed *Pinus radiata*, measured at 7 years of age, grew more slowly, had poorer stems and was less resistant to pests than outcrossed controls (Wilcox 1983). When compared at 19 years of age, selfed progeny of *Pinus rigida* were considerably smaller than outcrossed ones (Fig. 9.4) (Bush & Smouse 1991). Gymnosperms are generally outcrossing, and a large disadvantage to the progeny of selfed compared with outcrossed individuals often occurs in the fraction of seeds filled, the germination rate, plant size and survival (Charlesworth & Charlesworth 1987). Sakai *et al.* (1989) compared the progeny of selfed and outcrossed matings in *Schiedea salicaria*, a gynodioecious woody angiosperm shrub that is endemic to Hawaii. The progeny of selfing were inferior to those of outcrossing in number of seeds set, seed germination, seedling survival and number of flowers produced.

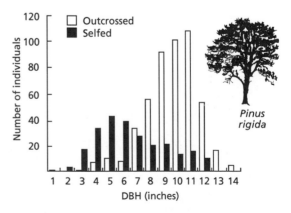

Fig. 9.4 The size distribution of selfed and outcrossed progeny of *Pinus rigida* at 19 years of age (from Bush & Smouse 1991).

Overall, inbred progeny had a relative fitness (W_i) of between 0.06–0.38. Inbreeding depression, d, is measured as $1 - W_i$, giving values 0.94–0.62 for this species. In the related dioecious species *Scheidea globosa* the progeny of crosses made within families were compared with those of between-family crosses. As one would expect, brother–sister matings in this species showed less severe inbreeding depression than selfing in *S. salicaria*, but this was still substantial ranging as high as $d = 0.49$.

Based on the 50% advantage of selfing, population genetic models have repeatedly shown that a gene that increases the selfing rate without altering pollen production should be able to invade an outcrossing population when $d < \frac{1}{2}k$, where k is the relative efficiency of fertilization by outcrossing, as compared to selfing (Maynard Smith 1989). When seed set in an outcrossing species is not limited by the availability of pollen $k = 1$, but if outcrossing is less efficient than selfing $k < 1$, and lower values of inbreeding depression than $d = 0.5$ will be sufficient to favour selfing (Schemske & Lande 1985). The way in which the details of different sexual systems alter the amount of inbreeding depression required to favour selfing has been determined by Lloyd (1979).

The traditional view of the evolution of plant mating systems has been that the selfing rate is fine tuned by selection to achieve an optimal value between 0 and 1 that depends upon local circumstances (e.g. Grant 1975). This view is amply corroborated by the great range of variation in selfing rates observed within many species. For example, Horovitz and Harding (1972) found that the selfing rate in different populations of the annual *Lupinus nanus* varied between 0 and 1 and was negatively correlated with the local abundance of pollinators. However, Schemske and Lande (1985; Lande & Schemske 1985) argue that the condition for selfing to spread constitutes a break point that should create **disruptive selection on the selfing rate**, causing a rapid increase in selfing when $d < \frac{1}{2}k$ and a rapid increase in outcrossing when $d > \frac{1}{2}k$. In fact most theoretical models of the evolution of mating systems have difficulty explaining how mixed mating can exist as an equilibrium state. Three solutions to this problem have been proposed.

The first, proposed by Schemske and Lande (1985), is that there really is no problem. They argue that mixed mating systems are actually relatively rare and that values of $\hat{t}$ are bimodally distributed, with most populations having a value near 1 or near 0. The evidence for this is disputed by Waller (1986) and Aide (1986). Schemske and Lande (1985) argue that species with mixed mating such as *Lupinus nanus* may not be demographically stable and are especially liable to pass through colonizing events or local catastrophes that create population bottlenecks (Chapter 3). Selfing induced by small population size may actually reduce subsequent inbreeding depression because there is strong selection against the deleterious genes that are expressed in homozygotes. A single generation of reduced population size would reduce inbreeding depression by the fraction $1/2N_e$, where N_e is the effective population size at the bottleneck (Lande & Schemske 1985). By lowering inbreeding depression, bottlenecks could therefore favour the subsequent spread of selfing. In three populations of the annual *Clarkia tembloriensis* with different outcrossing rates, Holtsford and Ellstrand (1990) found that inbreeding depression was negatively correlated with selfing rate (S). Differences in mating system between populations may reflect their different recent histories or time since foundation.

The second solution, proposed by Holsinger (1986), is that in addition to inbreeding depression, selfed progeny suffer another handicap when they disperse from the parent because they are only adapted to local conditions. Outcrossed progeny, having greater genetic variety, on average do better. In essence this is a restatement of the **lottery model** (Section 9.3), and there is some empirical evidence to support it in this context. Plants such as *Impatiens capensis*, *Viola* spp. and *Amphicarpaea bracteata* that produce chasmogamous and cleistogamous flowers on the same plant have mixed mating systems (depending on the outcrossing rate of chasmogamous flowers) and in *every case* selfed seeds are far less dispersed than outcrossed ones (Chapter 3). Schmitt and Gamble (1990) planted selfed and outcrossed progeny of *I. capensis* at different distances from the sites originally occupied by their maternal parents. They found that the fitness of inbred offspring declined significantly, and the magnitude of inbreeding depression increased with distance from the parental site, supporting the hypothesis that there had been local adaptation to microsites within the 40×40 m area of the experiment. This provides experimental support for one of the assumptions of Holsinger's model although another assumption, that selfing flowers produce pollen able to outcross, does not hold for cleistogamous flowers because these do not open.

The third solution to the problem of mixed mating systems also echoes an argument seen in the discussion of the evolution of sex. It proposes that the advantage of selfing is frequency-dependent. As we have seen in the context of bottlenecks, by decreasing d selfing in a mixed mating population should accelerate the evolution of an even higher selfing rate. However, selfing also has an effect on population genetic structure that may operate in the other direction as selfing becomes more common. Repeated selfing results in increased relatedness between neighbouring plants, and Uyenoyama (1986) argues that this will decrease the cost of outcrossing because offspring will carry more than half their mother's genes. In fact an 'outcrossing' plant in a population with this genetic structure is just inbreeding with relatives instead of with itself, but this **bi-parental inbreeding** could conceivably maintain a mixed mating system in equilibrium.

9.4.2 Reproductive costs for males and females

The costs associated with reproduction as a male or female are an important part of the ecology of sex. In evolutionary terms a plant should allocate its resources in such a way as to maximize its fitness. This is a complicated problem and involves not only a 'choice' between channelling resources between growth and reproduction, but also between male and female organs. In theory, in an outcrossing species, there should be equal investment in pollen and ovules in the population as a whole because every seed must have a mother and a father, so on average in the population,

transmission of genes through male and female function must be equal.

Fisher (1958) was the first to explain why, under natural selection, the two sexes are usually produced in approximately equal numbers. For the sake of argument suppose a population existed in which male births were less common than female, then male progeny would have better mating prospects than female progeny. Thus parents in this situation who are genetically predisposed to produce sons would tend to have greater than average numbers of grandchildren, and so the genes for male-producing tendencies would spread and male births would become more common. As the sex ratio approached unity the advantage associated with producing sons would be lost. A similar argument would obtain for females. Strictly, it is not the numbers which would be equal at equilibrium, but the parental expenditure on males and females, because selection acts upon the overall resource allocated. If the costs of producing sons and daughters are equal then equal expenditure implies a $1:1$ ratio in their numbers.

There are problems in testing this hypothesis, not least the one of knowing how to measure the cost of investment in male and female function and the fitness gained as a result. For example, it is clear that there are certain economies in making hermaphrodite flowers — after all the cosexual plant is free to act as either a maternal or paternal parent in a variety of matings, and a single set of petals and other decor is able to advertise both the male and female organs — an economic 'two-fer' (Charnov 1982; Charnov & Bull 1986). Hermaphrodite flowers are particularly economical if the flower is able to accept its own pollen for sexual reproduction and thus economize on the losses due, for example, to bees eating the pollen on the way to another flower, as well as the costs of providing nectar lunches (and sometimes narcotics!) for pollinators.

For hermaphrodites then, the question arises, how do we apportion the bill for petals, nectaries, etc. — how much to the male and how much to the female debit columns? Perhaps these costs should be divided fifty-fifty? In fact, experiments by Willson and Rathcke (1974) and Bell (1985) suggest that the attractive parts of the flower primarily promote the export of pollen, and have a lesser influence on the number of ovules fertilized. Female function may be satisfied by a single insect visit — but it requires several insect visits for a flower to disperse all of its pollen. The word for flower is female in nearly all languages we know of with gender-defined nouns; it is ironic to think that they *may* turn out to be essentially a male organ! However, there is a good evolutionary explanation for males being the showier of the sexes, as we shall see in Section 9.5.

Darwin (1877) reported that female, male and cosexual plants all occurred in the varieties of strawberries cultivated in his day. Flowers were smallest but the fruit crop heaviest in females, whereas cosexes had fewer and smaller berries but produced more runners. Because the cosexes and males produced more runners than the females Darwin commented that they tended to supplant the females and he concluded from this that female reproductive costs were greater than male — a conclusion that still held a century later when Lloyd and Webb (1977) reviewed the subject of secondary sexual characters in seed plants.

Virtually all studies indicate that the cost of being a mother are significantly greater than those of being a father. (Note that this is not the same as the relative costs to a mother of producing sons and daughters.) This constrains the ecology of females in a variety of ways, giving rise to a number of **secondary sexual differences** between males and females. For example, females of the double coconut *Lodoicea maldivica* are shorter and appear to have a greater risk of death than males because of the large crops of huge fruit they develop in their crowns which make them prone to decapitation by high winds (Silvertown 1987b). In the shrub staghorn sumac *Rhus typhina* it is the survivorship and rate of flowering of female ramets that are adversely affected by reproduction (Lovett Doust & Lovett Doust 1988).

Females of the herbs *Silene dioica* and *Mercurialis perennis* occur in predominantly better sites for growth than males, possibly because they have lower survival in poorer sites that males are able to tolerate (Freeman *et al.* 1976; Cox 1981). This niche differentiation between the sexes has been called the 'Jack Sprat effect', after the nursery rhyme:

Jack Sprat could eat no fat,
his wife could eat no lean,
And so between the two of them,
they licked the platter clean!

In Jack-in-the-pulpit (no relation) *Arisaema triphyllum* females are found in better sites than males, but this species is able to change sex so, in a sense, plants choose their sex according to the quality of the site (Lovett Doust & Cavers 1982b). In *Rumex acetosella* (Lovett Doust & Lovett Doust 1987) and spinach (Onyekwelu & Harper 1979) females remain alive longer each season, enabling them to cover their greater costs and even, in *R. acetosella*, to end up ahead of males in terms of clonal growth. However, in most species the heavier burden of reproductive costs that falls on females probably explains why sex ratios are more often male biased than female biased in established populations of dioecious species.

9.4.3 One gender or two?

A plant's gender is a function of how it divides its reproductive resources between male and female structures. In most plant species there is only a single gender class, but a significant minority of species have evolved two (Table 2.1). What conditions favour the extremes of gender specialization that produce gynodioecy and dioecy?

9.4.3.1 Gynodioecy

For a single nuclear gene that causes male sterility to spread in a cosexual population, the fitness of the females that carry this gene must exceed the fitness of cosexes. Because every sexual individual has a mother and a father, the fitness of a cosex must *on average* be acquired half through seed production and half through gene transmission by pollen. Lacking pollen, male-steriles must therefore produce more than twice as many seed progeny as a cosexual plant to achieve higher fitness (Lewis 1941; Lloyd 1975; Charlesworth & Charlesworth 1978). The principle of Fisher's sex-ratio argument applies to any population with two gender classes, so the fitness of male-steriles will be frequency-dependent, and an equilibrium between the two gender classes will be attained when their fitnesses reach equality. The highest equilibrium frequency of females that can be maintained is 50% (Gouyon & Couvet 1988) (Fig. 9.5).

Kohn (1988, 1989) studied gynodioecious populations of the perennial buffalo gourd *Cucurbita foetidissima* in Arizona and New Mexico. Females produced 1.5 times the number of seeds produced by cosexes, and seedling progeny from females survived their first year 2.8 times more often than cosex progeny, conferring a total fourfold advantage on females. Cosexual plants were highly selfing ($S = 0.9$) and crossing experiments showed that the poor survival of hermaphrodite progeny was due to inbreeding depression. A substantial part of the fitness advantage of females was therefore attributable to the benefits of outcrossing.

Gynodioecy in buffalo gourd qualitatively confirms simple theoretical expectations, but the situation in other gynodioecious populations tends to be much more complicated. For example, gynodioecy occurs in *Plantago lanceolata*, which is self-incompatible, so in this species females cannot attain an advantage over cosexes by release from inbreeding depression (van Damme 1984). Common thyme *Thymus vulgaris* is self-compat-

ible, but the frequency of females in gynodioecious populations in France varies between 5–95%, which cannot be explained if only nuclear genes for male sterility are involved (Gouyon & Couvet 1988). In both cases male sterility is inherited through **cytoplasmic genes** that are transmitted through seeds but not through pollen, and nuclear genes exist that restore male function. Antagonistic interactions between nuclear and cytoplasmic genes can cause the frequency of male-steriles to oscillate over many generations (Gouyon & Couvet 1988), so a wide range of female frequencies is to be expected in natural populations.

9.4.3.2 Dioecy

Natural selection operating on nuclear genes favours the *combination* of male and female allocation with the highest fitness. In certain circumstances an individual may achieve higher fitness by specializing as one sex than by being cosexual. One way this can evolve is via gynodioecy, because once male sterility has invaded a cosexual population the existence of pure females and inbreeding depression in cosexes can combine to favour outcrossing and the evolution of pure males (Charlesworth & Charlesworth 1978). This appears to have happened in the genus *Rumex* and in the genus *Silene*, which contains both gynodioecious species such as *S. vulgaris* and dioecious species such as *S. dioica* and *S. alba*.

We have seen that male and female fitnesses are frequency-dependent for genetic reasons (i.e. Fisher's sex-ratio argument), but they may also be interdependent in a variety of other ways that affect the fitness of cosexes as compared with males or females. Such interactions may be represented by plotting male fitness versus female fitness on a graph called a **fitness set** (Fig. 9.6). If *m* represents the fitness gained through male function, and *f* the fitness gained through female function, then the optimal allocation of resources for an outcrossed cosexual plant is that which produces the highest value of the product of male and female fitnesses, *mf* (Charnov 1982). When *mf* is at a maximum, any small change in allocation will cause the fitness loss via one sex to exceed the gain in fitness via the other sex.

Fig. 9.5 Equilibrium frequency of females for different values of female : hermaphrodite seed production in a gynodioecious population with male sterility determined by a nuclear gene. Female frequency = $(f-2)/(2f-2)$, where f = ratio of seed production by females : hermaphrodites.

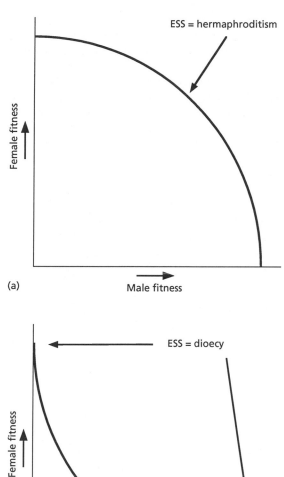

Fig. 9.6 Fitness sets showing the relationship between fitness returns from male and female function when: (a) hermaphroditism is favoured over dioecy, (b) when dioecy is favoured over hermaphroditism.

Because the allocation strategy that maximizes *mf* cannot be bettered, this ratio is termed an **evolutionarily stable strategy** (ESS).

When the fitness set is convex like that in Fig. 9.6a, *mf* is at a maximum at a point where male and female function contribute equally to fitness, and no gender type can have higher fitness than this cosexual plant. Curves of this shape may result when male and female functions can share costs of structures such as flowers (the 'two-fer'), or when male and female functions make their physiological demands upon the plant at different times, and so do not compete with one another (Charnov 1982). In experiments with a cultivated variety of gherkin cucumber, Silvertown (1987c) found that male flowers had no measurable cost in terms of female function, implying that the fitness set for this cosexual plant was indeed convex. In the wild species *Cucurbita foetidissima*, Kohn (1989) found that male flowers had a very significant cost in terms of seed production.

If the fitness set is concave, as in Fig. 9.6b, there are two maxima, representing plants of one sex or the other. Hence a concave fitness set favours dioecy. A fitness set of this shape could result from interference between male and female function, as for example when the proximity of male and female organs on the same plant leads to selfing that causes inbreeding depression in its progeny. Cox (1982) found an unusual form of interference between male and female function in *Freycinitia reineckei*, a dioecious liana in the Pandanaceae. Among other pollinators, *F. reineckei* was pollinated by a species of large bat that ate male flower spikes, but not female ones. Rarely, plants produced a cosexual flower spike and Cox found that female structures on such spikes were usually destroyed by bats as they licked the waxy material containing pollen off the male structures.

9.4.4 Sex change

Particularly in plants, the reproductive success of an individual is often dependent upon its size (chapters 1 and 4). If the curves of fitness plotted against size for male function and female function intersect each other (Fig. 9.7), the best strategy will be to specialize as one sex when small and the other when large, so natural selection ought to favour a change in sex expression as a plant grows (Ghiselin 1969). Sex change with size is a fundamental feature of the life history of a number of plants, for example in *Acer grandidentatum* (Barker *et al.* 1982), *Atriplex* spp. (Freeman *et al.* 1984) and *Arisaema triphyllum* (Policansky 1981; Lovett Doust & Cavers 1982b). In all these cases small plants are male and large ones are female or cosexual.

9.5 SEXUAL SELECTION

Charles Darwin distinguished between 'natural selection' in which individuals are selected according to their abilities to survive and reproduce, and 'sexual selection' in which they are selected according to their abilities to obtain more or better mates than other individuals. This is an artificial distinction but one which is customarily still made. According to Darwin (1871), sexual selection can occur in two fundamentally different ways: (i) competition between the members of one sex (usually males) for access to members of the second sex (usually females); or (ii) choice by the second sex of a mate from among the many potentially available. **Male–male competition** is the commonest example of the first of these mechanisms, and may regulate the *quantity* of offspring that males sire. The second mechanism is called **mate choice** and may be exercised by females, permitting them to regulate the *quality*

of offspring in which they invest. Oogamy lies at the bottom of this distinction between male–male competition for quantity and female choice of quality. Because pollen grains (like most male gametes) are cheap and plentiful, a male's reproductive success is limited by the number of ovules he can fertilize, rather than their quality. Conversely, because ovules are relatively few and costly and pollen is usually readily available, a female's fitness will be more closely related to which males she mates with than to the number of matings. (Beware of the temptation to extend this to human behaviour; the extended rearing period for human progeny means that both parents can do much to control the survival and quality of their offspring, and monogamy means that the male, as well as the female, may elect for quality.)

9.5.1 Paternity and male–male competition

Sexual selection between males can occur when the variance in male reproductive success is higher than that of female success (i.e. a few males father many offspring and many father very few or none at all), and when the mating success of males is associated with inherited traits that can be transmitted to their offspring. Competition between males may occur prior to pollen dispersal, when plants may compete to attract pollinators, or after pollen has reached its destination when pollen grains may compete within the style to reach the ovary first.

Darwin (1871) wrote at length about sexual selection in animals, but confined his remarks about plants to a footnote on floral protandry (see Table 2.1). In a considerable range of dioecious species, males begin growth and flowering earlier in the season than females (Lloyd & Webb 1977). Darwin suggested that the sexual precocity of male flowers in protandrous species results from competition among males for access to females. In a population of the monoecious wild cucumber *Echinocystis lobata*, for example, the first male flowers appeared a full 2 weeks before any female flowers were produced (Silvertown 1985).

Sexual selection provides a plausible explanation for the ostentatious colouring of many male birds and also for forms of ritualized combat between competing males. Bateman (1948) provided initial

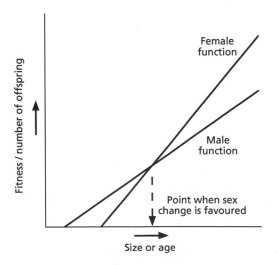

Fig. 9.7 Relationship between fitness returns of male and female function when these are age- or size-dependent and favour a change in sex.

evidence (from a study of reproductive success in fruit flies) that the relative success of males could be significantly more variable than that of females. This would mean that most females would breed successfully, while individual males would stand a greater chance of failure. Meagher (1986, 1991) found evidence of male–male competition in his study of *Chamaelirium luteum* (Chapter 3), where variance in the number of mates for males was nearly an order of magnitude greater than for females.

Willson (1979) suggested that the showy and often sophisticated decoration of angiosperm flowers enhances male function more than it does female function. This has been called the **male function hypothesis** and, if true, it would suggest that these floral traits are the product of sexual selection via male–male competition. Broyles and Wyatt (1990) tested the male function hypothesis using paternity analysis to determine the number of seeds sired by individual genotypes in a natural population of *Asclepias exalta*. Flowers of this species occur in umbels of varying size, and the pods that set seed are many fewer than the number pollinated. They found that male success (numbers of seeds sired) and female success (number of seeds produced) by a plant were both significantly correlated with the numbers of flowers per plant. Furthermore, the functional gender of a plant, as measured by the ratio of male to female success, was *not* correlated with the number of flowers per plant, indicating that large floral displays in this species increased overall fitness and were not dedicated simply to male function. This finding is contrary to the results of Willson and Rathcke (1974) and Bell (1985), mentioned in Section 9.4.2, who used indirect, non-genetic methods to estimate male and female reproductive success.

An intriguing example of male–male competition occurring after pollen dispersal is described by Uma Shaanker and Ganeshaiah (1989) in an Indian tree, *Kleinhovia hospita* (Sterculiaceae), in which they found that early-formed zygotes plug the style, so preventing the growth of any more pollen tubes and the fertilization of other ovules that would compete for maternal resources. Similar spoiling tactics are known in fish and insects.

9.5.2 Maternity and female choice

Willson and Burley (1983) suggested that, particularly in the case of mate choice, in order for sexual selection to operate in plants it is necessary that pollen should not be limiting. They reviewed the literature and concluded that female reproductive success in plants is often limited by resources rather than by pollen, and in such situations male reproductive success is likely to be limited by the number of fertilizations effected. Of course, seed production may be limited by both pollen supply and provisioning resources, and both may vary in time and space. Many temperate plants flowering in early spring have been shown to be pollen limited (e.g. Schemske *et al.* 1978; Bierzychudek 1981, 1982; Elmquist *et al.* 1988).

Most plants, particularly perennials, produce many more ovules than actually mature as seeds (Sutherland 1986a, b; Lee 1988; Marshall & Ellstrand 1988), which raises the possibility that plants may exercise mate choice by selective abortion. Depending upon the species, it has been found that patterns of fruit abortion can be influenced by choice of pollen parent, or pollination intensity (Bookman 1984; Lee 1988). Heavy pollination can result in competition between individual pollen tubes for access to ovules, so that rapidly growing pollen tubes are more likely to effect fertilization than slow ones. Willson (1979) has argued that it might be in the interests of the female parent to promote the deposition of excess pollen, as this could increase the average quality of her offspring. Schlichting *et al.* (1987) found evidence for this in controlled experiments using *Lotus corniculatus* and *Cucurbita pepo*. Offspring derived from conditions of more intense pollen competition were more vigorous and exhibited reduced genetic variation compared with offspring derived from conditions of little or no pollen competition. So, the outcome of increased competition in these species was clearly better quality offspring.

In experiments with *Delphinium nelsonii*, Waser *et al.* (1987) found that seed set was sensitive to the physical distance between pollen donor and recipient (Fig. 9.8). Reciprocal transplant experiments had shown that there was local adaptation in *D. nelsonii* populations, suggesting that the distance between populations was a useful index of their genetic difference. They found that the optimal separation between mating partners, measured in terms of seed set, was between 1–100 m. Follow-up experiments indicated that offspring of crosses from the optimal outcrossing distance survived better than did those from other distances (either greater or smaller), under field conditions. Several further experiments suggest an important physiological interaction between pollen and pistil. For example, in controlled crosses, pollen of donors 10 m from the recipient was significantly more likely to deliver a pollen tube to the ovary than was pollen from donors either 1 or 100 m away. Waser *et al.* (1987) interpret the pollen−pistil interaction as evidence of female discrimination among different mates that had adaptive value for the female parent and her offspring.

An *optimal* outcrossing distance (fitness rising and then falling with the distance between parents) has so far only been observed in *Delphinium nelsonii*, but straightforward distance-dependence has been observed in other species. Levin (1984) studied patterns of seed abortion in

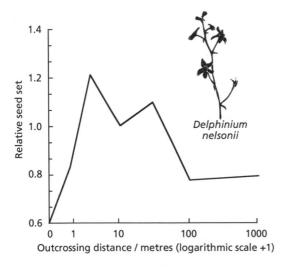

Fig. 9.8 Effects of outcrossing distance on relative seed set per flower in field populations of *Delphinium nelsonii* for plants that were hand pollinated with pollen collected from plants various distances away (from Waser *et al.* 1987).

Phlox drummondii as a function of the proximity of parents in natural populations. Seed abortion was negatively correlated with between-plant distance in each of the 12 populations considered. Because of limited pollen and seed dispersal, neighbouring plants would be more closely related to each other than to those further away. Levin postulated that proximity-dependent abortion was a consequence of inbreeding depression affecting the offspring of related neighbours. The studies of *Delphinium* and *Phlox* both describe female outcrossing − pollen is brought to a maternal parent from various pollen donors. The converse experiment of using a single pollen donor and females at different distances has not been done so far.

There is some experimental evidence that ovule abortion may be non-random, suggesting that female choice may play a role in this in some instances. For example, Stephenson and Winsor (1986) randomly thinned young ovaries in *Lotus corniculatus*, allowing the development of fruits that would otherwise abort. Compared with these random patterns of fruit abortion, natural patterns of abortion yielded mature fruits having significantly more seeds. Furthermore, these offspring had greater germinability, were more vigorous as seedlings, and had greater reproductive output as adults. Thus natural abortion tends to raise the average quality of surviving offspring. Casper (1988) tested the idea that ovules having more vigorous embryos are more likely to mature. She reported the results of hand thinning ovules in the herbaceous perennial *Cryptantha flava*, in each ovary of which three of the four ovules typically fail to develop. Following fertilization but before ovule abortion, three randomly chosen ovules of the four per flower were destroyed experimentally. The performance of offspring from these flowers was compared with that of offspring from control flowers. For a variety of measures of success, the survivors of 'natural' ovaries were more vigorous than those chosen randomly for survival. In *Cryptantha flava*, offspring seem to have been 'graded on a curve', and a fixed proportion of them get aborted, leaving the top 25%.

9.6 SUMMARY

Plant life histories are rich in variety. Fundamentally, this variety is due to the fact that limited resources cause **trade-offs** between different activities, so that there is no unique solution to the maximization of fitness. Individuals must allocate resources between competing demands (the **principle of allocation**). A trade-off between reproduction and other activities manifests itself as a **cost of reproduction**. A trade-off between **size** and **number** of offspring is also virtually universal. Trade-offs can be measured by **phenotypic correlation**, by **experimental manipulation**, or by **genetic correlation**.

The **evolution of sex** is paradoxical because natural selection favours genotypes that transmit the most copies of their genes to future generations, yet sexual reproduction produces offspring that are different from their parents. A sexually reproducing female has a **twofold disadvantage** to an asexually reproducing one because she transmits only half her genes to each offspring. This paradox may be resolved by **long-term** or **short-term advantages** of sex. The latter are more likely and include advantages based upon **frequency-dependent** fitness and **sib-competition**. **Apomicts** produce seeds asexually, but their evolutionary history suggests that this short-term advantage may be offset by longer-term disadvantages.

The **evolution of mating systems** is influenced by selection acting on the **quantity** and the **quality of offspring**. The detrimental effect of **inbreeding depression** seen in offspring produced by selfing is generally thought to be the main selective force behind the evolution of outcrossing mechanisms. It has been argued that the selfing rate is subject to disruptive selection, and that **mixed mating systems** should be rare. However, mixed mating systems may be maintained by a mechanism similar to the **lottery model** for the maintenance of sex, or by a local population genetic structure that leads to **bi-parental inbreeding**.

The equilibrium **sex ratio** in dioecious species should normally be 50 : 50, but the **costs of reproduction** tend to fall more heavily on females than males, which may bias the actual ratio in favour of males. A plant's **gender** refers to how it divides its reproductive resources between male and female structures. Most species have only one (cosexual) gender class, but a significant number have two and are **gynodioecious** (females and cosexes) or **dioecious** (females and males). Gynodioecy may be favoured when females have a higher fitness than cosexes, but this advantage will be **frequency-dependent** if male sterility is caused by nuclear genes. If **cytoplasmic genes** are involved, females may become much commoner than cosexes. **Dioecy** has sometimes evolved from gynodioecy. Dioecy may be favoured by trade-offs between male and female fitnesses that generate a **concave fitness set**.

Sexual selection is generated by differences between individuals in the number or quality of mates they are able to obtain. **Male–male competition** may regulate the quantity of offspring that a male sires and, according to the **male function hypothesis**, may account for the showiness of flowers. **Mate choice** exercised by females may regulate the quality of her offspring.

Chapter 10
Life History Evolution:
Birth, Growth and Death

10.1 INTRODUCTION

In this chapter we look at how evolution shapes a great range of the life history characteristics of plants: when they reproduce, how often they produce seed crops, how large seeds are, when seeds germinate, how plants grow, and when they die (see Table 9.1). Although this is a diverse set of traits, it is possible to say something about how each of them affects reproduction and survival, and therefore to compare the relative fitness of phenotypes with opposing life history characters. As we have already seen in Chapter 9, this phenotypic approach to selection makes it possible to analyse in some detail how ecological circumstances direct evolutionary change in heritable characters. However, note that it is particularly important to establish that the characters we are interested in here are heritable, because most life history traits will be closely correlated with fitness and may therefore have low heritability (Chapter 2).

10.2 REPRODUCTIVE VALUE AND REPRODUCTIVE MATURITY

Reproduction utilizes resources and meristems accumulated during prior periods of growth. This is why bigger individuals produce more seeds than smaller ones (Chapter 4). Because reproduction depends upon prior growth (and of course survival), trade-offs between reproduction and growth/survival also lead to trade-offs between reproduction now and reproduction later. Law (1979) found that in annual meadow grass *Poa annua*, high rates of reproduction early in life lead to lower rates of reproduction and smaller plant size thereafter. In short-lived plants such

trade-offs often arise from the fact that a limited number of meristems must be shared between reproduction and growth (Chapter 1). In the annual *Polygonum arenastrum*, Geber (1990) found that the rate of production of new meristems declined as soon as flowering began, and this was necessarily followed by reduced fecundity, growth and longevity. There were strong negative correlations with a genetic basis between age-specific growth and fecundity, reflecting genotypic differences between plants in the age at which they began flowering.

If reproduction incurs no costs, and a population is in a phase of increase, the earlier reproduction occurs the better for fitness. The reason for this is that when $\lambda > 1$, 10 offspring from an annual plant can have (say) 10 offspring each in the following year, these will in turn have 10 offspring each and in 6 years time the original annual plant will have given rise to 10^6 descendants. A semelparous plant delaying reproduction and taking 2 years to reach maturity will have only 10^3 descendants at the end of the same period. In such a scenario, a year's delay per generation reduces fitness measured at the end of 6 years by a factor of 10^3. In fact, of course, reproduction does have a cost and no population can be in a perpetual phase of increase, but this simple calculation demonstrates how important the length of the pre-reproductive period can be to fitness and that, other things being equal, delaying reproduction incurs a **demographic penalty**.

In reality, the optimum age at which an individual should begin reproduction depends upon how reproduction at a particular age would affect later survival and reproduction. The cost of reproduction in a plant reproducing at young age may slow its growth and increase the risk of death because small plants are more vulnerable (Chapter 4). Thus $\Sigma l_x m_x$ may be maximized by delaying reproduction until the plant reaches such a size as to be able to survive, or at least to complete, its first bout of reproduction. In general, the optimum age of reproductive maturity is reached when no further increase in $\Sigma l_x m_x$ can be obtained by any further delay. The contribution an average individual aged x will make to the next generation before it dies is its **reproductive value** V_x (Fisher 1958). In a population at demographic equilibrium, V_x may be calculated as:

$$V_x = m_x + \sum_{i=1}^{i=\infty} \frac{l_{x+i}}{l_x} m_{x+i} \qquad (10.1)$$

where l_x is the survivorship to age x and m_x is the average fecundity of plants aged x. These statistics are taken from the life table and fecundity schedule for a population (Chapter 1), but we will see later that the same ideas can be used to analyse a stage-classified population. Putting Eqn 10.1 into words: reproductive value (V_x) is the sum of the average number of offspring produced in the current age interval (m_x) plus the sum of the average number produced in later age intervals (m_{x+i}), allowing for the probability that an individual now of age x will survive to each of those intervals (l_{x+i}/l_x). Two examples of how reproductive value changes with age are shown in Fig. 10.1.

Equation 10.1 can be split into two components, **current fecundity** (m_x) and the expectations of reproduction, called the **residual reproductive value**, which is given by the second term on the right hand side of Eqn 10.1, or by $V_x - m_x$. The residual reproductive value of a plant which reproduces at age x is equivalent to the chances that remain to it to produce further offspring in following seasons. In other words it is the probability of living one more year (l_{x+1}/l_x) times the reproductive value of a plant one year older $(x + 1$ years old$)$ which we can denote by V_{x+1}. Therefore residual reproductive value can also be calculated as $(l_{x+1}/l_x)V_{x+1}$. This formulation can be useful in calculating reproductive values from life table data.

In a population that is not at demographic equilibrium, the residual reproductive value must

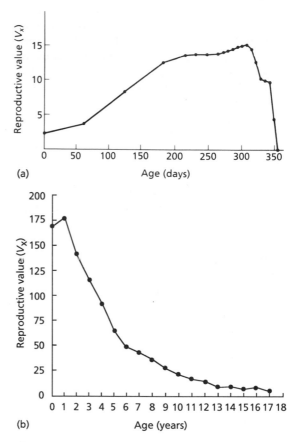

(a)

(b)

Fig. 10.1 Reproductive values for: (a) *Phlox drummondii*, and (b) *Acacia suaveolens*.

be scaled up or down, depending upon whether the population is decreasing or increasing, because the demographic penalty incurred by an offspring produced later in life compared with one produced without delay depends on the value of λ. When $\lambda > 1$, the greater its value, the greater the demographic penalty against plants which delay, and the more plants with precocious reproduction are favoured. Conversely, when $\lambda < 1$ there is an advantage to delaying reproduction. We can make allowance for the effect of λ on the balance between current fecundity and residual reproductive value by multiplying the term for residual reproductive value on the right hand side of Eqn 10.1 by N_x/N_{x+i}, where N_x is the size of the *entire* population when the cohort is x years of age, and N_{x+i} its size i years later. If the population is increasing $N_{x+i} > N_x$, so $N_x/N_{x+i} < 1$, and the residual reproductive value is *lower* than it would be in a stable population. Conversely, when $N_{x+i} > N_x$, $N_x/N_{x+i} > 1$ and the residual reproductive value will be *higher* than it would be in a stable population. When $\lambda = 1$, $N_x/N_{x+i} = 1$, and it can be ignored.

A high adult mortality risk favours earlier reproduction because it lowers residual reproductive value. Environmental causes of mortality (not directly related to the cost of reproduction) may determine where the upper limit of the optimum age of first reproduction lies. In unpredictable habitats, such as deserts, plants which delay reproduction experience a high risk of dying before reproductive maturity. This mortality risk lowers the value of l_{x+i}/l_x, and therefore also lowers the relative importance of residual reproductive value compared with current reproduction (Eqn 10.1). This is a plausible explanation for the high proportion of annuals found in desert floras (Fig. 10.2).

In a study of the annual woodland herb *Impatiens pallida* in Illinois, Schemske (1978, 1984) observed that a population in the interior of a wood flowered significantly earlier than a population only 64 m away at the woodland edge (Fig. 10.3). During 6 years of observation, the interior population was annually destroyed in July by a chrysomelid beetle, but the edge population escaped beetle attack. In fact plants in the woodland interior which delayed flowering produced drastically fewer seeds because late-flowering plants were eaten. Plants transplanted reciprocally between woodland interior and woodland edge had lower fitness than natives. Plants from the edge did less well in the interior than natives because they were eaten before they flowered, while interior plants did less well at the edge than natives because they flowered for a

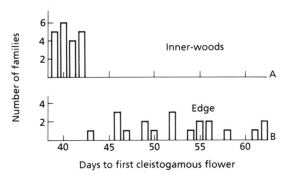

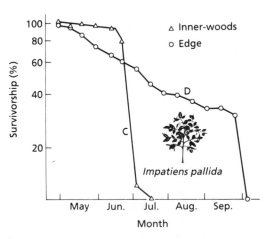

Fig. 10.3 The distribution of dates of first flowering in a greenhouse for *Impatiens pallida* collected from A, a site where it was attacked by a herbivore in July, and from B, a site where it was not attacked. C and D show survivorship of natural populations at the two sites (from Schemske 1984).

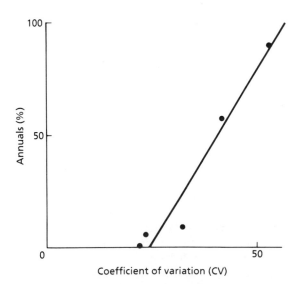

Fig. 10.2 Percentage of the herbaceous flora accounted for by annuals plotted against coefficient of variation in total annual rainfall (CV) in five North American desert habitats (from Schaffer & Gadgil 1975).

shorter period of time and consequently produced fewer seeds.

Kadereit and Briggs (1985) found that groundsel (*Senecio vulgaris*) sampled from the flower beds of the Cambridge Botanic Garden developed more quickly and flowered significantly earlier than plants of the same species sampled from habitats where gardeners do not rogue out this weed. Law *et al.* (1977) compared the life histories of *Poa annua* populations collected from disturbed environments such as building sites where mortality risks were high, and from relatively stable pasture environments. Populations of plants grown from seeds collected in these two types of environment were raised in uniform conditions and their age-specific survival (l_x) and reproduction (m_x) were determined at monthly intervals (x) (Fig. 10.4). The population from a disturbed environment showed early reproduction and early death, while reproduction in the pasture population was delayed so that only one major period of reproduction could be observed in the 18-month duration of the experiment. Although the short-lived population produced more inflorescences than the long-lived one in the first reproductive season (5–10 months), the situation was reversed in the second season (15–18 months). Delayed reproduction and increased survival in the pasture population allowed increased growth per plant and an increased average reproductive output.

In plant species that span a broad geographical range there is commonly a positive correlation between the latitude of a population and the age at which plants reach reproductive maturity. Studies by Bøcher (1949; Bøcher & Larsen 1958) of the herbs *Prunella vulgaris* and *Holcus lanatus* in Europe were among the first to show that northern populations flowered later than southern ones, and that the difference had a genetic basis. Similar trends with a genetic basis were found by Lacey (1986, 1988) in wild carrot *Daucus carota* and in *Verbascum thapsus* by Reinartz (1984) in North America. Both species are semelparous, but there was variation within and between populations in the length of the pre-reproductive period, which varied from 1 to 3 years. In *D. carota*, first-year survival in North Carolina populations (36°N) was double that in Ottawa (45°N), but survival in the second year showed the opposite trend and was 10 times greater in Ottawa than in North Carolina. This result strongly supports the hypothesis that differences in age-specific survival explain the differences in age at maturity between wild carrot populations at different latitudes. High mortality in the second year in North Carolina was thought to be caused by two fungal pathogens that are a particular hazard to cultivated carrots in Southeast USA. Why this mortality factor should be age-specific is not clear.

Flowering in *Daucus carota* and other semel-

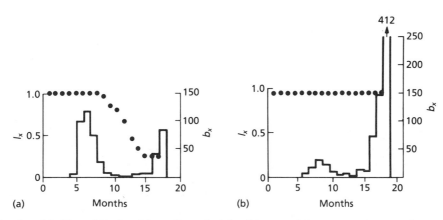

Fig. 10.4 Survivorship (dotted line) and fecundity schedules (histogram) for two populations of *Poa annua* derived from: (a) disturbed, (b) stable environments (from Law *et al.* 1977).

parous perennials typically occurs when plants reach a threshold size (e.g. Gross 1981). Latitudinal differences between *Daucus* populations in the age at maturity could therefore either result from latitudinal differences in the growth rate, which determines how quickly the reproductive threshold is reached, or from differences in the threshold size itself. In carrot there was some evidence for genetic differences in growth rate (Lacey 1986), but in *V. thapsus* Reinartz (1984) found that the threshold size at which rosettes flowered was significantly larger in Canadian populations than in populations in South USA, and that this difference was maintained in a common garden.

All of the principles that determine the optimum *age* of reproductive maturity also apply to determining the optimum *size* at which reproduction should begin in populations where size rather than age determines survival and fecundity. For practical purposes, the reproductive value for each size class may be determined by iteration of the population projection matrix to determine the left eigenvector of the matrix, which contains the relevant values (Caswell 1989).

10.3 BIRTH

Strictly speaking a plant is 'born' when it is released from the tissues of its mother, which in seed plants is the moment of germination. We will look at the evolution of germination timing later in this section, but two prior events merit discussion first: how often a plant produces seed crops, and the size of seeds themselves. All of these life history characters, including germination timing, are under some degree of maternal control, which is another way of saying that they are all *maternal traits*.

10.3.1 Seed crop frequency

Total crop size per plant often fluctuates from year to year. The magnitude of this variation can be very different in different species. It is most extreme in tree species such as beech *Fagus* spp., oak *Quercus* spp. and pine *Pinus* spp. in the temperate zone and in dipterocarp trees in Southeast Asia. These species produce vast crops of

seed, called **mast**, in some years, but few or no seeds in the intervening periods between mast years. Typically, seed production by different individuals in such populations is synchronized and mast years are correlated with climatic variables. In Europe, for instance, the beech (*F. sylvatica*) may mast in the year following a hot summer but rarely in the year following a cold one. In Southeast Asia, mast years in dipterocarp trees appear to be triggered by slight climatic changes in the region that are associated with El Niño, an oceanic event that occurs many thousands of miles away in the Pacific, but which has global climatic repercussions (Ashton *et al.* 1988).

Why do some species miss opportunities to reproduce, and then produce large crops in mast years? The null hypothesis to explain this phenomenon must be that climatic conditions simply suit seed production better in some years than in others and that barren periods result because trees take time to recover from the effort of reproduction (see Fig. 9.1). This hypothesis would certainly seem to explain why fruit trees such as pear and apple only bear large fruit crops on a biennial cycle. An alternative hypothesis, prompted by the observation that masting seems to waste opportunities for reproduction, is that these lost opportunities are in some way compensated because the habit increases the fitness of individual trees. One argument is that synchronized fruiting improves the chance of cross-pollination in wind-pollinated species (Smith *et al.* 1990). There is some evidence for this in trees such as *Fagus sylvatica* and *Pinus sylvestris* in which there is a smaller percentage of unfilled seeds in mast years than in years when crops are small (Sarvas 1968; Nilsson & Wästljung 1987), but the hypothesis cannot apply to dipterocarps because they are insect-pollinated. One thing all large trees, and indeed all plants, have in common is predation on their seeds by animals. This provides a more general explanation for the possible advantage of the masting habit.

The argument is that seed predators consume a large proportion of small seed crops but that they cannot consume a tree's entire crop in a mast year. Hence the probability of a seed escaping predation is greatest when crops are large. However, it would be disastrous for the tree if large crops were produced regularly because predators would simply build up their numbers from one year to the next on succeeding bumper crops. This is known as the **predator satiation** hypothesis. It predicts that there should be a negative relationship between the probability of a seed being eaten and the size of the current seed crop in a masting species. This prediction is supported by information collected by foresters on a number of species (Silvertown 1980b). The effect of two predators on seed survival in ponderosa pine is shown in Fig. 10.5.

The predator satiation hypothesis might be strengthened if it could be shown that the masting habit is most pronounced in those tree populations where seed predation is strongest. Using data on variation in the annual seed production of trees and data on seed predation in crops of different sizes (e.g. Fig. 10.5) for a range of species, it is possible to test the idea that masting species are attacked more severely than non-masting species when they produce small seed crops. Comparing seed production and seed predation data for 15 species in this way, Silvertown (1980b) found that five of the seven most heavily preyed upon species showed the masting habit. Among the eight species which suffered lower seed predation, only two showed very variable seed production.

Another intriguing piece of evidence which supports the idea that masting is a defensive strategy which protects trees from seed predation comes from a comparison of the seeding behaviour of two populations of the tropical tree *Hymenaea coubaril*. This tree occurs both on the island of Puerto Rico where one of its major insect seed predators is absent, and on mainland Costa Rica where these predators are present. The mainland tree population shows the masting habit but the island one does not. Other morphological features of *H. coubaril* fruit which help deter predators in Costa Rica are also absent from populations in Puerto Rico (Janzen 1975).

It follows from the argument that masting prevents animals consuming an entire tree crop, that trees with fleshy fruits and animal-dispersed seeds should not mast but should produce fruit regularly. The seeds in such fruits generally pass through the gut of the dispersal agent intact, so

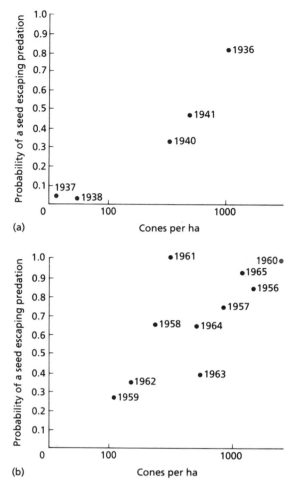

(a)

(b)

Fig. 10.5 The relationship between annual cone crop size and the probability of a seed of ponderosa pine escaping predation by (a) chalcid wasps, and (b) abert squirrels. Study (a) was done in California by Fowells and Schubert (1956), study (b) in Arizona by Larson and Schubert (1970).

that the only effect of masting in a fleshy-fruited species would be to prevent seed dispersal. The hypothesis that fleshy-fruited species do not mast may be tested by using a method of between-species comparison as before. A comparison of this kind, using species from the North American sylva, confirms the hypothesis and shows that most trees with non-fleshy dispersal units mast to some degree, while most of those with fleshy dispersal units do not (Silvertown 1980b).

10.3.2 Seed size

The principle of allocation dictates that a plant can package its reproductive effort into a few large seeds or many small ones. This produces a **size–number trade-off** in seed output. The resources available to a plant vary with its size, its age and a host of environmental factors. Should a plant respond to this by keeping seed size constant and varying seed number, or by the reverse strategy? A variety of phenotypic selection models point to the conclusion that there ought to be an optimum seed size within any particular population, and that this size should be conserved at the expense of seed number when resources are scarce (e.g. Lloyd 1987). Seeds much smaller than the optimum will be non-viable or have low fitness, seeds much bigger than the optimum waste resources that could be put to better use by provisioning more seeds, and are therefore a fitness cost to the parent. It may be to the individual advantage of the *embryo* in a seed to grow as large as possible, but the maternal parent controls the flow of resources to its seeds, so fitness costs to the parent should place an upper limit on the profitability of large seeds (Casper 1990). These theoretical considerations give rise to the sigmoidal fitness curve shown in Fig. 10.6. The optimum seed size on this curve is given by the point at which increasing investment by the parent brings diminishing returns in fitness (Lloyd 1987). This theory is yet to be experimentally tested in plants, but it is consistent with the pattern of variation in seed size between species, which shows significant correlations between seed size and habitat that suggest optima do exist. The picture for variation within species is more complicated, as we shall see.

10.3.2.1 Variation between species

An enormous range in seed size occurs between species, from the tiny seeds weighing 10^{-6} g produced by the orchid *Goodyera repens* to the monstrous seed of the double coconut (*Lodoicea maldivica*) which weighs in at over 10^4 g (18–27 kg) (Harper *et al.* 1970). A number of comparative studies of seed size have been made and these suggest that this character has been adjusted by

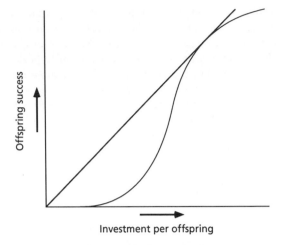

Fig. 10.6 The theoretical relationship between a parent's investment in a seed, measured by seed size, and offspring fitness. The optimum seed size occurs at the point where the diagonal line, which represents constant returns in fitness for constant increments in seed size, meets the curve at a tangent (after Lloyd 1987).

natural selection in various ways, depending upon the life history and habitat of species. The mean weight of seeds increases progressively through herb, shrub and tree species in the floras of the British Isles, California and neotropical forest. It also follows the same trend on a world scale (Table 10.1). Within the herb group in California, the mean seed weight of annuals is significantly less than that of perennials (Baker 1972). This difference does not occur in Britain (Salisbury 1942;

Table 10.1 Seed weight in relation to plant habit. British, Californian and worldwide data are from Levin and Kerster (1974) and sources therein. Neotropical data are for 54 tree, 55 shrub and 17 herb species in the Rubiaceae. Other families and neotropical floras as a whole follow the same trend (Rockwood 1985)

Habit	Mean seed weight (mg)			
	Britain	California	Neotropics	Worldwide
Herbs	2.0	5.7	1.69	7.0
Shrubs	85.4	7.5	8.82	69.1
Trees	653.4	9.6	29.46	327.9

Hart 1977) when the floras of all habitats (woodland, grassland, etc.) are lumped together. Trends of seed size with variations in life history are likely to be obscured when a whole flora is compared because of the stronger association of seed size with habitat.

A comparison of seed size for annuals and perennials which occur in the same habitat enables habitat to be eliminated as a variable. When such a comparison is made for plants in the flora of British calcareous grasslands for instance, annuals are found to have significantly smaller seeds than perennial herbs (Silvertown 1981). The annuals in this grassland sample were semelparous while most of the perennials were iteroparous. It seems likely that natural selection has increased seed number at the expense of seed size in these annuals. In the most detailed study yet, Mazer (1989) correlated ecological variables with seed weight in a sample of 648 angiosperm species from the Indiana Dunes region of the USA. This study confirmed the patterns found by previous studies of other floras, but was able to apportion the variance in seed weight between species among different variables. Plant family was most important (30% of variance in seed weight), life history was next (22%), and habitat accounted for 8% of the variance in seed mass. It appears that, at least in the Indiana Dunes flora, much of the association between habitat and seed size is due to a correlation between family and seed size and a tendency for family, life history and habitat to be correlated with one another.

Neotropical forest trees that are able to regenerate in small gaps or in shade have significantly larger seeds than pioneer trees and those which need large light gaps (Foster & Janson 1985). This difference is presumably the result of the increased food reserves required for seedling establishment in shade, which is reflected in higher seedling mortality among species with smaller seeds in experimental shade conditions (Fig. 10.7). In tropical forest species from Peru, Foster and Janson (1985) found that taller plants had significantly larger seeds than shorter ones of the same life form. Relationships between plant stature, shade tolerance and seed size in modern floras have been applied to an analysis of seeds in palaeofloras by Tiffney (1984) who found that Cretaceous

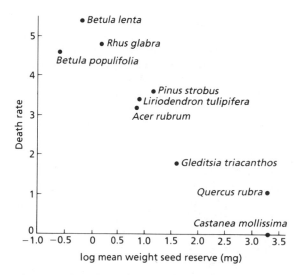

Fig. 10.7 The relation between death rate in shade conditions and log mean seed weight in nine tree species (from Grime & Jeffrey 1965).

angiosperms had smaller seeds than those in later Tertiary floras. On this basis, he suggests that the early flowering plants may have been mostly shrubs and small trees which occupied gaps in vegetation dominated by gymnosperms.

Variations in seed weight between species in the Californian flora appear to be more strongly related to the risk of seedling mortality due to drought than due to shade, and a positive relationship between seed weight and the dryness of the habitat occurs both among herb species within the same genus and for whole herb communities. A similar relationship between moisture availability and seed weight is also found in Californian trees. The explanation offered for these relationships is that a larger seed enables a seedling to produce a more extensive root system and thus obtain water more rapidly and efficiently than a small seed in a dry environment. California has a large number of introduced species which fit the same patterns of seed weight and environmental conditions shown by native species (Baker 1972).

This correspondence between the behaviour of introduced and native species has some interesting and far-reaching implications for how we view the adaptive fit between seed size and environment (e.g. species with large seeds occurring in shaded habitats). Two hypotheses for this

fit are possible: (i) species have evolved a seed
size characteristic of a particular habitat largely
under selective forces operating within that habitat; or (ii) species with seeds ill-suited to regeneration in a particular habitat suffer ecological
displacement. In other words, species which
happen to have large seeds become woodland
plants when they invade an area. Invading species
with small seeds cannot occupy woodland but can
become weeds or grassland plants. The process of
ecological displacement probably explains the
distribution of most of the introduced plants of
California. For how many native species could
this also apply? The answer must depend upon
the availability of heritable variation for seed size
within species.

10.3.2.2 Variation within species

Well-established cases of adaptive variation in
seed size between populations of the same species
are relatively few, though this may be mainly for
want of investigation. Probably the best example
is the weed *Camelina sativa* whose seeds have
diverged in size from those of its wild progenitor,
and have ballistic properties better resembling
the seeds of flax, a crop which it infests. Selection
on the size of *Camelina* seeds was brought about
by the use of winnowing machines that separated
flax seed from the husk. *Camelina* seeds that
separated out with flax seeds were selectively
favoured by being resown with the crop (Barrett
1986).

Particularly in competitive situations and when
germinating in shade, the size of a seed is closely
correlated with fitness. For example, Black (1958)
planted large and small seeds of subterranean
clover (*Trifolium subterraneum*) in a mixture and
found that the percentage share of light intercep-
tion by seedlings from small seeds was reduced
virtually to zero after 80 days (Fig. 10.8). Wulff
(1986b) compared the growth of plants from large
and small seeds of another legume, *Desmodium
paniculatum*, when raised on their own, with
their performance when competing in mixtures
with each other. Pure stands grown from small
seeds produced the same yield per plant as pure
stands grown from large seeds, but in mixtures
the seed production of plants from small seeds

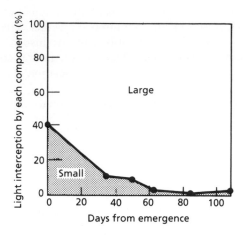

Fig. 10.8 The percentage of light intercepted by plants
from large and small seeds of *Trifolium subterraneum*
when grown in mixtures with each other (from Black
1958).

was less than half that of plants from large ones.
In an experiment with the wild oat *Avena fatua*,
whose seeds varied fivefold in weight, large seeds
produced more than twice the number of offspring
of small ones when growing in competition with
barley (Peters 1985).

When selection for seed size can be so strong, it
is not surprising that the size of some seeds is so
constant that those of the carob tree (*Ceratonia
siliqua*) have been used to define a unit of weight
(the carat) for trading in gold. But clearly, not all
species are as reliable as this and anyone who
chose the seeds of *T. subterraneum* as a unit of
measurement would find they had a standard that
could vary seventeenfold in weight! (Black 1959).
As one might expect of a character closely related
to fitness, the heritability of seed size in wild
plants tends to be low (e.g. Schaal 1980; Primack
& Antonovics 1981; Waller 1982; Mazer 1987;
Roach 1987), but what is puzzling is that phenotypic variation in seed size is still so great in many
species. In a survey of seed size variation in 39
North American plants, Michaels *et al.* (1988)
found significant variation in seed weight within
populations of 37 of them, with most of the
variance arising from differences between seeds
on the same plant. Other studies of individual
species have found much the same thing, with
most of the observed variation related to position

on the plant (e.g. Waller 1982; Stanton 1984; Wulff 1986a), to plant size (e.g. Hendrix 1984) or to season (e.g. Cavers & Steele 1984). Why is there such phenotypic plasticity in a character so important to fitness?

There are two possible answers. First is the null hypothesis that **developmental constraints** limit the degree to which plants are able to control the size of individual seeds (Silvertown 1989b). This would explain why *mean* seed size is usually quite constant between plants in the same population, but individual seed size varies with position on the plant. It is also consistent with the existence of trade-offs between seed size and seed number that have been detected in several species (Wolf *et al.* 1986; Lalonde & Roitberg 1989). For example, in *Primula vulgaris* seed number per capsule varies with pollination success, and when this is good, seeds are smaller and more numerous than when pollination is poor (Boyd *et al.* 1990). Where such a trade-off exists, a plant whose seed set is pollinator limited will have no direct control over the size of its seeds.

Alternative, **adaptive hypotheses** propose that variation in seed size raises the parents' fitness in an environment that is spatially or temporally heterogeneous (remember that seed size is a *maternal*, not a progeny, trait). There are a variety of ways in which this could work, although in all cases the inherent disadvantage of small seeds must be compensated by some other kind of advantage over larger ones. A trade-off between seed size and number within a plant's progeny could be adaptive. Seed abortion is extremely common (Chapter 9), so it is clear that many plants can and do control seed number independently of pollination. In plants that are pollinator limited, seed size and number might be negatively correlated because the optimum seed size depends upon seed number. When pollination is sufficient to produce them, many small seeds might be more valuable than fewer large ones, but when pollen is scarce, resources may be best used by increasing seed size.

McGinley *et al.* (1987) used a phenotypic selection model to examine the hypothesis that seed size variation could be favoured in a spatially heterogeneous environment if the optimum seed size was different in different micro-environments.

They found that this strategy would only be favoured if seeds could be selectively dispersed to their optimum environments. This may be possible when size variation is correlated with variation in dispersal syndrome, such as in the case of cleistogamously and chasmogamously produced seeds of *Impatiens capensis* (Chapter 9), but it is not likely in most species. A half-way house between this kind of adaptive hypothesis and the operation of evolutionary constraints is suggested by a study of seed size variation in *Prunella vulgaris*. Winn (1985) found that plants reciprocally transplanted between natural populations in a deciduous wood and an old-field had larger seeds in the wood than in the old-field, regardless of their source. In the *absence* of such experimental evidence, one would normally (and wrongly) suppose that large seed size was the product of selection in a shaded habitat. Winn's experiment showed that it was *phenotypic plasticity* of seed size, not seed size itself, that was adaptive. If selection operates on the phenotypic plasticity of a trait rather than the trait itself, it does not seem so surprising that there is substantial phenotypic variance associated with the trait (Silvertown 1989b).

To return to purely adaptive hypotheses, the most obvious advantage a small seed has is that it is cheap to produce, so it may pay a plant to put only some of its resources into a few large seeds and any 'spare' resources above a threshold into as many small seeds as possible (Capinera 1979). This **bet-hedging strategy** is favoured when there is uncertainty (i.e. temporal heterogeneity) about which kind of seed will produce the greatest fitness returns for the resources invested (i.e. what the optimum seed size will be). The annual *Amphicarpum purshii* follows this strategy, producing a fairly constant number of large seeds early in life, and increasing numbers of small ones later as the plant grows larger, if resources and time allow (Cheplick & Quinn 1982, 1983). The large seeds are produced from subterranean flowers and the small ones from aerial ones, a syndrome called **amphicarpy**, which is rare, but independently evolved in at least nine different families (Cheplick 1987). The bet-hedging allocation strategy in amphicarps is an easy one to spot because the seed types are so different from each other. More subtle variation among the seeds of other plants could function in a similar manner, but the allocation strategy would be more difficult to identify. The likeliest way in which seeds of different size could be part of a bet-hedging strategy is if they have different germination requirements.

10.3.3 Germination

We saw in Chapter 5 that the time a seed germinates can crucially affect its fitness because the hazards of the physical and biotic environment vary with season and priority of emergence. Setting to one side the seasonal factors that favour short-term dormancy, dormancy should otherwise be brief because of the demographic penalty against delayed reproduction that occurs when $\lambda > 1$ (Section 10.2). As we have now come to expect of characters closely correlated with fitness, the heritability of germination timing (or of dormancy) is often low. However, differences in the germination behaviour of seeds produced by the same plant, called **germination heteromorphism**, appear to be quite common. The similarities between germination heteromorphism and seed size variation are difficult to miss and, in fact, variation in seed size and germination requirements are often correlated with one another within a plant (Silvertown 1984). For example, Hendrix (1984) found that 45% of autumn emerging seedlings of wild parsnip *Pastinaca sativa* derived from the small seeds produced on tertiary umbels and that 35% came from large seeds of primary umbels. In the spring when most germination takes place, 26% of seedlings were from tertiary and 45% from primary umbels. If such variation is of adaptive value, the fitness of plants producing a clutch of seeds with uniform behaviour should be lower than the fitness of plants producing more than one seed type.

Consider the common situation of a habitat which varies in conditions for successful seedling establishment from year to year. Say, in some years, autumn germinating seeds do better, and in others spring ones. Simplistically, we might expect plants to produce the single kind of seed that does best in the commonest kind of year. However, this is incorrect because it implicitly

assumes that long-term fitness is the arithmetic average of each year's seed production and survival. Because the rate of increase in a population from year to year is multiplicative, the mean fitness of a phenotype whose seed production and survival vary through time must be obtained by taking the **geometric average** (Venable 1985). This point is fundamental to adaptive explanations of germination heteromorphism and other bet-hedging strategies. To see why, imagine two kinds of plant: a 'monomorphic producer' which produces 10 seeds of a single type (seed type I) per year and a 'dimorphic producer' which produces five seeds of type I and five of type II per year. The germination success of the two types of seeds is perfectly negatively correlated: i.e. when type I does well, type II does badly and vice versa. Now, compare in Table 10.2 the fitness of the two types of parent when calculated as the arithmetic mean and as the geometric mean over 5 years. Although the arithmetic calculation reveals no difference, the correct calculation using the geometric mean shows that the relative fitness of the monomorphic seed producer is $5.2 : 6 = 0.87 : 1$ and plants producing heteromorphic seeds are favoured over those producing only one seed type.

The **bet-hedging hypothesis**, that year-to-year variation in the relative success of different seed germination morphs favours germination heteromorphism, can be tested by calculating the fitness of each morph from demographic field data recorded over a period of years. For example, Mack and Pyke (1983) found that seeds of the annual grass *Bromus tectorum* growing in Washington State germinated over a prolonged period between autumn and spring. At a dry study site which cohorts did best depended on the year. Taking this variation into account, and calculating geometric mean fitness over a run of years, plants with heteromorphic seeds turn out to have higher fitness than monomorphic producers at this site (Fig. 10.9). Differences in morphology and germination between seeds in the same seed head are very common in the Asteraceae (Venable and Lawlor 1980). Venable *et al.* (1987) studied the species *Heterosperma pinnatum* in Mexico and found that seeds (more properly called 'achenes' in the Asteraceae) from the centre of the head germinated with the early spring rains.

Table 10.2 The number of successful offspring produced annually by plants producing the same total number of monomorphic or dimorphic seeds, and the 6-year average number of offspring of each type of parent, calculated as arithmetic and geometric means. The germination success of the two types of seed is exactly negatively correlated (see text for further explanation)

Parental strategy	Number of successful offspring							Arithmetic mean	Product	Geometric mean
	Year	A	B	C	D	E	Σ			
Monomorphic seeds										
Type I		10	8	6	4	2	30	6	3840	5.2
Dimorphic seeds										
Type I		5	4	3	2	1				
Type II		1	2	3	4	5				
Types I + II		6	6	6	6	6	30	6	7776	6

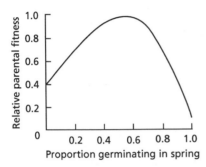

Fig. 10.9 The expected fitness of plants producing seed crops with different proportions of spring-germinating seeds, estimated from annual variability in the fitness of different cohorts of *Bromus tectorum*. Phenotypes producing a mixture of seed germination morphs are favoured (from Silvertown 1988).

These seedlings risked being caught by a later drought, but survivors could reach large size. Achenes from the margin of the fruiting head were slow to germinate and consequently did not risk drought, but produced smaller plants with fewer seeds. Venable and Búrquez (1989) found that the ratio of morphs in *H. pinnatum* heads varied between populations and had a high broad-sense heritability. This is consistent with a character whose fitness is affected by the weather, because selection of variable direction will maintain genetic variation (Chapter 3).

10.4 GROWTH

Because of their modular construction, it is not fanciful to argue that clonal growth comes naturally to plants. A tree adds modules to its branches and so extends vertically, a clonal plant adds modules at its base and extends horizontally. A few clonal shrubs and trees do both. The difference between these growth habits is mostly one of scale and it has been said that a clonal herb is just a tree lying on its side (Harper 1981) (see Fig. 1.7a), though of course the essential difference is that the branches of a clonal herb are ramets capable of an independent existence. Clonal growth is evolutionarily ancient and widespread among plant taxa (Mogie & Hutchings 1990). Salisbury (1942) estimated that more than two-thirds of the commonest perennials in the British flora show pronounced clonal growth. Even greater proportions of woodland and aquatic herbs show the habit.

10.4.1 Costs and benefits of clonal growth

The advantages of clonal growth are easy to see.
1 Rapid increases in size are possible and this enhances later survival and reproductive potential.
2 Movement is possible, which means bad environments can be left for better ones.
3 Space can be occupied and resources captured.
4 Young ramets have much lower mortality than young seedlings.
5 Competitors can be invaded and displaced.

6 Clonal (vegetative) reproduction incurs no costs of sex.

Given these substantial advantages and the fact that plants are predisposed to a clonal pattern of growth by their modular structure and evolutionary history, the most fundamental question to answer about growth is why a significant number of plants, including most trees, are *not* clonal. This question has received very little attention, so we will consider the possible disadvantages of clonal growth only briefly before looking at different modes of clonal growth.

Like all other features of life history, clonal growth has costs. Although clonal growth may ultimately increase reproduction because it increases the size of a genet, there are trade-offs between sexual reproduction and clonal growth in the short term. For example, in water hyacinth *Eichhornia crassipes*, Watson (1984) found that rapid population increase by clonal growth was slowed up when meristems began to produce flowers instead of ramets. Two genotypes differed in their relative allocation to flowering and clonal growth. Sutherland and Vickery (1988) found a trade-off between sexual reproduction and clonal growth in five closely related North American species of *Mimulus*. All five species produced rooted stolons or rhizomes as well as flowers that were chiefly pollinated by humming-birds. Across the five species there was a perfect negative correlation between the number of ramets per unit of plant weight and all measures of investment in sexual reproduction, such as nectar volume, fruit set and seed production. In a taxonomically broader comparison of 26 clonal species, Silvertown *et al.* (1993) found a negative correlation between the importance to λ of sexual reproduction and the importance of clonal growth.

To the extent that the reproductive costs of clonal growth delay flowering, clonal growth must increase residual reproductive value if its costs are to be repaid. We should therefore expect clonal growth to be absent from anywhere where selection favours early reproduction, such as in highly disturbed habitats where the risk of adult mortality is high (Section 10.2). Disruptive selection related to the degree of disturbance in different habitats is presumably responsible for the fact that several grasses and sedges show intraspecific variation for the clonal habit and have forms with and without stolons or rhizomes (Mogie & Hutchings 1990). Selection to reduce the cost of clonal growth on sexual reproduction may be the explanation for an interesting phenomenon seen in rosettes in the genus *Hieracium*, which produce stolons only after they have flowered (Bishop *et al.* 1978).

From a genetic point of view, clonal growth provides a mode of reproduction that can circumvent sex altogether though, as we saw in Chapter 9, sex is seldom lost entirely. Because clonal growth can be thought of as *asexual* reproduction, its advantages and disadvantages should be the converse of those of sexual reproduction (Chapter 9). So, for example, one might expect the genetic uniformity of clonal populations to make them particularly susceptible to disease. When elm disease appeared in the Netherlands 95% of the elms belonged to one susceptible clone, and some cities lost 70% of them (Burdon & Shattock 1980). There has been no systematic survey of wild clonal plants to determine whether they are more or less susceptible to disease than other plants, but studies of the clonal woodland herbs *Anemone nemorosa* (Ernst 1983), *Arisaema triphyllum* (Parker 1987), *Lactuca sibirica* (Wennström & Ericson 1992), *Mercurialis perennis* (Hutchings 1983) and *Podophyllum peltatum* (Sohn & Policansky 1977) all report the occurrence of significant levels of disease in wild populations.

10.4.2 Modes of clonal growth

Depending upon the length of connections and their frequency of branching, clonal growth can produce linear arrangements, networks or clumped distributions of ramets (see Fig. 1.7). There is a continuum of growth form from the solid advancing front of ramets, called a **phalanx**, to the production of widely spaced ramets that infiltrate the surrounding vegetation in **guerrilla** mode (Lovett Doust 1981a). It is not at all clear why there should be such a range of clonal behaviour. For example, among forest herbs, in *Medeola virginiana* the clone regularly fragments, but in *Aralia nudicaulis* the connections between ramets are permanent (Cook 1983; Edwards 1984;

Hutchings & Bradbury 1986). The answer probably lies in the benefits and costs associated with two functional aspects of the clonal growth habit: the degree to which ramets are physiologically dependent upon one another, and the degree to which clonal growth helps plants forage for resources.

10.4.2.1 Physiological integration

In most clonal species in which the rhizome, stolon or other connection remains intact, the genet behaves as a physiologically integrated unit. Radio-tracer studies, in which CO_2 labelled with the isotope ^{14}C has been fed to leaves, have shown for most species investigated that young ramets are supported by a flow of carbohydrates from older ones. Depending on the species, these carbon flows generally cease as young ramets mature, but they may be re-established if a ramet is shaded or defoliated (Marshall 1990). In the arctic herbs *Carex bigelowii* and *Lycopodium antoninum*, which possess the guerrilla structure, clones are physiologically integrated throughout life. Only the youngest tillers or shoots are photosynthetically active, and these export carbohydrate to an extensive rhizome system that returns nutrients captured from a wide area (Carlsson *et al.* 1990). Water and mineral transport between mature ramets appears to be commonplace in many clonal species (Marshall 1990).

A number of experiments have shown that the ability to translocate carbohydrates, water and nutrients from ramet to ramet allows a clone to even out spatial heterogeneity, and may benefit the clone as a whole. Disconnected ramets of three old-field species of *Solidago* and *Aster lanceolatus* suffered more severely from experimental defoliation than did connected ones (Schmid *et al.* 1988). Hartnett and Bazzaz (1985) found that *S. canadensis* ramets growing in backgrounds of other species produced fewer leaves with some neighbour species than others. When interconnected ramets were grown so that each ramet had a different neighbour species, ramets did not show a specific response to neighbours but appeared to average out the effect of different neighbours between them. However, to be sure that physiological integration is respon-sible for this result, the experiment would have to be repeated with unconnected ramets because an averaging effect could also result from the direct influence of a variety of neighbours upon the *aerial* part of each *individual* ramet.

In an experiment with *Ambrosia psilostachya*, only the below-ground environment of ramets was altered, thus avoiding the confounding of aerial effects with subterranean ones. The plant is a rhizomatous perennial that occurs in saline and non-saline habitats in the Great Plains of western USA. Salzman and Parker (1985) grew *A. psilostachya* in pots with two compartments. For a two-environment treatment, one compartment contained saline soil and a ramet connected by a rhizome to a neighbour growing in salt-free soil in the other compartment. Ramets in the saline compartment grew more than twice as well when connected with a salt-free neighbour than when connected with a saline neighbour. The average combined performance of connected ramets growing in different environments was better than the average performance of plants growing in the two kinds of environment connected to neighbours in a like environment. This surprising result suggests that physiological integration could cause some clonal plants to perform better in heterogeneous environments than in uniform ones.

Slade and Hutchings (1987a, c) grew connected ramets of the woodland herb *Glechoma hederacea* in individual pots that differed in nutrient supply and light environment. Older ramets in good conditions supported the growth of younger ones in poor environments without suffering any detectable cost compared with controls, but the relationship was not reversible when younger ramets were in the better environment and older ones in the poorer. Physiological integration in this species appears to be constrained by the direction assimilates can travel (Price *et al.* 1992). In comparable experiments with the beach strawberry *Fragaria chiloensis*, bi-directional transfers of water, nitrogen and carbohydrate were possible. Connections were capable of keeping two ramets alive in conditions where both would have died if disconnected (Alpert & Mooney 1986). Nitrogen transfer showed a particularly interesting pattern. The nitrogen a ramet received from its own roots promoted its own growth and the production of new stolons, but when nitrogen was acquired from another ramet, this went only into stolon production, not growth (Alpert 1991). This would seem to be an ideal foraging strategy.

10.4.2.2 Foraging

Foraging is a behaviour that increases resource capture in an environment that is spatially heterogeneous for the resource. Cook (1986) suggested that stolons of *Viola blanda* may respond to local pockets of nutrients, initiating new ramets in the vicinity of rotting wood. Rhizomes of couch grass *Elymus repens* may behave the same way in response to patches of soil rich in nitrogen (Mortimer 1984). In *Aralia nudicaulis* and other forest floor species, clonal growth may aid plants in locating light gaps, as the understorey runners of the tropical vine *Ipomoea phillomega* (see Fig. 1.8) evidently do. The behaviour of *Fragaria chiloensis* is another example, because ramets growing in resource-poor environments do not make any further investment in the spot they occupy, but allocate resources to stolons that will have a chance of rooting in a more nutrient-rich site.

In general, plants growing in resource-rich microsites should consolidate their hold over the site, while those in poor sites should place ramets elsewhere. Sutherland and Stillman (1988) tested this hypothesis by computer simulation and against the results of published experiments. *Glechoma hederacea* behaved exactly as expected. When growing in nutrient-rich conditions it branched profusely and produced many ramets on short internodes, but in poor conditions it branched less and produced fewer ramets on longer stolons (Slade & Hutchings 1987b). Other species were apparently less well behaved, and although all branched more in richer conditions, many had longer stolons in these sites than in poor ones. This may aid escape from the shade of large ramets that develop in nutrient-rich conditions and may also reflect the greater availability to such plants of nutrients to put into stolons.

10.5 DEATH

Since at least some plants appear to be capable of immortality, why do any die? Can death itself be adaptive? In a sense, the answer is yes, although it would be more accurate to say that there are circumstances when the trade-off between reproduction and survival is so heavily on the side of reproduction that death is an unavoidable consequence of maximizing fitness.

10.5.1 Annuals

Consider a hypothetical population of annual plants that produce three seeds each, and then die at the end of the year (e.g. *Vulpia fasciculata* would fit the bill, Chapter 5). Next year each seed germinates, there is no seedling mortality and each plant produces three seeds. Each of the original plants now has $3 \times 3 = 9$ descendants (Table 10.3). What advantage in fitness would a mutant individual in this population gain if it became perennial and iteroparous? It would start the second year with the three seeds produced the previous year, *plus* such resources as remained after producing them. Annual plants expend most of their resources on reproduction, so a perennial mutant would have little left if it produced as many seeds. It is possibly generous to assume that the mutant has enough resources left to reproduce as well at the end of its second year as it did in its first year, but making this convenient assumption for the sake of argument, the perennial mutant starts the second generation with three seeds that will produce three seeds each, plus a surviving plant that will also produce three seeds. At the end of the second year there will be 16 mutants (Table 10.3). In fact an annual that produced just one extra seed could do just as well as this, because it would have $4 \times 4 = 16$ descendants (Table 10.3a). It seems reasonable to suppose that the mutant would have to sacrifice more than one seed in order to switch to perenniality in the first place, so that the situation as it stands in our model is heavily weighted in favour of annual plants. Cole (1954), who derived this result, pointed out that it paradoxically predicts that all organisms should die after reproduction!

We have plainly left something important out

Table 10.3 Comparison of the rates of population increase over 2 years for a semelparous annual and an iteroparous perennial. (a) No pre-reproductive mortality, (b) pre-reproductive mortality of 33% (see text for further explanation)

Life history	Year 1						Year 2					
	Seeds at start	Seedling l_x	Plants reproducing	m_x	Seeds produced	+survivors	Seedling l_x	Plants reproducing	m_x	Seeds produced	+survivors	Total
(a) Semelparous	1	1	1	3	3	0	1	3	3	9	0	9
Semelparous	1	1	1	4	4	0	1	4	4	16	0	16
Iteroparous	1	1	1	3	3	1	1	4	3	12	4	16
(b) Semelparous	1	0.67	0.67	3	2	0	0.67	1.33	3	4	0	4
Semelparous	1	0.67	0.67	4	2.66	0	0.67	1.77	4	7	0	7
Iteroparous	1	0.67	0.67	3	2	0.67	0.67	2	3	6	2	8

of our model so far. In fact it is the second component of fitness — survival. Heavy seed and seedling mortality is commonplace in natural populations and values of more than 90% for pre-reproductive mortality are not unusual (Chapter 5). For arithmetic simplicity let us assume a modest pre-reproductive mortality risk of one-third. The mean number of seeds produced per original annual plant at the end of 2 years is now $(\frac{2}{3} \times 3)^2 = 4$. If perennial mutants that reach reproductive age all survive to flower again in following seasons, there will be eight of them at the end of 2 years (Table 10.3b). Now, perenniality is favoured and annual plants will have to produce more than one more seed to match the fitness of iteroparous perennials and in fact about 12 more when pre-reproductive mortality is 90%.

This reasoning depends upon the assumption that adult plants have a zero risk of mortality. Though this is clearly untrue, the actual advantage of an iteroparous perennial life history over an annual one depends upon the ratio of adult survival (p) to pre-reproductive survival (c) (Schaffer & Gadgil 1975). An annual population will increase at a rate λ_a, given by the product of annual seed and seedling survival (c) and mean seed production per individual (m_a):

$$\lambda_a = cm_a \qquad (10.2)$$

The equivalent expression for an iteroparous perennial is:

$$\lambda_p = cm_p + p \qquad (10.3)$$

An annual will then reproduce faster than a perennial when:

$$m_a > m_p + \frac{p}{c}$$

For example, in a demographic study of the grassland perennial *Ranunculus bulbosus*, Sarukhán and Harper (1973) found values of $c = 0.05$, $m_p \approx 30$, $p \approx 0.8$. An annual mutant in this population would have to increase its seed production by 53% over the perennial phenotype in order to just match its fitness (Schaffer & Gadgil 1975). This model predicts that populations which experience relatively heavy juvenile mortality should be iteroparous perennials, while populations with shallower survivorship curves in the

pre-reproductive phase should tend towards the annual life history. Although the shape of survivorship curves is greatly influenced by density and other aspects of spatial heterogeneity that make generalizations about different species difficult, the expected pattern does seem to occur when annuals and iteroparous perennials are compared (Silvertown 1987a).

10.5.2 Semelparous perennials

Using the same kind of reasoning with which we compared the fitness of an annual and an iteroparous perennial, we can estimate the fitness of a semelparous perennial (or 'biennial') with the equation:

$$\lambda_n = (lm_n)^{1/n} \qquad (10.4)$$

where λ_n is the yearly rate of increase of a semelparous plant living n years before flowering, l is the probability of survival until year n and m_n is the seed production of the plant. Notice that this equation is virtually the same as the one for an annual, but raised to the power $1/n$. An annual produces cm_a offspring every year but the semelparous perennial produces its lm_n offspring only every n years; cm_a offspring produced by the annual become $(cm_a)^n$ descendants after n years, so to measure the rate of increase of the semelparous plant on the same *per year* basis as the annual we must take the nth root. Equation 10.4 is equivalent to Eqn 10.2 when n = 1. To compare the fitness of an annual and a semelparous perennial living 2 years (n = 2) let us conservatively assume that $cm_a = lm_n = 16$. The equations then show us that the relative fitness of the two life history types is 16 : 4 or 1 : 0.25 in favour of the annual. If the semelparous perennial lives 3 years and the other variables keep the same values, relative fitnesses are 16 : 2 or 1 : 0.125.

Quite obviously delaying reproduction causes a serious decrease in fitness for a semelparous plant unless it has substantially higher seedling survival and/or much higher seed production than an annual (Hart 1977). In fact semelparous perennials do tend to have higher seed production than annuals (Salisbury 1942) but this is not the whole story. The advantage of the annual life history depends upon seeds being able to germinate every year so that its high potential geometric rate of increase can be realized (i.e. $\lambda > 1$). Conditions in many habitats make this impossible because perennial plants tend to fill in the vegetation gaps which annuals require to establish successfully. An annual that has to wait for x years for a gap, with its seeds lying dormant in the soil, will have a per year rate of increase:

$$\lambda_a = (cm_a)^{1/x} \qquad (10.5)$$

Notice that the annual is now in the same position as a biennial (Eqn 10.4), but with the disadvantage that it can only turn one season's assimilates into seeds. A habitat in which germination is only possible in intermittent gaps does also have a disadvantage for a semelparous perennial when compared with a completely open habitat, but the disparity in fitness between this life history and the annual one narrows dramatically as x (the time between gaps) increases (Silvertown 1983b, 1986). It is therefore significant that semelparous perennials are actually most abundant in those types of habitat where gaps occur intermittently. Perennial semelparity is a relatively rare kind of life history among angiosperms as a whole but, in Europe at least, it is unusually common in families such as the Asteraceae and Apiaceae which have a tap root that forms a storage organ and a branching structure that can produce very large numbers of seeds from stored reserves (Silvertown 1983b; Schat *et al.* 1989). These simple models of life histories indicate the kinds of ecological factors and developmental constraints which may have operated in the evolution of semelparous perennial herbs.

Looked at from a different point of view, the models also tell us something more general about when it is better for a plant to cue its reproduction by size rather than by age. By definition, annuals reproduce at a particular age ($\leqslant 12$ months), irrespective of how large they are. In some situations, plants like *Erophila verna* (Chapter 5) will flower when they are less than 10 mm high. We have seen that this behaviour can be advantageous in open habitats but not in others where the opportunities for seedling establishment in the year after the birth of the parent annual are low or chancy; or in other words not in sites where the local habitat *deteriorates*. In these habitats, plants which cue

reproduction by size lose nothing by the delay but, on the contrary, are able to grow for several seasons and hence produce more seeds when eventually they flower (Hirose 1983; Kachi & Hirose 1985). Cueing reproduction by size, rather than by age, is the rule in both semelparous and iteroparous perennials (Chapter 1) that occur in relatively closed habitats.

Equations 10.2–10.5 are rather crude, though instructive, models of some simple types of life history, but they do not allow for the fact that reproduction and survival trade-off against one another. An analysis based on using reproductive value (V_x) as an estimate of fitness can rectify this, because V_x is given by the sum of current fecundity and residual reproductive value (Eqn 10.1). We will define **reproductive allocation** (RA) as the proportion of resources devoted to reproduction in one season. We know that fecundity will increase with RA (Fig. 10.10a), and from the principle of allocation, we also know that there should be a negative relationship between residual reproductive value and RA (Fig. 10.10b). The sum of these two curves shows how V_x, and therefore fitness, will vary with RA (Fig. 10.10c). As Fig. 10.10a–c shows, when both curves are concave, V_x is at a maximum when RA is zero or 100%. These are therefore the conditions that favour semelparity. If the two curves are linear or convex, V_x is at a maximum at an intermediate value of RA (Fig. 10.10d–f). These are the conditions that favour iteroparity.

What are the ecological conditions which will result in curves of these shapes? Generally speaking, any circumstances which reduce the resource cost per seed (number of seeds per unit of RA) as the size of a seed crop increases, or which increase the probability that a seed will itself survive to reproduce as the size of the seed or seedling cohort increases, will produce a concave fecundity curve as in Fig. 10.10a. Seed predators can exercise this kind of effect because, even if a small plant devotes all its resources to reproduction, it can only produce a relatively small crop of seeds. As we have already seen, small crops generally suffer proportionately more losses to seed predators than large ones, and this creates a situation in which a plant may gain a disproportionate release from predation by delaying reproduction until it is big

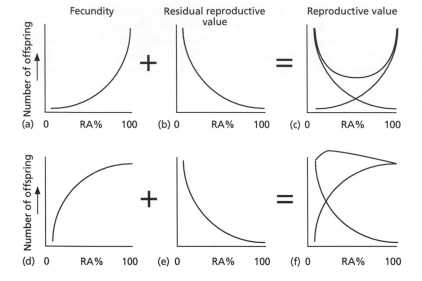

Fig. 10.10 Relationships between fecundity and reproductive allocation (RA) and between residual reproductive value and RA: (a–c) favour semelparity and (d–f) favour iteroparity.

enough to increase its crop size and swamp its seed predators. This is the explanation which has been put forward for the most spectacular delays of reproductive maturity which are found in species of semelparous bamboo, among which a 20-year pre-reproductive period is common. Before they reproduce, some Indian semelparous bamboos appear to grow at an exponential rate, causing clonal genets to expand at 10% a year until they flower (Gadgil & Prasad 1984). A Japanese species, *Phyllostachys bambusoides*, waits 120 years to flower and die. A few semelparous bamboos also synchronize reproduction within cohorts of the same age (Campbell 1985). This synchrony is probably important because it reduces the risk that predator populations will move from one bamboo population to another as they fruit (Janzen 1976).

Density-dependent seedling mortality between seeds from the same mother will produce the opposite effect and result in a convex fecundity curve that favours iteroparity (Fig. 10.10d). One might expect the huge crops of seed that semelparous perennials produce would compete with one another as seedlings and that this might alter the shape of the fecundity curve towards iteroparity (Young & Augspurger 1991), but presumably this effect is not as strong as the forces operating in the other direction.

Reproductive efficiency, or economies of scale

in seed production, seem to explain semelparity in a number of long-lived semelparous species. The genus *Agave* contains a number of semelparous perennials which grow in the deserts and chaparral of western North America. Some *Agave* spp. produce vegetative bulbils and consequently the genet is to be considered iteroparous. On the other hand, individual rosettes of most of these plants are semelparous. A typical species is the century plant (*A. deserti*) which delays reproduction for many years, storing water and carbohydrates in its rosette leaves. When it finally flowers, a rosette of only 60 cm in diameter is capable of producing an inflorescence up to 4 m tall. This feat can only be achieved by a massive translocation of water and assimilates from the rosette into the growing inflorescence which obtains 70% of its carbon from this source (Tissue & Nobel 1990).

The *Agave* has a similar habitat and morphology to plants in the genus *Yucca*, but rosettes of these species are mostly iteroparous. Rosettes of yuccas and agaves both have stiff xeromorphic leaves and bear a central spike of insect-pollinated flowers. In an attempt to explain the difference in reproductive habit between species in these two genera, Schaffer and Schaffer (1977, 1979) compared the life history of seven *Agave* and five *Yucca* species. One yucca (*Y. whipplei*) proved to have semelparous rosettes and one of the agaves had iteroparous ones (*A. parviflora*). This demonstrated that differences in reproductive habit between the two genera were not simply a result of different evolutionary descent, but had evolved separately in each genus. Schaffer and Schaffer suggested that semelparity in these plants was favoured by the selective behaviour of pollinating insects, which they showed visited the larger inflorescences disproportionately more often than smaller ones. Larger inflorescences also produced more fruits per flower than smaller ones in the semelparous species of both genera, but this pattern was absent in iteroparous species.

A number of semelparous giant rosette herbs occur in the alpine zone of tropical mountains in South America and Africa (Smith & Young 1987). Young (1990) studied populations of a semelparous species *Lobelia telekii* and compared them with populations of an iteroparous species *Lobelia keniensis* on Mount Kenya in Africa. Both species are pollinated by birds, and the semelparous species produced inflorescences up to two and a half times as long as the iteroparous species. As in the semelparous agaves and yuccas, the relationship between RA and fecundity in semelparous *Lobelia telekii*, but not *L. keniensis*, was concave, and there were more seeds per unit length of inflorescence in large flower spikes than small ones. Strangely enough, seed set in *L. telekii* was apparently *not* limited by pollination, so this was not the cause of the disproportionate seed set on large inflorescences. However, *L. telekii* and *L. keniensis* occupied different habitats, with the semelparous species occurring in drier sites than the iteroparous one. Young concluded that residual reproductive value was lower in dry conditions, and that this was decisive in favouring semelparity at these sites.

In the last two chapters of this book we hope to have demonstrated how the topic of life history evolution brings together the many strands of population biology. Textbook authors tend to dislike loose ends, but we freely admit that there are many. Find them and tie them!

10.6 SUMMARY

Phenotypic selection models may be used to analyse a wide variety of life history characters that have predictable consequences for survival and reproduction. Because reproduction and growth utilize the same plant resources, there tends to be a **trade-off** between early flowering and later growth and reproduction. On the other hand there is a **demographic penalty** to delayed reproduction when $\lambda > 1$. The costs and benefits of reproduction are therefore both age-specific. The relative contribution an average individual aged x will make to the next generation before it dies is its **reproductive value** V_x, which may be adjusted for the value of λ. V_x can be split into a component for **current fecundity** (m_x) and the **residual reproductive value**. A high adult mortality risk favours earlier reproduction because it lowers residual reproductive value.

Many plants, particularly forest trees, produce irregular large crops of seed called **mast**. Mast years usually correlate with climatic variables, but masting behaviour may be of advantage to plants because it **satiates predators**. A smaller proportion of seeds is lost to predators when crops are large than when crops are small. The **size of seeds** is governed by a **size–number trade-off** which suggests that there should be an **optimal seed size** that maximizes parental fitness. This idea is consistent with the observation that, between species, seed size is correlated with habitat. Plants of shaded habitats and arid conditions tend to have larger seeds than those of more open or mesic habitats. Seed size tends to have low heritability, presumably because it is a character closely correlated with fitness, but it still has a high phenotypic variance in some species. This may be because **developmental constraints** limit a plant's ability to control seed size, or because seed size variation is a **bet-hedging strategy**.

The timing of seed **germination** may be crucial to survival and future reproduction and it also appears to be a character that often has low heritability and a large phenotypic variance. There is good evidence in favour of the bet-hedging hypothesis for **germination heteromorphism**. **Clonal growth** is a common, but by no means universal, character in plants. The benefits of clonal growth are evident, but there also appear to be costs due to trade-offs with sexual reproduction. Two modes of clonal growth are commonly recognized: **phalanx** and **guerrilla**. The degree of **physiological integration** between ramets and transport of solutes between them vary with species and with the age and physiological state of connected ramets. One of the major advantages of the guerrilla mode of clonal growth is that it allows plants to **forage** for resources in environments where these are patchily distributed.

Semelparity may be favoured when the trade-off between reproduction and survival is so heavily on the side of reproduction that death is an unavoidable consequence of maximizing fitness. This may occur when adult mortality from environmental causes tends to be high. **Reproductive allocation** (RA) is the proportion of resources devoted to reproduction in one season. In semelparous plants RA has just two alternate values, zero during the vegetative phase and 100% when flowering occurs. This pattern of reproduction is favoured by ecological situations that produce a concave relationship between RA and fecundity.

References

Aarssen, L.W. & Turkington, R. (1985) Biotic specialization between neighbouring genotypes in *Lolium perenne* and *Trifolium repens* from a permanent pasture. *Journal of Ecology*, **73**, 605–614.

Abrams, P. (1983) The theory of limiting similarity. *Annual Review of Ecology and Systematics*, **14**, 359–376.

Abul-Fatih, H.A. & Bazzaz, F.A. (1979) The biology of *Ambrosia trifida* L.I. Influence of species removals on the organization of the plant community. *New Phytologist*, **83**, 813–816.

Ågren, J. & Eriksson, O. (1990) Age and size structure of *Pinus sylvestris* populations on mires in central and northern Sweden. *Journal of Ecology*, **78**, 1049–1062.

Aide, T.M. (1986) The influence of wind and animal pollination on variation in outcrossing rates. *Evolution*, **40**, 434–435.

Allan, M. (1977) *Darwin and His Flowers*. Faber and Faber, London.

Allard, R.W. (1990) Future directions in plant population genetics, evolution, and breeding. In Brown, A.H.D., Clegg, M.T., Kahler, A.L. & Weir, B.S. (eds) *Plant Population Genetics, Breeding, and Genetic Resources*, pp. 1–19. Sinauer, Sunderland, MA.

Allard, R.W. & Adams, J. (1969) Population studies in predominantly self-pollinating species XII. Intergenotypic competition and population structure in barley and wheat. *American Naturalist*, **103**, 621–645.

Allen, E.B. & Allen, M.F. (1990) The mediation of competition by mycorrhizae in successional and patchy environments. In Grace, J.B. & Tilman, D. (eds) *Perspectives on Plant Competition*, pp. 367–389. Academic Press, San Diego.

Allen, E.B. & Forman, R.T.T. (1976) Plant species removals and old-field community structure and stability. *Ecology*, **57**, 1233–1243.

Allen, M.F. (1991) *The Ecology of Mycorrhizae*. Cambridge University Press, Cambridge.

Alpert, P. (1991) Nitrogen sharing among ramets increases clonal growth in *Fragaria chiloensis*. *Ecology*, **72**, 69–80.

Alpert, P. & Mooney, H.A. (1986) Resource sharing among ramets in the clonal herb *Fragaria chiloensis*. *Oecologia*, **70**, 227–233.

Alvarez-Buylla, E.R. & García-Barrios, R. (1991) Seed and forest dynamics: a theoretical framework and an example from the neotropics. *American Naturalist*, **137**, 100–154.

Alvarez-Buylla, E.R. & Martinez-Ramos, M. (1990) Seed bank versus seed rain in the regeneration of a tropical pioneer tree. *Oecologia*, **84**, 314–325.

Antlfinger, A.E. (1981) The genetic basis of microdifferentiation in natural and experimental populations of *Borrichia frutescens* in relation to salinity. *Evolution*, **35**, 1056–1068.

Antlfinger, A.E., Curtis, W.F. & Solbrig, O.T. (1985) Environmental and genetic determinants of plant size in *Viola sororia*. *Evolution*, **39**, 1053–1064.

Antonovics, J. & Fowler, N.L. (1985) Analysis of frequency and density effects on growth in mixtures of *Salvia splendens* and *Linum grandiflorum* using hexagonal fan designs. *Journal of Ecology*, **73**, 219–234.

Antonovics, J. & Primack, R.B. (1982) Experimental ecological genetics in *Plantago* VI. The demography of seedling transplants of *P. lanceolata*. *Journal of Ecology*, **70**, 55–75.

Arthur, A.E., Gale, J.S. & Lawrence, K.J. (1973) Variation in wild populations of *Papaver dubium*: VII. Germination time. *Heredity*, **30**, 189–197.

Ashton, P.S. (1988) Dipterocarp biology as a window to the understanding of tropical forest structure. *Annual Review of Ecology and Systematics*, **19**, 347–370.

Ashton, P.S., Givnish, T.J. & Appanah, S. (1988) Staggered flowering in the Dipterocarpaceae: new insights into floral induction and the evolution of mast fruiting in the aseasonal tropics. *American Naturalist*, **132**, 44–66.

Atkinson, W.D. & Shorrocks, B. (1981) Competition on a divided and ephemeral resource: a simulation model. *Journal of Animal Ecology*, **50**, 461–471.

Augspurger, C.K. (1983) Seed dispersal of the tropical tree *Platypodium elegans*, and the escape of its seedlings from fungal pathogens. *Journal of Ecology*,

71, 759–771.

Augspurger, C.K. (1984) Seedling survival of tropical trees: interactions of dispersal distance, light gaps, and pathogens. *Ecology*, **65**, 1705–1712.

Auld, T.D. (1986a) Population dynamics of the shrub *Acacia suaveolens* (Sm.) Willd.: dispersal and the dynamics of the soil seed-bank. *Australian Journal of Ecology*, **11**, 235–254.

Auld, T.D. (1986b) Population dynamics of the shrub *Acacia suaveolens* (Sm.) Willd.: fire and the transition to seedlings. *Australian Journal of Ecology*, **11**, 373–385.

Auld, T.D. (1987) Population dynamics of the shrub *Acacia suaveolens* (Sm.) Willd: survivorship throughout the life cycle, a synthesis. *Australian Journal of Ecology*, **12**, 139–151.

Auld, T.D. & Myerscough, P.J. (1986) Population dynamics of the shrub *Acacia suaveolens* (Sm.) Willd.: seed production and predispersal seed predation. *Australian Journal of Ecology*, **11**, 219–234.

Austin, M.P. (1985) Continuum concept, ordination methods and niche theory. *Annual Review of Ecology and Systematics*, **16**, 39–61.

Baker, H.G. (1972) Seed weight in relation to environmental conditions in California. *Ecology*, **53**, 997–1010.

Baker, H.G. (1989) Some aspects of the natural history of seed banks. In Leck, M.A., Parker, V.T. & Simpson, R.L. (eds) *Ecology of Soil Seed Banks*, pp. 9–21. Academic Press, San Diego.

Ballaré, C.L., Scopel, A.L. & Sanchez, R.A. (1990) Far-red radiation reflected from adjacent leaves: an early signal of competition in plant canopies. *Science*, **247**, 329–332.

Barker, P., Freeman, D.C. & Harper, K.T. (1982) Variation in the breeding system of *Acer grandidentatum*. *Forest Science*, **28**, 563–572.

Barrett, J.A. (1988) Frequency-dependent selection in plant–fungal interactions. *Philosophical Transactions of the Royal Society of London B*, **319**, 473–483.

Barrett, J.W., Knowles, P. & Cheliak, W.M. (1987) The mating system in a black spruce clonal seed orchard. *Canadian Journal of Forest Research*, **17**, 379–382.

Barrett, S.C.H. (1986) Mimicry in plants. *Scientific American*, **257**, 68–75.

Barrett, S.C.H. (1988) The evolution, maintenance, and loss of self-incompatibility systems. In Lovett Doust, J. & Lovett Doust, L. (eds) *Plant Reproductive Ecology: Patterns and Strategies*, pp. 98–124. Oxford University Press, New York.

Barrett, S.C.H. & Kohn, J.R. (1991) Genetic and evolutionary consequences of small population size in plants: implications for conservation. In Falk, D.A. &

Holsinger K.E. (eds) *Genetics and Conservation of Rare Plants*, pp. 3–30. Oxford University Press, New York.

Baskin, J.M. & Baskin, C.C. (1972) Influence of germination date on survival and seed production in a natural population of *Leavenworthia stylosa*. *American Midland Naturalist*, **88**, 318–323.

Baskin, J.M. & Baskin, C.C. (1980) Ecophysiology of secondary dormancy in seeds of *Ambrosia artemisifolia*. *Ecology*, **61**, 475–480.

Baskin, J.M. & Baskin, C.C. (1983) Seasonal changes in the germination response of buried seeds of *Arabidopsis thaliana* and ecological interpretation. *Botanical Gazette*, **144**, 540–543.

Baskin, J.M. & Baskin, C.C. (1985) The annual dormancy cycle in buried weed seeds: a continuum. *Bioscience*, **35**, 492–498.

Baskin, J.M. & Baskin, C.C. (1987) Environmentally induced changes in the dormancy states of buried weed seeds. *British Crop Protection Conference*, **C-2**, 695–706.

Bateman, A.J. (1948) Intra-sexual selection in *Drosophila*. *Heredity*, **2**, 349–368.

Bayer, R.J. (1987) Evolution and phylogenetic relationships of the *Antennaria* (Asteraceae : Inuleae) polyploid agamic complexes. *Biologisches Zentralblatt*, **106**, 683–698.

Bayer, R.J. (1989) Patterns of isozyme variation in Western North American *Antennaria* (Asteraceae : Inuleae) II. Diploid and polyploid species of section Alpinae. *American Journal of Botany*, **76**, 679–691.

Bayer, R.J. (1990) Investigations into the evolutionary history of the *Antennaria rosea* (Asteraceae : Inuleae) polyploid complex. *Plant Systematics and Evolution*, **169**, 97–110.

Bayer, R.J. & Stebbins, G.L. (1983) Distribution of sexual and apomictic populations of *Antennaria parlinii*. *Evolution*, **37**, 555–561.

Bayer, R.J. & Stebbins, G.L. (1987) Chromosome numbers, patterns of distribution, and apomixis in *Antennaria* (Asteraceae : Inuleae). *Systematic Botany*, **12**, 305–319.

Bazzaz, F.A., Levin, D.A. & Schmierbach, M.R. (1982) Differential survival of genetic variants in crowded populations of *Phlox*. *Journal of Applied Ecology*, **19**, 891–900.

Beatley, J.C. (1970) Perennation in *Astragalus lentiginosus* and *Tridens pulchellus* in relation to rainfall. *Madroño*, **20**, 326–332.

Beattie, A.J. & Culver, D.C. (1979) Neighborhood size in *Viola*. *Evolution*, **33**, 1226–1229.

Becker, P. & Wong, M. (1985) Seed dispersal, seed predation, and juvenile mortality of *Aglaia* sp. (Meliaceae) in lowland dipterocarp rainforest. *Biotropica*, **17**,

230–237.

Bell, A.D. (1991) *Plant Form*. Oxford University Press, Oxford.

Bell, G. (1982) *The Masterpiece of Nature: the Evolution and Genetics of Sexuality*. University of California Press, Berkeley.

Bell, G. (1985) On the function of flowers. *Proceedings of the Royal Society of London B*, **224**, 223–265.

Bender, E.A., Case, T.J. & Gilpin, M.E. (1984) Perturbation experiments in community ecology: theory and practice. *Ecology*, **65**, 1–13.

Benjamin, L.R. (1984) Role of foliage habit in the competition between differently sized plants in carrot crops. *Annals of Botany*, **53**, 549–557.

Benjamin, L.R. & Hardwick, R.C. (1986) Sources of variation and measures of variability in even-aged stands of plants. *Annals of Botany*, **58**, 757–778.

Bennett, K.D. (1983) Postglacial population expansion of forest trees in Norfolk, UK. *Nature*, **303**, 164–167.

Berendse, F. (1981) Competition between plant populations with different rooting depths II. Pot experiments. *Oecologia*, **48**, 334–341.

Berendse, F. (1982) Competition between plant populations with different rooting depths III. Field experiments. *Oecologia*, **53**, 50–55.

Bergelson, J.M. & Crawley, M.J. (1988) Mycorrhizal infection and plant species diversity. *Nature*, **334**, 202.

Bergmann, F. (1978) The allelic distribution at an acid phosphatase locus in Norway spruce (*Picea abies*) along similar climatic gradients. *Theoretical and Applied Genetics*, **52**, 57–64.

Bernstein, H., Hopf, F.A. & Michod, R.E. (1988) Is meiotic recombination an adaptation for repairing DNA, producing genetic variation, or both? In Michod, R.E. & Levin, B.R. (eds) *The Evolution of Sex. An Examination of Current Ideas*, pp. 139–160. Sinauer, Sunderland, MA.

Bierzychudek, P. (1981) Pollination limitation of plant reproductive effort. *American Naturalist*, **117**, 838–840.

Bierzychudek, P. (1982) Life histories and demography of shade-tolerant temperate forest herbs: a review. *New Phytologist*, **90**, 757–776.

Bierzychudek, P. (1987a) Resolving the paradox of sexual reproduction: a review of experimental tests. In Stearns, S.C. (ed.) *The Evolution of Sex and its Consequences*, pp. 163–174. Birkhäuser, Basle.

Bierzychudek, P. (1987b) Patterns in plant parthenogenesis. In Stearns, S.C (ed.) *The Evolution of Sex and its Consequences*, pp. 197–217. Birkhäuser Basle.

Bierzychudek, P. (1987c) Pollinators increase the cost of sex by avoiding female flowers. *Ecology*, **68**, 444–447.

Bierzychudek, P. (1989) Environmental sensitivity of sexual and apomictic *Antennaria*: do apomicts have general-purpose genotypes? *Evolution*, **43**, 1456–1466.

Bigwood, D.W. & Inouye, D.W. (1988) Spatial pattern analysis of seed banks: an improved method and optimized sampling. *Ecology*, **69**, 497–507.

Birks, H.J.B. (1989) Holocene isochrone maps and patterns of tree-spreading in the British Isles. *Journal of Biogeography*, **16**, 503–540.

Bishop, G.F., Davy, A.J. & Jeffries, R.L. (1978) Demography of *Hieracium pilosella* in a Breck grassland. *Journal of Ecology*, **66**, 615–629.

Black, J.N. (1958) Competition between plants of different initial seed sizes in swards of subterranean clover (*Trifolium subterraneum* L.) with particular reference to leaf area and the microclimate. *Australian Journal of Agricultural Research*, **9**, 299–318.

Black, J.N. (1959) Seed size in herbage legumes. *Herbage Abstracts*, **29**, 235–241.

Bocher, T.W. (1949) Racial divergences in *Prunella vulgaris* in relation to habitat and climate. *New Phytologist*, **48**, 285–314.

Bocher, T.W. & Larsen, K. (1958) Geographical distribution of initiation of flowering, growth habit and other factors in *Holcus lanatus*. *Botaniska Notiser*, **3**, 289–300.

Bonan, G.B. (1988) The size structure of theoretical plant populations: spatial patterns and neighbourhood effects. *Ecology*, **69**, 1721–1730.

Bond, W.J. (1985) Canopy-stored seed reserves (serotiny) in Cape Proteaceae. *South African Journal of Botany*, **51**, 181–186.

Bookman, S.S. (1984) Evidence for selective fruit production in Asclepias. *Evolution*, **38**, 72–86.

Borchert, M.I. (1985) Serotiny and cone-habit variation in populations of *Pinus coulteri* (Pinaceae) in the southern coastal ranges of California. *Madroño*, **32**, 29–48.

Bossema, I. (1979) Jays and oaks: an eco-ethological study of a symbiosis. *Behaviour*, **70**, 1–117.

Bousquet, J., Cheliac, W.M. & Lalonde, M. (1987) Allozyme variability in natural populations of green alder (*Alnus crispa*) in Quebec. *Genome*, **29**, 345–352.

Boyd, M., Silvertown, J. & Tucker, C. (1990) Population ecology of heterostyle and homostyle *Primula vulgaris*: growth, survival and reproduction in field populations. *Journal of Ecology*, **78**, 799–813.

Bradshaw, A.D. (1965) Evolutionary significance of phenotypic plasticity in plants. *Advances in Genetics*, **13**, 115–155.

Bradshaw, A.D. & McNeilly, T. (1981) *Evolution and Pollution*. Edward Arnold, London.

Bradstock, R.A. & Myerscough, P.J. (1981) Fire effects on seed release and the emergence and establishment of seedlings of *Banksia ericifolia* L.f. *Australian*

Journal of Botany, **29**, 521–532.

Brand, D.G. & Magnussen, S. (1988) Asymmetric, two-sided competition in even-aged monocultures of red pine. *Canadian Journal of Forestry Research*, **18**, 901–910.

Briggs, D. & Walters, S.M. (1984) *Plant Variation and Evolution*, 2nd edn. Cambridge University Press, Cambridge.

Brokaw, N.V.L. (1982) The definition of treefall gap and its effect on measures of forest dynamics. *Biotropica*, **14**, 158–160.

Brokaw, N.V.L. (1985) Gap-phase regeneration in a tropical forest. *Ecology*, **66**, 682–687.

Brown, A.H.D. (1984) Multilocus organization of plant populations. In Wohrmann, K. & Loeschcke, V. (eds) *Population Biology and Evolution*, pp. 159–169. Springer Verlag, Berlin.

Brown, A.H.D. (1989) Genetic characterization of plant mating systems. In Brown, A.H.D., Clegg, M.T., Kahler, A.L. & Weir, B.S. (eds) *Plant Population Genetics, Breeding, and Genetic Resources*, pp. 145–162. Sinauer, Sunderland, MA.

Brown, A.H.D. & Feldman, M.W. (1981) Population structure of multilocus associations. *Proceedings of the National Academy of Science USA*, **78**, 5913–5916.

Brown, J.H., Reichman, O.J. & Davidson, D.W. (1979) Granivory in desert ecosystems. *Annual Review of Ecology and Systematics*, **10**, 201–227.

Broyles, S.B. & Wyatt, R. (1990) Plant parenthood in milkweeds: a direct test of the pollen donation hypothesis. *Plant Species Biology*, **5**, 131–142.

Buchanan, G.A., Crowley, R.H., Street, J.E. & McGuire, J.A. (1980) Competition of sicklepod (*Cassia obtusifolia*) and redroot pigweed (*Amaranthus retroflexus*) with cotton (*Gossypium hirsutum*). *Weed Science*, **28**, 258–262.

Burdon, J.J. (1987) *Disease and Plant Population Biology*. Cambridge University Press, Cambridge.

Burdon, J.J. & Chilvers, J.A. (1975) Epidemiology of damping-off disease (*Pythium irregulare*) in relation to density of *Lepidium sativum* seedlings. *Annals of Applied Biology*, **81**, 135–143.

Burdon, J.J. & Chilvers, J.A. (1982) Host density as a factor in plant disease ecology. *Annual Review of Phytopathology*, **20**, 143–166.

Burdon, J.J., Groves, R.H., Kaye, P.E. & Speer, S.S. (1984) Competition in mixtures of susceptible and resistant genotypes of *Chondrilla juncea* differentially infected with rust. *Oecologia*, **64**, 199–203.

Burdon, J.J. & Jarosz, A.M. (1988) The ecological genetics of plant–pathogen interactions in natural communities. *Philosophical Transactions of the Royal Society of London B*, **321**, 349–363.

Burdon, J.J. & Shattock, R.C. (1980) Disease in plant communities. *Applied Biology*, **5**, 145–220.

Burrows, F.M. (1991) Biomass production, structural deformation, self-thinning and thinning mechanisms in monocultures. *Philosophical Transactions of the Royal Society of London B*, **333**, 119–245.

Bush, R.M. & Smouse, P.E. (1991) The impact of electrophoretic genotype on life history traits in *Pinus taeda*. *Evolution*, **45**, 481–498.

Buss, L. (1987) *The Evolution of Individuality*. Princeton University Press, Princeton.

Campbell, J.J.N. (1985) Bamboo flowering patterns: a global view with special reference to East Asia. *Journal of the American Bamboo Society*, **6**, 17–35.

Canham, C.D. (1985) Suppression and release during canopy recruitment in *Acer saccharum*. *Bulletin of the Torrey Botany Club*, **112**, 134–145.

Cannell, M.G.R., Rothery, P. & Ford, E.D. (1984) Competition within stands of *Picea sitchensis* and *Pinus contorta*. *Annals of Botany*, **53**, 349–362.

Capinera, J.L. (1979) Qualitative variation in plants and insects: effect of propagule size on ecological plasticity. *American Naturalist*, **114**, 350–361.

Carlquist, S. (1974) *Island Biology*. Columbia University Press, New York.

Carlsson, B.A., Jónsdóttir, B.M., Svensson, B.M. & Callaghan, T.V. (1990) Aspects of clonality in the arctic: a comparison between *Lycopodium antoninum* and *Carex bigelowii*. In van Groenendael, J. & de Kroon, H. (eds) *Clonal Growth in Plants*, pp. 131–151. SPB Academic Publishing, The Hague.

Carter, R.N. & Prince, S.D. (1981) Epidemic models used to explain biogeographical distribution limits. *Nature*, **293**, 644–645.

Carter, R.N. & Prince, S.D. (1988) Distribution limits from a demographic viewpoint. In Davy, A.J., Hutchings, M.J. & Watkinson, A.R. (eds) *Plant Population Ecology*, pp. 165–184. Blackwell Scientific Publications, Oxford.

Casper, B.B. (1988) Evidence for selective embryo abortion in *Cryptantha flava*. *American Naturalist*, **132**, 318–326.

Casper, B.B. (1990) Seedling establishment from one- and two-seeded fruits of *Cryptantha flava*: a test of parent–offspring conflict. *American Naturalist*, **136**, 167–177.

Caswell, H. (1989) *Matrix Population Models: Construction, Analysis, and Interpretation*. Sinauer, Sunderland, MA.

Cavers, P.B. & Benoit, D.L. (1989) Management and soil seed banks. In Leck, M.A., Parker, V.T. & Simpson, R.L. (eds) *Ecology of Soil Seed Banks*, pp. 309–328. Academic Press, San Diego.

Cavers, P.B. & Steele, M.G. (1984) Patterns of change in seed weight over time on individual plants. *American*

Naturalist, **124**, 324–335.

Chanway, C.P., Holl, F.B. & Turkington, R. (1989) Effect of *Rhizobium leguminosarum* biovar *trifolii* genotype on specificity between *Trifolium repens* and *Lolium perenne*. *Journal of Ecology*, **77**, 1150–1160.

Charlesworth, B. (1980) The cost of sex in relation to mating system. *Journal of Theoretical Biology*, **84**, 655–671.

Charlesworth, B. & Charlesworth, D. (1978) A model for the evolution of dioecy and gynodioecy. *American Naturalist*, **112**, 975–997.

Charlesworth, D. & Charlesworth, B. (1987) Inbreeding depression and its evolutionary consequences. *Annual Review of Ecology and Systematics*, **18**, 237–268.

Charlesworth, D. & Morgan, M.T. (1991) Allocation of resources to sex functions in flowering plants. *Philosophical Transactions of the Royal Society of London B*, **332**, 91–102.

Charnov, E.L. (1982) *The Theory of Sex Allocation*. Princeton University Press, Princeton.

Charnov, E.L. & Bull, J.J. (1986) Sex allocation, pollinator attraction and fruit dispersal in cosexual plants, *Journal of Theoretical Biology*, **118**, 321–325.

Cheplick, G.P. (1987) The ecology of amphicarpic plants. *Trends in Ecology and Evolution*, **2**, 97–101.

Cheplick, G.P. & Quinn, J.A. (1982) *Amphicarpum purshii* and the 'pessimistic strategy' in amphicarpic annuals with subterranean fruit. *Oecologia*, **52**, 327–332.

Cheplick, G.P. & Quinn, J.A. (1983) The shift in aerial/subterranean fruit ratio in *Amphicarpum purshii*: causes and significance. *Oecologia*, **57**, 374–379.

Cheplick, G.P., Clay, K. & Marks, S. (1989) Interactions between infection by endophytic fungi and nutrient limitation in the grasses *Lolium perenne* and *Festuca arundinacea*. *New Phytologist*, **111**, 89–97.

Christy, E.J. & Mack, R.N. (1984) Variation in demography of juvenile *Tsuga heterophylla* across the substratum mosaic. *Journal of Ecology*, **72**, 75–91.

Clark, D.A. & Clark, D.B. (1984) Spacing dynamics of a tropical forest tree: evaluation of the Janzen–Connell model. *American Naturalist*, **124**, 769–788.

Clausen, J., Keck, D.D. & Heisey, W.M. (1948) Experimental studies on the nature of species III. Environmental responses of climatic races of *Achillea*. *Carnegie Institution of Washington Publication*, **581**.

Clay, K. (1990) Fungal endophytes of grasses. *Annual Review of Ecology and Systematics*, **21**, 275–297.

Clegg, M.T. & Allard, R.W. (1973) Viability vs. fecundity selection in the slender wild oat species, *Avena barbata*. *Science*, **181**, 667–668.

Cody, M.L. (1978) Distribution ecology of *Haplopappus* and *Chrysothamnus* in the Mojave Desert I. Niche position and niche shifts on north-facing granitic

slopes. *American Journal of Botany*, **65**, 1107–1116.

Cody, M.L. (1986) Structural niches in plant communities. In Diamond, J. & Case, T.J. (eds) *Community Ecology*, pp. 381–405. Harper Row, New York.

Cody, M.L. (1989) Growth-form diversity and community structure in desert plants. *Journal of Arid Environments*, **17**, 199–209.

Cole, L.C. (1954) The population consequences of life history phenomena. *Quarterly Review of Biology*, **29**, 103–1337.

Condit, R., Hubbell, S.P. & Foster, R.B. (1992) The recruitment of conspecific and heterospecific saplings near adults and the maintenance of tree diversity in a tropical forest. *American Naturalist*, **140**, 261–286.

Connell, H.H., Tracey, J.G. & Webb, L.J. (1984) Compensatory recruitment, growth, and mortality as factors maintaining rain forest tree diversity. *Ecological Monographs*, **54**, 141–164.

Connell, J.H. (1971) On the role of natural enemies in preventing competitive exclusion in some marine animals and in rain forests. In den Boer, P.J. & Gradwell, G.R. (eds) *Dynamics of Populations*, pp. 298–310. PUDOC, Wageningen.

Connell, J.H. (1978) Diversity in tropical rainforests and coral reefs. *Science*, **199**, 1302–1310.

Connell, J.H. (1990) Apparent versus 'real' competition in plants. In Grace, J.B. & Tilman, D. (eds) *Perspectives on Plant Competition*, pp. 9–26. Academic Press, San Diego.

Connolly, J. (1986) On difficulties with replacement series methodology in mixture experiments. *Journal of Applied Ecology*, **23**, 125–137.

Connolly, J. (1987) On the use of response models in mixture experiments. *Oecologia*, **72**, 95–103.

Connolly, J., Wayne, P. & Murray, R. (1990) Time course of plant–plant interactions in experimental mixtures of annuals: density, frequency, and nutrient effects. *Oecologia*, **82**, 513–526.

Cook, R.E. (1980) Germination and size-dependent mortality in *Viola blanda*. *Oecologia*, **47**, 115–117.

Cook, R.E. (1983) Clonal plant populations. *American Scientist*, **71**, 244–253.

Cook, R.E. (1986) Growth and demography in clonal plants. In Jackson, J., Buss, L. & Cook, R.E. (eds) *Population Biology and Evolution of Clonal Organisms*, pp. 259–296. Yale University Press, New Haven, CT.

Cottam, D.A., Whittaker, J.B. & Malloch, A.J.C. (1986) The effects of chrysomelid beetle grazing and plant competition on the growth of *Rumex obtusifolius*. *Oecologia*, **70**, 452–456.

Courtney, A.D. (1968) Seed dormancy and field emergence in *Polygonum aviculare*. *Journal of Applied Ecology*, **5**, 675–684.

Cousens, R. (1985) A simple model relating yield loss to weed density. *Annals of Applied Biology*, **107**, 239–252.

Cox, P.A. (1981) Niche partitioning between sexes of dioecious plants. *American Naturalist*, **117**, 295–307.

Cox, P.A. (1982) Vertebrate pollination and the maintenance of dioecism in *Freycinetia*. *American Naturalist*, **120**, 65–80.

Crawford, T.J. (1984a) The estimation of neighbourhood parameters in plant populations. *Heredity*, **52**, 272–283.

Crawford, T.J. (1984b) What is a population? In Shorrocks, B. (ed.) *Evolutionary Ecology*, pp. 135–173. Blackwell Scientific Publications, Oxford.

Crawley, M.J. (1987) What makes a community invasible? In Gray, A.J., Crawley, M.J. & Edwards, P.J. (eds) *Colonization, Succession and Stability*, pp. 429–454. Blackwell Scientific Publications, Oxford.

Croat, T.B. (1978) *Flora of Barro Colorado Island*. Stanford University Press, Stanford, CA.

Crow, J.F. (1986) *Basic Concepts in Population, Quantitative, and Evolutionary Genetics*. W.H. Freeman, New York.

Cwynar, L.C. & MacDonald, G.M. (1987) Geographical variation of lodgepole pine in relation to population history. *American Naturalist*, **129**, 463–469.

Darwin, C. (1859) *The Origin of Species*. Penguin edn (1968). Penguin Books, Harmondsworth.

Darwin, C. (1871) *The Descent of Man and Selection in Relation to Sex*. J. Murray, London.

Darwin, C. (1876) *On the Effects of Cross- and Self-Fertilization in the Vegetable Kingdom*. J. Murray, London.

Darwin, C. (1877) *The Different Forms of Flowers on Plants of the Same Species*. J. Murray, London.

Davidson, D.W., Samson, D.A. & Inouye, R.S. (1985) Granivory in the Chihuahua Desert: interactions within and between trophic levels. *Ecology*, **66**, 486–502.

Davy, A.J. & Smith, H. (1985) Population differentiation in the life-history characteristics of salt-marsh annuals. *Vegetatio*, **61**, 117–125.

Davy, A.J. & Smith, H. (1988) Life-history variation and environment. In Davy, A.J., Hutchings, M.J. & Watkinson, A.R. (eds) *Plant Population Ecology*, pp. 1–22. Blackwell Scientific Publications, Oxford.

Dawkins, R. (1976) *The Selfish Gene*. Oxford University Press, Oxford.

Debaeke, P. (1988) Population dynamics of some broad-leaved weeds in cereal. I Relation between standing vegetation and soil seed bank. *Weed Research*, **28**, 251–263. (In French.)

de Kroon, H., Plaiser, A. & van Groenendael, J.M. (1986)

Elasticity: the relative contribution of demographic parameters to population growth rate. *Ecology*, **67**, 1427–1431.

del Moral, R. (1983) Competition as a control mechanism in subalpine meadows. *American Journal of Botany*, **70**, 232–245.

Devlin, B. & Ellstrand, N.C. (1990) The development and application of a refined method for estimating gene flow from angiosperm paternity analysis. *Evolution*, **44**, 248–259.

Devlin, B., Roeder, K. & Ellstrand, N.C. (1988) Fractional paternity assignment: theoretical development and comparison to other methods. *Theoretical and Applied Genetics*, **76**, 369–380.

De Vries, H. (1905) *Species and Varieties, Their Origin by Mutation*. Open Court Publishing, Chicago.

Eckert, C.G. & Barrett, S.C.H. (1992) Stochastic loss of style morphs from populations of tristylous *Lythrum salicaria* and *Decodon verticillatus* (Lythraceae). *Evolution*, **46**, 1014–1029.

Edwards, J. (1984) Spatial pattern and clone structure of the perennial herb, *Aralia nudicaulis* L. (Araliaceae). *Bulletin of the Torrey Botany Club*, **111**, 28–33.

Edwards, M. (1980) Aspects of the population ecology of charlock. *Journal of Applied Ecology*, **17**, 151–171.

Eis, S., Garman, E.H. & Ebel, L.F. (1965) Relation between cone production and diameter increment of Douglas fir (*Pseudtsuga menzesii* (Mirb.) Franco), grand fir (*Abies grandis* Dougl.), and western white pine (*Pinus monticola* Dougl.). *Canadian Journal of Botany*, **43**, 1553–1559.

Eissenstat, D.M. & Caldwell, M.M. (1988) Competitive ability is linked to rates of water extraction. A field study of two aridland tussock grasses. *Oecologia*, **75**, 1–7.

Eissenstat, D.M. & Newman, E.I. (1990) Seedling establishment near large plants: effects of vesicular-arbuscular mycorrhizas on the intensity of plant competition. *Functional Ecology*, **4**, 95–99.

Ellenberg, H. (1978) *Vegetation Mitteleuropas mit den Alpen*, 2nd edn. Ulmer, Stuttgart.

Ellison, A.M. (1987) Density-dependent dynamics of *Salicornia europaea* monocultures. *Ecology*, **68**, 737–741.

Ellner, S. & Shmida, A. (1981) Why are adaptations for long-range seed dispersal rare in desert plants? *Oecologia*, **51**, 133–144.

Ellstrand, N.C. & Antonovics, J. (1985) Experimental studies of the evolutionary significance of sexual reproduction. II. A test of the density-dependent hypothesis. *Evolution*, **39**, 657–666.

Ellstrand, N.C. & Roose, M.L. (1987) Patterns of genotypic diversity in clonal plant species. *American Journal of Botany*, **74**, 123–131.

Ellstrand, N.C., Devlin, B. & Marshall, D.L. (1989) Gene flow by pollen into small populations: data from experimental and natural stands of wild radish. *Proceedings of the National Academy of Science*, **86**, 9044–9047.

Elmquist, T., Ågren, J. & Tunlid, A. (1988) Sexual dimorphism and between-year variation in flowering, fruit set and pollinator behaviour in a boreal willow. *Oikos*, **53**, 58–66.

Endler, J.A. (1986) *Natural Selection in the Wild*. Princeton University Press, Princeton, NJ.

Ennos, R. (1985) The significance of genetic variation for root growth within a natural population of white clover (*Trifolium repens*). *Journal of Ecology*, **73**, 615–624.

Ennos, R.A. (1981) Detection of selection in populations of white clover (*Trifolium repens* L.). *Biological Journal of the Linnean Society*, **15**, 75–82.

Enti, A.A. (1968) Distribution and ecology of *Hildegardia barteri* (Mast.) Kosterin. *Bulletin de l'IFAN*, **30**, 881–895.

Epling, C., Lewis, H. & Ball, E.M. (1960) The breeding group and seed storage: a study in population dynamics. *Evolution*, **14**, 238–255.

Erickson, R.O. (1943) Population size and geographical distribution of *Clematis fremontii* var. *riehlii*. *Annals of the Missouri Botanical Garden*, **30**, 63–68.

Erickson, R.O. (1945) The *Clematis fremontii* var. *riehlii* population in the Ozarks. *Annals of the Missouri Botanical Garden*, **32**, 413–460.

Eriksson, O. (1989) Seedling dynamics and life histories in clonal plants. *Oikos*, **55**, 231–238.

Eriksson, O. & Jerling, L. (1990) Hierarchical selection and risk spreading in clonal plants. In van Groenendael, J. & de Kroon, H. (eds) *Clonal Growth in Plants*, pp. 79–94. SPB Academic Publishing, The Hague.

Ernst, W.H.O. (1983) Population biology and mineral nutrition of *Anemone nemorosa* with emphasis on its parasitic fungi. *Flora*, **173**, 335–348.

Evans, D.R., Hill, J., Williams, T.A. & Rhodes, I. (1985) Effects of coexistence on the performance of white clover–perennial ryegrass mixtures. *Oecologia*, **66**, 536–539.

Ewell, J.J. (1986) Invasibility: lessons from south Florida. In Mooney, H.A. & Drake, J.A. (eds) *Ecology of Biological Invasions of North America and Hawaii*, pp. 214–230. Springer, New York.

Faille, A., Lemée, G. & Pontailler, J.Y. (1984) Dynamique des clairières d'une fôret inexploitée (réserves biologiques de la fôret de Fontainbleau) I. Origine et état actuel des ouvertures. *Acta Oecologica Oecologica Generalis*, **5**, 35–51.

Falconer, D.S. (1989) *Introduction to Quantitative Genetics*, 3rd edn. Longman, New York.

Farris, M.A. (1987) Natural selection on the plant-water relations of *Cleome serrulata* growing along natural moisture gradients. *Oecologia*, **72**, 434–439.

Fenner, M. (1980) Germination tests on thirty-two East African weed species. *Weed Research*, **20**, 135–138.

Fenster, C.B. (1991a) Gene flow in *Chamaecrista fasciculata* (Legunosae) I. Gene dispersal. *Evolution*, **45**, 398–409.

Fenster, C.B. (1991b) Gene flow in *Chamaecrista fasciculata* (Legunosae) II. Gene establishment. *Evolution*, **45**, 410–422.

Finlay, K.W. & Wilkinson, G.N. (1963) The analysis of adaptation in a plant breeding programme. *Australian Journal of Agricultural Research*, **14**, 742–754.

Firbank, L.G. & Watkinson, A.R. (1985) On the analysis of competition within two-species mixtures of plants. *Journal of Applied Ecology*, **22**, 503–517.

Firbank, L.G. & Watkinson, A.R. (1987) On the analysis of competition at the level of the individual plant. *Oecologia*, **71**, 308–317.

Firbank, L.G. & Watkinson, A.R. (1990) On the effects of competition: from monocultures to mixtures. In Grace, J.B. & Tilman, D. (eds) *Perspectives on Plant Competition*, pp. 165–192. Academic Press, San Diego.

Fisher, R.A. (1958) *The Genetical Theory of Natural Selection*, 2nd edn. Dover, New York.

Fleming, T.H. & Williams, C.F. (1990) Phenology, seed dispersal, and recruitment in *Cecropia peltata* (Moraceae) in Costa Rican tropical dry forest. *Journal of Tropical Ecology*, **6**, 163–178.

Flor, H.H. (1956) The complementary genetic systems in flax and flax rust. *Advances in Genetics*, **8**, 29–54.

Flor, H.H. (1971) Current status of the gene-for-gene concept. *Annual Review of Phytopathology*, **9**, 275–296.

Forcella, F. & Harvey, S.J. (1988) Patterns of weed migration in northwestern USA. *Weed Science*, **36**, 194–201.

Ford, E.D. (1975) Competition and stand structure in some even-aged plant monocultures. *Journal of Ecology*, **63**, 311–333.

Foster, S.A. & Janson, C.H. (1985) The relationship between seed size and establishment conditions in tropical woody plants. *Ecology*, **66**, 773–780.

Fowells, H.A. & Schubert, G.H. (1956) Seed crops of forest trees in the pine region of California. *USDA Technical Bulletin*, **1150**, 48 pp.

Fowler, N. (1981) Competition and coexistence in a North Carolina grassland II. The effects of the removal of species. *Journal of Ecology*, **69**, 843–854.

Fowler, N.L. (1986a) Density-dependent population regulation in a Texas grassland. *Ecology*, **67**, 545–554.

Fowler, N.L. (1986b) The role of competition in plant communities in arid and semiarid regions. *Annual*

Review of Ecology and Systematics, **17**, 89–110.

Fox, J.F. (1981) Intermediate levels of disturbance maximize alpine plant species diversity. *Nature,* **293**, 564–565.

Franco, M. & Harper, J.L. (1988) Competition and the formation of spatial pattern in spacing gradients: an example using *Kochia scoparia. Journal of Ecology,* **76**, 959–974.

Frank, S.A. & Slatkin, M. (1992) Fisher's fundamental theorem of natural selection. *Trends in Ecology and Evolution,* **7**, 92–95.

Freeman, D.C., Klikoff, L.G. & Harper, K.T. (1976) Differential resource utilization by the sexes of dioecious plants. *Science,* **193**, 597–599.

Freeman, D.C., McArthur, E.D. & Harper, K.T. (1984) The adaptive significance of sexual lability in plants, using *Atriplex canescens* as a principal example. *Annals of the Missouri Botanical Garden,* **71**, 265–277.

Frissell, S.S. (1973) The importance of fire as a natural ecological factor in Itasca State Park. *Quaternary Research,* **3**, 397–407.

Fryxell, P.A. (1957) Mode of reproduction of higher plants. *Botanical Review,* **23**, 135–233.

Furnier, G.R., Knowles, P., Clyde, M.A. & Dancik, B.P. (1987) Effects of avian seed dispersal on the genetic structure of whitebark pine populations. *Evolution,* **41**, 607–612.

Futuyma, D.J. & Wasserman, S.S. (1980) Resource concentration and herbivory in oak forests. *Science,* **210**, 920–922.

Gadgil, M. & Prasad, S.N. (1984) Ecological determinants of life history evolution of two Indian bamboo species. *Biotropica,* **16**, 161–172.

Gagnon, P.S., Vadas, R.L., Burdick, D.B. & May, B. (1980) Genetic identity of annual and perennial forms of *Zostera marina* L. *Aquatic Botany,* **18**, 157–162.

Galen, C., Shore, J.S. & Deyoe, H. (1991) Ecotypic divergence in alpine *Polemonium viscosum*: genetic structure, quantitative variation, and local adaptation. *Evolution,* **45**, 1218–1226.

Galen, C., Zunmer, K.A. & Newport, M.E. (1987) Pollination in floral scent morphs of *Polemonium viscosum*: a mechanism for disruptive selection on flower size. *Evolution,* **41**, 599–606.

Garbutt, K. & Bazzaz, F.A. (1987) Population niche structure. Differential response of *Abutilon theophrasti* progeny to resource gradients. *Oecologia,* **72**, 291–296.

Garbutt, K., Bazzaz, F.A. & Levin, D.A. (1985) Population and genotype niche width in clonal *Phlox paniculata. American Journal of Botany,* **72**, 640–648.

Garbutt, K. & Whitcombe, J. (1986) The inheritance of seed dormancy in *Sinapis arvensis* L. *Heredity,* **56**, 25–31.

Garwood, N.C. (1982) Seasonal rhythm of seed germination in a semideciduous tropical forest. In Leigh, E.G. Jr, Rand, A.S. & Windsor, D.M. (eds) *The Ecology of a Tropical Forest: Seasonal Rhythms and Long-term Changes,* pp. 173–185. Smithsonian Institution Press, Washington, DC.

Garwood, N.C. (1983) Seed germination in a seasonal tropical forest in Panama: a community study. *Ecological Monographs,* **53**, 159–181.

Geber, M.A. (1990) The cost of meristem limitation in *Polygonum arenastrum*: negative genetic correlations between fecundity and growth. *Evolution,* **44**, 799–819.

Ghiselin, M. (1974) *The Economy of Nature and the Evolution of Sex.* University of California Press, Berkeley.

Ghiselin, M.T. (1969) The evolution of hermaphroditism among animals. *Quarterly Review of Biology,* **44**, 189–208.

Givnish, T.J. (1981) Serotiny, geography and fire in the pine barrens of New Jersey. *Evolution,* **35**, 101–123.

Gliddon, C. & Saleem, M. (1985) Gene-flow in *Trifolium repens* — an expanding genetic neighbourhood. In Jacquard, P., Heim, G. & Antonovics, J. (eds) *Genetic Differentiation and Dispersal in Plants,* pp. 293–309. Springer-Verlag, Heidelberg.

Gliddon, C., Belhassen, E. & Gouyon, P.-H. (1987) Genetic neighbourhoods in plants with diverse systems of mating and different patterns of growth. *Heredity,* **59**, 29–32.

Go, M. (1981) Correlation of DNA exonic regions with protein structural units in haemoglobin. *Nature,* **291**, 90–92.

Goldberg, D.E. (1987) Neighbourhood competition in an old-field plant community. *Ecology,* **68**, 1211–1223.

Goldberg, D.E. (1990) Components of resource competition in plant communities. In Grace, J.B. & Tilman, D. (eds) *Perspectives on Plant Competition,* pp. 27–49. Academic Press, San Diego.

Goldberg, D.E. & Barton, A.M. (1992) Patterns and consequences of interspecific competition in natural communities: a review of field experiments with plants. *American Naturalist,* **139**, 771–801.

Goldberg, D.E. & Fleetwood, L. (1987) Competitive effect and response in four annual plants. *Journal of Ecology,* **75**, 1131–1143.

Goldberg, D.E. & Werner, P.A. (1983) Equivalence of competitors in plant communities: a null hypothesis and a field experimental approach. *American Journal of Botany,* **70**, 1098–1104.

Gottlieb, L.D. (1977) Genotypic similarity of large and small individuals in a natural population of the annual plant *Stephanomeria exigua* ssp. *coronaria*

(Compositae). *Journal of Ecology*, **65**, 127–134.

Gottlieb, L.D. (1981) Electrophoretic evidence and plant populations. *Progress in Phytochemistry*, **7**, 1–46.

Gouyon, P.H. & Couvet, D. (1988) A conflict between two sexes, female and hermaphrodite. In Stearns, S.C. (ed.) *Evolution of Sex and its Consequences*, pp. 245–261. Birkhäuser Verlag, Basel.

Gouyon, P.H., Gliddon, C.J. & Couvet, D. (1988) The evolution of reproductive systems: a hierarchy of causes. In Davy, A.J., Hutchings, M.J. & Watkinson, A.R. (eds) *Plant Population Ecology*, pp. 23–33. Blackwell Scientific Publications, Oxford.

Govindaraju, D.R. (1988) Relationship between dispersal ability and levels of gene flow in plants. *Oikos*, **52**, 31–35.

Grace, J.B. (1985) Juvenile vs adult competitive abilities in plants: size-dependence in cattails (TYpha). *Ecology*, **66**, 1630–1638.

Grace, J.B. & Tilman, D. (1990) *Perspectives in Plant Competition*. Academic Press, London. San Diego.

Graham, F.B. & Bormann, F.H. (1966) Natural root grafts. *Botanical Review*, **32**, 255–292.

Grant, J.D. (1983) The activities of earthworms and the fates of seeds. In Satchwell, J.E. (ed.) *Earthworm Ecology*, pp. 107–122. Chapman & Hall, London.

Grant, V. (1975) *Genetics of Flowering Plants*. Columbia University Press, New York.

Gray, A.J., Benham, P.E.M. & Raybould, A.F. (1990) *Spartina anglica* – the evolutionary and ecological background. In Gray, A.J. & Benham, P.E.M. (eds) *Spartina anglica – a Research Review*, pp. 5–10. HMSO, London.

Gray, A.J., Marshall, D.F. & Raybould, A.F. (1991) A century of evolution in *Spartina anglica*. *Advances in Ecological Research*, **21**, 1–62.

Grime, J.P. (1979) *Plant Strategies and Vegetation Processes*. Wiley, Chichester.

Grime, J.P. & Jeffrey, D.W. (1965) Seedling establishment in vertical gradients of sunlight. *Journal of Ecology*, **53**, 621–642.

Grime, J.P., Mackey, J.M.L., Hillier, S.H. & Read, D.J. (1987) Floristic diversity in a model system using experimental microcosms. *Nature*, **328**, 420–422.

Gross, K.L. (1981) Predictions of fate from rosette size in four 'biennial' plant species: *Verbascum thapsus*, *Oenothera biennis*, *Daucus carota*, and *Tragopogon dubius*. *Oecologia*, **48**, 209–213.

Groves, R.H. & Williams, J.D. (1975) Growth of skeleton weeds (*Chondrilla juncea* L.) as affected by growth of subterranean clover (*Trifolium subterraneum* L.) and infection by *Puccinia chondrilla* Bubak and Syd. *Australian Journal of Agricultural Research*, **26**, 975–983.

Grubb, P.J. (1977) The maintenance of species richness in plant communities: the importance of the regeneration niche. *Biological Reviews*, **52**, 107–145.

Grubb, P.J. (1986) Problems posed by spares and patchily distributed species in species-rich plant communities. In Diamond, J. & Case, T.J. (eds) *Community Ecology*, pp. 207–225, Harper & Row, New York.

Grubb, P.J. (1988) The uncoupling of disturbance and recruitment, two kinds of seed bank, and persistence of plant populations at the regional and local scales. *Annales Zoologici Fennci*, **25**, 23–36.

Hamrick, J.L. & Godt, M.J. (1990) Allozyme diversity in plant species. In Brown, A.D.D., Clegg, M.T., Kahler, A.L. & Weir, B.S. (eds) *Plant Population Genetics, Breeding, and Genetic Resources*, pp. 43–63. Sinauer, Sunderland, MA.

Hamrick, J.L. & Loveless, M.D. (1986) The influence of seed dispersal mechanisms on the genetic structure of plant populations. In Estrada, A. & Fleming, T.H. (eds) *Frugivores and Seed Dispersal*, pp. 211–223, W. Junk, Dordrecht.

Handel, S.N. (1976) Dispersal ecology of *Carex pedunculata* (Cyperaceae), a new North American mymecochore. *American Journal of Botany*, **63**, 1071–1079.

Hanzawa, F.M., Beattie, A.J. & Culver, D.C. (1988) Directed dispersal: demographic analysis of an ant–seed mutualism. *American Naturalist*, **131**, 1–13.

Harberd, D.J. (1961) Observations on population structure and longevity of *Festuca rubra*. *New Phytologist*, **61**, 85–100.

Harberd, D.J. (1962) Some observations on natural clones of *Festuca ovina*. *New Phytologist*, **61**, 85–100.

Harlan, J.R. (1976) Diseases as a factor in plant evolution. *Annual Review of Phytopathology*, **14**, 31–51.

Harper, J.L. (1981) The concept of population in modular organisms. In May, R.M. (ed.) *Theoretical Ecology*, 2nd edn, pp. 53–77. Sinauer, Sunderland, MA.

Harper, J.L., Jones, M. & Sackville-Hamilton, N.R. (1991) The evolution of roots and the problems of analysing their behaviour. In Atkinson, D. (ed.) *Plant Root Growth, an Ecological Perspective*, pp. 3–22. Blackwell Scientific Publications, Oxford.

Harper, J.L., Lovell, P.H. & Moore, K.G. (1970) The shapes and size of seeds. *Annual Review of Ecology and Systematics*, **1**, 327–356.

Harper, J.L., Williams, J.T. & Sagar, G.R. (1965) The behaviour of seeds in the soil: I. The heterogeneity of soil surfaces and its role in determining the establishment of plants. *Journal of Ecology*, **53**, 273–286.

Hart, R. (1977) Why are biennials so few? *American Naturalist*, **111**, 792–799.

Hartgerink, A.P. & Bazzaz, F.A. (1984) Seedling-scale environmental heterogeneity influences individual fitness and population structure. *Ecology*, **65**, 198–206.

Hartnett, D.C. & Bazzaz, F.A. (1985) The integration of neighbourhood effects by clonal genets in *Solidago canadensis*. *Journal of Ecology*, **73**, 415–427.

Hassell, M.P., Lawton, J.H. & May, R.M. (1976) Patterns of dynamical behaviour in single-species populations. *Journal of Animal Ecology*, **45**, 471–486.

Hendrix, S.D. (1984) Variation in seed weight and its effects on germination in *Pastinaca sativa* L. (Umbelliferae). *American Journal of Botany*, **71**, 795–802.

Hengveld, R. & Haeck, J. (1982) The distribution of abundance. I. Measurements. *Journal of Biogeography*, **9**, 303–316.

Henry, J.D. & Swan, J.M.A. (1974) Reconstructing forest history from live and dead plant material — an approach to the study of forest succession in southwestern New Hampshire. *Ecology*, **55**, 772–783.

Hetrick, B.A.D., Wilson, G.W.T. & Hartnett, D.C. (1989) Relationship between mycorrhizal dependence and competitive ability of two tallgrass prairie grasses. *Canadian Journal of Botany*, **67**, 2608–2615.

Hett, J.M. (1971) A dynamic analysis of age in sugar maple seedlings. *Ecology*, **52**, 1071–1074.

Heywood, J.S. (1986) The effect of plant size variation on genetic drift in populations of annuals. *American Naturalist*, **127**, 851–861.

Hils, M.H. & Vankat, J.L. (1982) Species removals from a first-year old-field plant community. *Ecology*, **63**, 705–711.

Hirose, T. (1983) A graphical analysis of life history evolution in biennial plants. *Botanical Magazine of Tokyo*, **96**, 37–47.

Hobbs, R.J. & Mooney, H.A. (1985) Community and population dynamics of serpentine grassland annuals in relation to gopher disturbance. *Oecologia*, **67**, 342–351.

Holland, R.F. & Jain, S.K. (1981) Insular biogeography of vernal pools in the Central Valley of California. *American Naturalist*, **117**, 24–37.

Holmsgaard, E. (1956) Effect of seed-bearing on the increment of European beech (*Fagus sylvatica* L.) and Norway spruce (*Picea abies* (L) Karst). *Proc. Int. Univ. For. Res. Org.*, 12th Congress, Oxford, 158–161.

Holsinger, K.E. (1986) Dispersal and plant mating systems: the evolution of self-fertilization in subdivided populations. *Evolution*, **40**, 405–413.

Holtsford, T.P. & Ellstrand, N.C. (1990) Inbreeding effects in *Clarkia tembloriensis* (Onagraceae) populations with different natural outcrossing rates. *Evolution*, **44**, 2031–2046.

Horovitz, A. & Harding, J. (1972) Genetics of *Lupinus* V. Intraspecific variability for the reproductive traits in *Lupinus nanus*. *Botanical Gazette*, **133**, 155–165.

Horvitz, C.C. & Schemske, D.W. (1986) Seed dispersal and environmental heterogeneity in a neotropical herb: a model of population and patch dynamics. In Estrada, A. & Fleming, T.H. (eds) *Frugivores and Seed Dispersal*, pp. 170–186. W. Junk, Dordrecht.

Horvitz, C.C. & Schemske, D.W. (1988) Demographic cost of reproduction in a neotropical herb: an experimental study. *Ecology*, **69**, 1741–1745.

Howe, H.F. & Smallwood, J. (1982) Ecology of seed dispersal. *Annual Review of Ecology and Systematics*, **13**, 201–228.

Hubbell, S.P. (1979) Tree dispersion, abundance, and diversity in a tropical dry forest. *Science*, **203**, 1299–1309.

Hubbell, S.P. (1980) Seed predation and coexistence of tree species in tropical forests. *Oikos*, **35**, 214–229.

Hubbell, S.P., Condit, R. & Foster, R.B. (1990) Presence and absence of density dependence in a neotropical tree community. *Philosophical Transactions of the Royal Society of London B*, **330**, 269–281.

Hubbell, S.P. & Foster, R.B. (1986) Canopy gaps and the dynamics of a neotropical forest. In Crawley, M.J. (ed.) *Plant Ecology*, pp. 77–96. Blackwell Scientific Publications, Oxford.

Hubbell, S.P. & Foster, R.B. (1990a) Structure, dynamics, and equilibrium status of old-growth forest on Barro Colorado Island. In Gentry, A.H. (ed.) *Four Neotropical Forests*, pp. 522–541. Yale University Press, New Haven, CT.

Hubbell, S.P. & Foster, R.B. (1990b) The fate of juvenile trees in a neotropical forest: implications for the natural maintenance of tropical tree diversity. In Bawa, K.S. & Hadley, M. (eds) *Reproductive Ecology of Tropical Forest Plants*, pp. 325–349. UNESCO/IUBS.

Hubbell, S.P. & Werner, P.A. (1979) On measuring the intrinsic rate of increase of populations with heterogeneous life histories. *American Naturalist*, **113**, 277–293.

Hunter, A.F. & Aarssen, L.W. (1988) Plants helping plants. New evidence indicates that beneficence is important in vegetation. *BioScience*, **38**, 34–40.

Huntley, B. & Birks, H.J.B. (1983) *An Atlas of Past and Present Pollen Maps of Europe 0–13 000 Years Ago*. Cambridge University Press, Cambridge.

Hurka, H. & Benneweg, M. (1979) Patterns of seed size variation in populations of the common weed *Capsella bursa pastoris* (Brassicaceae). *Biologisches Zentralblatt*, **98**, 699–709.

Husband, B.C. & Barrett, S.C.H. (1992) Effective population size and genetic drift in tristylous *Eichhornia paniculata* (Pontederiaceae). *Evolution*, **46**, 1875–1890.

Huston, M. (1979) A general hypothesis of species

diversity. *American Naturalist*, **113**, 81–101.

Hutchings, M.J. (1979) Weight–density relationships in ramet populations of clonal perennial herbs, with special reference to the −3/2 power law. *Journal of Ecology*, **67**, 21–33.

Hutchings, M.J. (1983) Shoot performance and population structure in pure stands of *Mercurialis perennis* L., a rhizomatous perennial herb. *Oecologia*, **58**, 260–264.

Hutchings, M.J. & Bradbury, I.K. (1986) Ecological perspectives on clonal perennial herbs. *Bioscience*, **36**, 178–182.

Hutchinson, G.E. (1978) *An Introduction to Population Ecology*. Yale University Press, New Haven, CT.

Inghe, O. & Tamm, C.O. (1985) Survival and flowering of perennial herbs IV. The behaviour of *Hepatica nobilis* and *Sanicula europaea* on permanent plots during 1943–1981. *Oikos*, **45**, 400–420.

Inghe, O. & Tamm, C.O. (1988) Survival and flowering of perennial herbs. V. Patterns of flowering. *Oikos*, **51**, 203–219.

Inouye, R.S. & Schaffer, W.M. (1981) On the ecological meaning of ratio (de Wit) diagrams in plant ecology. *Ecology*, **62**, 1679–1681.

Istock, C.A. (1983) Boundaries to life history variation and evolution. In King, C.E. & Dawson, P.S. (eds) *Population Biology: Retrospect and Prospect*. Columbia University Press, New York.

Jackson, L.E. (1985) Ecological origins of California's mediterranean grasses. *Journal of Biogeography*, **12**, 349–361.

Jain, S.K. (1976) Patterns of survival and microevolution in plant populations. In Karlin, S. & Nevo, E. (eds) *Population Genetics and Ecology*, pp. 49–89. Academic Press, New York.

Jalloq, M. (1975) The invasion of molehills by weeds as a possible factor in the degeneration of reseeded pasture. *Journal of Applied Ecology*, **12**, 643–657.

Janzen, D.H. (1970) Herbivores and the number of tree species in tropical forest. *American Naturalist*, **104**, 501–528.

Janzen, D.H. (1975) Behaviour of *Hymanaea coubaril* when its predispersal seed predator is absent. *Science*, **189**, 145–147.

Janzen, D.H. (1976) Why bamboos wait so long to flower. *Annual Review of Ecology and Systematics*, **7**, 347–391.

Jefferies, R.L., Davy, A.J. & Rudmik, T. (1981) Population biology of the salt marsh annual *Salicornia europaea* agg. *Journal of Ecology*, **69**, 17–31.

Jefferies, R.L., Jensen, A. & Bazely, D. (1983) The biology of the annual, *Salicornia europaea* agg. at the limit of its range in Hudson Bay. *Canadian Journal of Botany*, **61**, 762–773.

Johnson, W.C. & Webb, T. III (1989) The role of blue jays (*Cyanocitta cristata* L.) in the postglacial dispersal of fagaceous trees in eastern North America. *Journal of Biogeography*, **16**, 561–571.

Kachi, N. & Hirose, T. (1985) Population dynamics of *Oenothera glazioviana* in a sand-dune system with special reference to the adaptive significance of size-dependent reproduction. *Journal of Ecology*, **73**, 887–901.

Kadereit, J.W. & Briggs, D. (1985) Speed of development of radiate and non-radiate plants of *Senecio vulgaris* L. from habitats subject to different degrees of weeding pressure. *New Phytologist*, **99**, 155–169.

Kadmon, R. & Shmida, A. (1990) Patterns and causes of spatial variation in the reproductive success of a desert annual. *Oecologia*, **83**, 139–144.

Kalisz, S. (1991) Experimental determination of seed bank age structure in the winter annual *Collinsa verna*. *Ecology*, **72**, 575–585.

Kanzaki, M. (1984) Regeneration in subalpine coniferous forests: I. Mosaic structure and regeneration process in *Tsuga diversifolia* forest. *Botanical Magazine of Tokyo*, **97**, 297–311.

Kay, Q.O.N. (1978) The role of preferential and assortative pollination in the maintenance of flower colour polymorphisms. In Richards, A.J. (ed.) *Pollination of Flowers by Insects*, pp. 175–190. *Linnaean Society Symposium, Series 6*. Academic Press, London.

Kay, Q.O.N. (1982) Intraspecific discrimination by pollinators and its role in evolution. In Armstrong, J.A., Powell, J.M. & Richards, A.J. (eds) *Pollination and Evolution*, pp. 9–28. Publications of the Royal Botanic Gardens, Sydney.

Kays, S. & Harper, J.L. (1974) The regulation of plant and tiller density in a grass sward. *Journal of Ecology*, **62**, 97–105.

Keddy, P.A. (1981) Experimental demography of the sand-dune annual, *Cakile edentula*, growing along an environmental gradient in Nova Scotia. *Journal of Ecology*, **69**, 615–630.

Keddy, P.A. (1982) Population ecology on an environmental gradient: *Cakile edentula* on a sand dune. *Oecologia*, **52**, 348–355.

Keddy, P.A. (1989) *Competition*. Chapman and Hall, London.

Keddy, P.A. & Constabel, P. (1986) Germination of ten shoreline plants in relation to seed size, soil particle size and water level: an experimental study. *Journal of Ecology*, **74**, 133–141.

Keeler, K. (1978) Intra-population differentiation in annual plants II. Electrophoretic variation in *Veronica peregrina*. *Evolution*, **32**, 638–645.

Kelley, S.E. (1989) Experimental studies of the evolutionary significance of sexual reproduction. V. A field test of the sibcompetition lottery hypothesis. *Evolution*, **43**, 1054–1065.

Kelley, S.E. & Clay, K. (1987) Interspecific competitive interactions and the maintenance of genotypic variation within the populations of two perennial grasses. *Evolution*, **41**, 92–103.

Kelley, S.E., Antonovics, J. & Schmitt, J. (1988) A test of the short-term advantage of sexual reproduction. *Nature*, **331**, 714–716.

Kemp, P.R. (1989) Seed banks and vegetation processes in desert. In Leck, M.A., Parker, V.T. & Simpson, R.L. (eds) *Ecology of Soil Seed Banks*, pp. 257–281. Academic Press, San Diego.

Kenkel, N.C. (1988) Pattern of self-thinning in Jack pine: testing the random mortality hypothesis. *Ecology*, **69**, 1017–1024.

Kimura, M. (1983) *The Neutral Theory of Molecular Evolution*. Cambridge University Press, Cambridge.

Kimura, M. (1991a) Recent development of the neutral theory viewed from the Wrightian tradition of theoretical population genetics. *Proceedings of the National Academy of Sciences USA*, **88**, 5969–5973.

Kimura, M. (1991b) The neutral theory of molecular evolution: a review of recent evidence. *Japanese Journal of Genetics*, **66**, 367–386.

King, L.M. & Schaal, B.A. (1990) Genotypic variation within asexual lineages of *Taraxacum officinale*. *Proceedings of the National Academy of Sciences USA*, **87**, 998–1002.

King, T.J. (1977) The plant ecology of ant-hills in calcareous grasslands I. Patterns of species in relation to ant-hills in southern England. *Journal of Ecology*, **65**, 235–256.

Kirkpatrick, K.J. & Wilson, H.D. (1988) Interspecific gene flow in *Cucurbita*: *Cucurbita texana* vs *Cucurbita pepo*. *American Journal of Botany*, **75**, 519–527.

Klekowski, E.J. Jr (1988) *Mutation, Developmental Selection, and Plant Evolution*. Columbia University Press, New York.

Kohn, J.R. (1988) Why be female? *Nature*, **335**, 431–433.

Kohn, J.R. (1989) Sex ratio, seed production, biomass allocation, and the cost of male function in *Cucurbita foetidissima* HBK (Cucurbitaceae). *Evolution*, **43**, 1424–1434.

Lacey, E.P. (1986) The genetic and environmental control of reproductive timing in a short-lived monocarp, *Daucus carota*. *Ecology*, **69**, 220–232.

Lacey, E.P. (1988) Latitudinal variation in reproductive timing of a short-lived monocarpic species, *Daucus carota*. *Journal of Ecology*, **74**, 73–86.

Lack, A.J. & Kay, Q.O.N. (1988) Allele frequencies, genetic relationships and heterozygosity in *Polygala vulgaris* populations from contrasting habitats in southern Britain. *Biological Journal of the Linnaean Society*, **34**, 119–147.

Lagercrantz, U. & Ryman, N. (1990) Genetic structure of Norway spruce (*Picea abies*): concordance of morphological and allozymic variation. *Evolution*, **44**, 38–53.

Lalonde, R.G. & Roitberg, B.D. (1989) Resource limitation and offspring size and number trade-offs in *Cirsium arvense* (Asteraceae). *American Journal of Botany*, **76**, 1107–1113R.

Lande, R. & Arnold, S.J. (1983) The measurement of selection on correlated characters. *Evolution*, **36**, 1210–1226.

Lande, R. & Schemske, D.W. (1985) The evolution of self-fertilization and inbreeding depression in plants. 1. Genetic models. *Evolution*, **39**, 24–40.

Larson, M.M. & Schubert, G.H. (1970) Cone crops of ponderosa pine in central Arizona, including the influence of Abert squirrels. *USDA Forest Service Research Paper*, **RM 58**, 15 pp.

Law, R. (1979) The cost of reproduction in annual meadow grass. *American Naturalist*, **113**, 3–16.

Law, R. (1983) A model for the dynamics of a plant population containing individuals classified by age and size. *Ecology*, **64**, 224–230.

Law, R., Bradshaw, A.D. & Putwain, P.D. (1977) Life history variation in *Poa annua*. *Evolution*, **31**, 233–246.

Law, R. & Watkinson, A.R. (1987) Response-surface analysis of two-species competition: an experiment on *Phleum arenarium* and *Vulpia fasciculata*. *Journal of Ecology*, **75**, 871–886.

Lee, T.D. (1988) Patterns of fruit and seed production. In Lovett Doust, J. & Lovett Doust, L. (eds) *Plant Reproductive Ecology: Patterns and Strategies*, pp. 179–202. Oxford University Press, New York.

Levin, D.A. (1975) Pest pressure and recombination systems in plants. *American Naturalist*, **109**, 437–451.

Levin, D.A. (1977) The organization of genetic variability in *Phlox drummondii*. *Evolution*, **31**, 477–494.

Levin, D.A. (1981) Dispersal *versus* gene flow in plants. *Annals of the Missouri Botanical Garden*, **68**, 233–253.

Levin, D.A. (1983) Polyploidy and novelty in flowering plants. *American Naturalist*, **122**, 1–25.

Levin, D.A. (1984) Inbreeding depression and proximity-dependent crossing success in *Phlox drummondii*. *Evolution*, **38**, 116–127.

Levin, D.A. (1988) Local differentiation and the breeding structure of plant populations. In Gottlieb, L.D. & Jain, S.K. (eds) *Plant Evolutionary Biology*, pp. 305–329. Chapman and Hall, London.

Levin, D.A. & Clay, K. (1984) Dynamics of synthetic *Phlox drummondii* populations at the species margin. *American Journal of Botany*, **71**, 1040–1050.

Levin, D.A. & Kerster, H.W. (1968) Local gene dispersal in *Phlox pilosa*. *Evolution*, **22**, 130–139.

Levin, D.A. & Kerster, H.W. (1969a) The dependence of bee-mediated pollen and gene dispersal upon plant density. *Evolution*, **23**, 560–571.

Levin, D.A. & Kerster, H.W. (1969b) Density-dependent gene dispersal in *Liatris*. *American Naturalist*, **103**, 61–74.

Levin, D.A. & Kerster, H.W. (1973) Assortative pollination for stature in *Lythuum salicaria*. *Evolution*, **27**, 144–152.

Levin, D.A. & Kerster, H.W. (1974) Gene flow in seed plants. *Evolutionary Biology*, **7**, 139–220.

Levin, D.A. & Wilson, J.B. (1978) The genetic implications of ecological adaptations in plants. In Freysen A.H. & Woldendorp J.W. (eds) *Structure and Functioning of Plant Communities*, pp. 75–100. PUDOC, Wageningen.

Levins, R. (1963) Theory of fitness in a heterogeneous environment. II. Developmental flexibility and niche selection. *American Naturalist*, **97**, 75–90.

Levins, R. (1970) Extinction. In Gerstenhaber, M. (ed.) *Lectures on Mathematics in the Life Sciences, 2*, pp. 77–107. American Mathematical Society, Providence, RI.

Levins, R. & Culver, D. (1971) Regional coexistence of species and competition between rare species. *Proceedings of the National Academy of Sciences USA*, **68**, 1246–1248.

Lewis, D. (1941) Male-sterility in a natural population of hermaphrodite plants. *New Phytologist*, **40**, 56–63.

Lewis, D. & Jones, D.A. (1992) The genetics of heterostyly. In Barrett, S.C.H. (ed.) *Evolution and Function of Heterostyly*, pp. 129–150. Springer Verlag, Berlin.

Lewontin, R. (1974) *The Genetic Basis of Evolutionary Change*. Columbia University Press, New York.

Lieberman, D. & Lieberman, M. (1987) Forest tree growth and dynamics at La Selva, Costa Rica (1969–1982). *Journal of Tropical Ecology*, **3**, 347–358.

Lieberman, D., Lieberman, M., Hartshorn, G. & Peralta, R. (1985b) Growth rates and age–size relationships of tropical wet forest trees in Costa Rica. *Journal of Tropical Ecology*, **1**, 97–109.

Lieberman, D., Lieberman, M., Peralta, R. & Hartshorn, G.S. (1985a) Mortality patterns and stand turnover rates in a wet tropical forest in Costa Rica. *Journal of Ecology*, **73**, 915–924.

Linhart, Y.B. (1988) Intrapopulation differentiation in annual plants III. The contrasting effects of intra and interspecific competition. *Evolution*, **42**, 1047–1064.

Linhart, Y.B. & Baker, I. (1973) Intra-population differentiation in response to flooding in a population of *Veronica peregrina*, L. *Nature*, **242**, 275–276.

Linnaeus, C. (1737) *Genera plantarum*. Tumba, Sweden, Microfiche, International Documentation Centre.

Liston, A., Riesberg, L.H. & Elias, T.S. (1990) Functional androdioecy in the flowering plant *Datisca glomerata*. *Nature*, **343**, 641–642.

Lloyd, D.G. (1975) The maintenance of gynodioecy and androdioecy in angiosperms. *Genetica*, **45**, 325–339.

Lloyd, D.G. (1979) Some reproductive factors affecting the selection of self-fertilization in plants. *American Naturalist*, **113**, 67–97.

Lloyd, D.G. (1980) Sexual strategies in plants. III. A quantitative method for describing the gender of plants. *New Zealand Journal of Botany*, **18**, 103–108.

Lloyd, D.G. (1982) Selection of combined versus separate sexes in seed plants. *American Naturalist*, **120**, 571–585.

Lloyd, D.G. (1987) Selection of offspring size at independence and other size-versus-number strategies. *American Naturalist*, **129**, 800–817.

Lloyd, D.G. & Webb, C.J. (1977) Secondary sex characters in seed plants. *Botanical Review*, **43**, 177–216.

Lonsdale, W.M. (1990) The self-thinning rule: dead or alive? *Ecology*, **71**, 1373–1388.

Lord, E.M. (1981) Cleistogamy: a tool for the study of floral morphogenesis, function and evolution. *Botanical Review*, **47**, 421–449.

Lovett Doust, J. (1990) Botany agonistes: on phytocentrism and plant sociobiology. *Evolutionary Trends in Plants*, **4**, 121–133.

Lovett Doust, J. & Cavers, P.B. (1982a) Biomass allocation in hermaphrodite flowers. *Canadian Journal of Botany*, **60**, 2530–2534.

Lovett Doust, J. & Cavers, P.B. (1982b) Sex and gender dynamics in Jack-in-the-pulpit *Arisaema triphyllum* (Araceae). *Ecology*, **63**, 797–808.

Lovett Doust, J. & Eaton, G.W. (1982) Demographic aspects of flowering and fruit production in bean plants, *Phaseolus vulgaris* L. *American Journal of Botany*, **69**, 1156–1164.

Lovett Doust, J. & Lovett Doust, L. (1988) Modules of production and reproduction in a dioecious clonal shrub, *Rhus typhina*. *Ecology*, **69**, 741–750.

Lovett Doust, L. (1981a) Population dynamics and local specialization in a clonal perennial (*Ranunculus repens*). I. The dynamics of ramets in contrasting habitats. *Journal of Ecology*, **69**, 743–755.

Lovett Doust, L. (1981b) Population dynamics and local specialization in a clonal perennial (*Ranunculus repens*). II. The dynamics of leaves and a reciprocal transplant experiment. *Journal of Ecology*, **69**, 757–768.

Lovett Doust, L. & Lovett Doust, J. (1987) Leaf demography and clonal growth in female and male *Rumex acetosella. Ecology*, **68**, 2056–2058.

Lyman, J.C. & Ellstrand, N.C. (1984) Clonal diversity in *Taraxacum officinale* (Compositae), an apomict. *Heredity*, **53**, 1–10.

MacArthur, R.H. (1972) *Geographical Ecology*. Princeton University Press, Princeton, NJ.

MacDonald, G.M. & Cwynar, L.C. (1991) Post-glacial population growth rates of *Pinus contorta* ssp. *latifolia* in western Canada. *Journal of Ecology*, **79**, 417–429.

MacDonald, S.E. & Chinnappa, C.C. (1989) Population differentiation for phenotypic plasticity in the *Stellaria longipes* complex. *American Journal of Botany*, **76**, 1627–1637.

Mack, R.N. (1981) Invasion of *Bromus tectorum* L. into western North America: an ecological chronicle. *Agroecosystems*, **7**, 145–165.

Mack, R.N. & Harper, J.L. (1977) Interference in dune annuals: spatial pattern and neighbourhood effects. *Journal of Ecology*, **65**, 345–363.

Mack, R.N. & Pyke, D.A. (1983) The demography of *Bromus tectorum*: variation in time and space. *Journal of Ecology*, **71**, 69–93.

Maddox, G.D. & Cappuccino, N. (1986) Genetic determination of plant susceptibility to an herbivorous insect depends on environmental context. *Evolution*, **40**, 863–866.

Maddox, G.D., Cook, R.E., Wimberger, P.H. & Gardescu, S. (1989) Clone structure in four *Solidago altissima* (Asteraceae) populations: rhizome connections within genotypes. *American Journal of Botany*, **76**, 318–326.

Mahdi, R., Law, R. & Willis, A.J. (1989) Large niche overlaps among coexisting plant species in a limestone grassland community. *Journal of Ecology*, **77**, 386–400.

Marks, M. & Prince, S. (1981) Influence of germination date on survival and fecundity in wild lettuce *Lactuca serriola. Oikos*, **36**, 326–330.

Marshall, C. (1990) Source-sink relations of interconnected ramets. In van Groenendael, J. & de Kroon, H. (eds) *Clonal Growth in Plants*, pp. 23–41. SPB Academic Publishing, The Hague.

Marshall, D.L. & Ellstrand, N.C. (1988) Effective mate choice in wild radish: evidence for selective seed abortion and its mechanism. *American Naturalist*, **131**, 739–756.

Marshall, D.R. & Jain, S.K. (1968) Phenotypic plasticity of *Avena fatua* and *A. barbata. American Naturalist*, **102**, 457–467.

Marshall, D.R. & Jain, S.K. (1969) Interference in pure and mixed populations of *Avena fatua* and *A.barbata. Journal of Ecology*, **57**, 251–270.

Martin, M.M. & Harding, J. (1981) Evidence for the evolution of competition between two species of annual plants. *Evolution*, **35**, 975–987.

Martin, M.P.L.D. & Field, R.J. (1984) The nature of competition between perennial ryegrass and white clover. *Grass and Forage Science*, **39**, 247–253.

Martínez-Ramos, M., Alvarez-Buylla, E. & Piñero, D. (1988) Treefall age determination and gap dynamics in a tropical forest. *Journal of Ecology*, **76**, 700–716.

Martínez-Ramos, M., Alvarez-Buylla, E. & Sarukhán, J. (1989) Tree demography and gap dynamics in a tropical rain forest. *Ecology*, **70**, 555–558.

Maynard Smith, J. (1978) *The Evolution of Sex*. Cambridge University Press, Cambridge.

Maynard Smith, J. (1989) *Evolutionary Genetics*. Oxford University Press, New York.

Mazer, S.J. (1987) Parental effects on seed development and seed yield in *Raphanus raphanistrum*: implications for natural and sexual selection. *Evolution*, **41**, 340–354.

Mazer, S.J. (1989) Ecological, taxonomic and life history correlates of seed mass among Indiana dune angiosperms. *Ecological Monographs*, **59**, 153–175.

McCall, C., Mitchell-Olds, T. & Waller, D.M. (1989) Fitness consequences of outcrossing in *Impatiens capensis*: tests of the frequency-dependent and sib-competition models. *Evolution*, **43**, 1075–1084.

McCauley, D.E. (1991) Genetic consequences of local population extinction and recolonization. *Trends in Ecology and Evolution*, **6**, 5–8.

McDonald, B.A., McDermott, J.M., Allard, R.W. & Webster, R.K. (1989b) Coevolution of host and pathogen populations in the *Hordeum vulgare–Rhynchosporium secalis* pathosystem. *Proceedings of the National Academy of Sciences USA*, **86**, 3924–3927.

McDonald, B.A., McDermott, J.M., Goodwin, S.B. & Allard, R.W. (1989a) The population biology of host–pathogen interactions. *Annual Review of Phytopathology*, **27**, 77–94.

McGinley, M.A., Temme, D.H. & Geber, M.A. (1987) Parental investment in offspring in variable environments: theoretical and empirical considerations. *American Naturalist*, **130**, 370–398.

McGraw, J.B. & Antonovics, J. (1983) Experimental ecology of *Dryas octopetala* ecotypes 1. Ecotypic differentiation and life-cycle stages of selection. *Journal of Ecology*, **71**, 879–897.

McMaster, G.S. & Zedler, P.H. (1981) Delayed seed dispersal in *Pinus torreyana* (Torrey pine). *Oecologia*, **51**, 62–66.

Meagher, T.R. (1986) Analysis of paternity within a natural population of *Chamaelirium luteum* I. Identification of most-likely male parents. *American Naturalist*, **127**, 199–215.

Meagher, T.R. (1991) Analysis of paternity within a natural population of *Chamaelirium luteum* II. Patterns of male reproductive success. *American Naturalist*, **137**, 738–752.

Meagher, T.R. & Thompson, E. (1987) Analysis of parentage for naturally established seedlings of *Chamaelirium luteum* (Liliaceae). *Ecology*, **68**, 803–812.

Meeuse, B.J.D. & Morris, S. (1984) *The Sex Life of Flowers*. Facts on File, New York.

Mettler, L.E., Gregg, T.G. & Schaffer, H.E. (1988) *Population Genetics and Evolution*, 2nd edn. Prentice Hall, Englewood Cliffs, NJ.

Michaels, H.J. & Bazzaz, F.A. (1986) Resource allocation and demography of sexual and apomictic *Antennaria parlinii*. *Ecology*, **67**, 27–36.

Michaels, H.J., Benner, B., Hartgerink, A.P. *et al.* (1988) Seed size variation: magnitude, distribution, and ecological correlates. *Evolutionary Ecology*, **2**, 157–166.

Michod, R.E. & Levin, B.R. (1988) *The Evolution of Sex*. Sinauer, Sunderland, MA.

Miller, T.E. (1987) Effects of emergence time on survival and growth in an early old-field plant community. *Oecologia*, **72**, 272–278.

Miller, T.E. & Werner, P.A. (1987) Competitive effects and responses between plant species in a first-year old field community. *Ecology*, **68**, 1201–1210.

Mitchell-Olds, T. & Bergelson, J. (1990a) Statistical genetics of an annual plant, *Impatiens capensis*. I. Genetic basis of quantitative variation. *Genetics*, **124**, 407–415.

Mitchell-Olds, T. & Bergelson, J. (1990b) Statistical genetics of an annual plant, *Impatiens capensis*. II. Natural selection. *Genetics*, **124**, 417–421.

Mitchell-Olds, T. & Shaw, R.G. (1987) Regression analysis and natural selection: statistical inference and biological interpretation. *Evolution*, **41**, 1149–1161.

Mitchley, J. (1987) Diffuse competition in plant communities. *Trends in Ecology and Evolution*, **2**, 104–106.

Mithen, R., Harper, J.L. & Weiner, J. (1984) Growth and mortality of individual plants as a function of 'available area'. *Oecologia*, **57**, 57–60.

Mogie, M. & Hutchings, M.J. (1990) Phylogeny, ontogeny and clonal growth in vascular plants. In Groenendael, J. van & Kroon, D. de (eds) *Clonal Growth in Plants*, pp. 3–22. SPB Academic Publishing, The Hague.

Mohler, C.L. (1990) Co-occurrence of oak sub-genera: implications for niche differentiation. *Bulletin of the Torrey Botanical Club*, **117**, 247–255.

Mohler, C.L., Marks, P.L. & Sprugel, D.G. (1978) Stand structure and allometry of trees during self-thinning of pure stands. *Journal of Ecology*, **66**, 599–614.

Moloney, K.A. (1988) Fine-scale spatial and temporal variation in the demography of a perennial bunchgrass. *Ecology*, **69**, 1588–1598.

Moloney, K.A. (1990) Shifting demographic control of a perennial bunchgrass along a natural habitat gradient. *Ecology*, **71**, 1133–1143.

Moody, M.E. & Mack, R.N. (1988) Controlling the spread of plant invasions: the importance of nascent foci. *Journal of Applied Ecology*, **25**, 1009–1021.

Mopper, S., Mitton, J.B., Whitham, T.G., Cobb, N.S. & Christensen, K.M. (1991) Genetic differentiation and heterozygosity in Pinyon pine associated with resistance to herbivory and environmental stress. *Evolution*, **45**, 989–999.

Mortimer, A.M. (1984) Population ecology and weed science. In Dirzo, R. & Sarukhán, J. (eds) *Perspectives on Plant Population Ecology*, pp. 363–388. Sinauer, Sunderland, MA.

Mousseau, T.A. & Roff, D.A. (1987) Natural selection and the heritability of fitness components. *Heredity*, **59**, 181–197.

Muir, P.S. & Lotan, J.E. (1984) Serotiny and life history of *Pinus contorta* var. latifolia. *Canadian Journal of Botany*, **63**, 938–945.

Nei, M., Maruyama, T. & Chakraborty, R. (1975) The bottleneck effect and genetic variability in populations. *Evolution*, **29**, 1–10.

Nevo, E., Beiles, A., Kaplan, D., Golenberg, E.M., Olsvig-Whittaker, L.S. & Naveh, Z. (1986) Natural selection of allozyme polymorphisms: a microsite test revealing ecological genetic differentiation. *Evolution*, **40**, 13–20.

Nevo, E., Beiles, A. & Krugman, T. (1988) Natural selection of allozyme polymorphisms: a microgeographical differentiation by edaphic, topographical and temporal factors in wild emmer wheat (*Triticum dicoccoides*). *Theoretical & Applied Genetics*, **76**, 737–752.

New, J.K. (1958) A population study of *Spergula arvensis* I. Two clines and their significance. *Annals of Botany*, **22**, 457–477.

New, J.K. (1978) Change and stability of clines in *Spergula arvensis* L. (corn spurrey) after 20 years. *Watsonia*, **12**, 137–143.

New, J.K. & Herriott, J.C. (1981) Moisture for germination as a factor affecting the distribution of the seedcoat morphs of *Spergula arvensis* L. *Watsonia*, **13**, 323–324.

Newman, E.I. (1988) Mycorrhizal links between plants: their functioning and ecological significance. *Advances in Ecological Research*, **18**, 243–270.

Ng, F.S.P. (1983) Ecological principles of tropical lowland rain forest conservation. In Sutton, S.L., Whitmore, T.C. & Chadwick, A.C. (eds) *Tropical Rain Forest Ecology and Management*, pp. 359–375.

Blackwell Scientific Publications, Oxford.

Nilsson, S.G. & Wästljung, U. (1987) Seed predation and cross-pollination in mast-seeding beech (*Fagus sylvatica*) patches. *Ecology*, **68**, 260–265.

Novoplansky, A., Cohen, D. & Sachs, T. (1990) How *Portulaca* seedlings avoid their neighbours. *Oecologia*, **82**, 490–493.

Nuñez-Farfán, J. & Dirzo, R. (1988) Within-gap spatial heterogeneity and seedling performance in a Mexican tropical rainforest. *Oikos*, **51**, 274–284.

Nunney, L. (1989) The maintenance of sex by group selection. *Evolution*, **43**, 245–257.

Nygren, A. (1967) Apomixis in the angiosperms. *Handbuch der Pflanzenphysiologie*, **18**, 551–596.

Nyquist, W.E. (1991) Estimation of heritability and prediction of selection response in plant populations. *Critical Reviews in Plant Sciences*, **10**, 235–322.

Olivieri, I. & Gouyon, P.H. (1985) Seed dimorphism for dispersal: theory and observations. In Haeck, J. & Woldendorp, J.W. (eds) *Structure and Functioning of Plant Populations*, pp. 77–90. North Holland, Amsterdam.

Olivieri, I., Couvert, D. & Gouyon, P.-H. (1990) The genetics of transient populations: research at the metapopulation level. *Trends in Ecology and Evolution*, **5**, 207–210.

Onyekwelu, S.S. & Harper, J.L. (1979) Sex ratio and niche differentiation in spinach (*Spinaca oleracea* L.). *Nature*, **282**, 609–611.

Osawa, A. & Sugita, S. (1989) The self-thinning rule: another interpretation of Weller's results. *Ecology*, **70**, 279–283.

Pacala, S.W. & Silander, J.A. Jr (1985) Neighbourhood models of plant population dynamics 1. Single-species models of annuals. *American Naturalist*, **125**, 385–411.

Pacala, S.W. & Silander, J.A. Jr (1987) Neighborhood interference among velvet leaf, *Abutilon theophrasti*, and pigweed, *Amaranthus retroflexus*. *Oikos*, **48**, 217–224.

Pacala, S.W. & Silander, J.A. Jr (1990) Tests of neighbourhood population dynamic models in field communities of two annual weed species. *Ecological Monographs*, **60**, 113–134.

Paige, K.N. & Whitham, T.G. (1985) Individual and population shifts in flower colour by scarlet gilia: a mechanism for pollinator tracking. *Science*, **227**, 315–317.

Palmblad, I.G. (1968) Competition studies on experimental populations of weed with emphasis on the regulation of population size. *Ecology*, **49**, 26–34.

Parish, R. & Turkington, R. (1990a) The influence of dung pats and molehills on pasture composition. *Canadian Journal of Botany*, **68**, 1698–1705.

Parish, R. & Turkington, R. (1990b) The colonization of dung pats and molehills in permanent pastures. *Canadian Journal of Botany*, **68**, 1706–1711.

Parker, M.A. (1987) Pathogen impact on sexual versus asexual reproductive success in *Arisaema triphyllum*. *American Journal of Botany*, **74**, 1758–1763.

Partridge, L. & Harvey, P.H. (1988) The ecological context of life-history evolution. *Science*, **241**, 1449–1455.

Pavone, L.V. & Reader, R.J. (1982) The dynamics of seed bank size and seed state of *Medicago lupulina*. *Journal of Ecology*, **70**, 537–547.

Pease, C.M. & Bull, J.J. (1988) A critique of methods for measuring life history trade-offs. *Journal of Evolutionary Biology*, **1**, 293–304.

Peñalosa, J. (1983) Shoot dynamics and adaptive morphology of *Ipomoea phillomega* (Vell.) House (Convolulaceae), a tropical rainforest liana. *Annals of Botany*, **52**, 737–754.

Peters, N.C.B. (1985) Competitive effects of *Avena fatua* L. plants derived from seeds of different weights. *Weed Research*, **25**, 67–77.

Phillips, D.L. & MacMahon, J.A. (1981) Competition and spacing patterns of desert shrubs. *Journal of Ecology*, **69**, 97–115.

Pickett, S.T.A. & White, P.S. (1985) *The Ecology of Natural Disturbance and Patch Dynamics*. Academic Press, New York.

Pinder, I.J.E. (1975) Effects of species removals on an old-field plant community. *Ecology*, **56**, 747–751.

Piñero, D., Martínez-Ramos, M. & Sarukhán, J. (1984) A population model of *Astrocaryum mexicanum* and a sensitivity analysis of its finite rate of increase. *Journal of Ecology*, **72**, 977–991.

Piñero, D., Sarukhán, J. & Alberdi, P. (1982) The costs of reproduction in a tropical palm, *Astrocaryum mexicanum*. *Journal of Ecology*, **70**, 473–481.

Pitelka, L.F. (1984) Application of the −3/2 power law to clonal herbs. *American Naturalist*, **123**, 442–449.

Plantenkamp, G.A.J. (1990) Phenotypic plasticity and genetic differentiation in the demography of the grass *Anthoxanthum odoratum*. *Journal of Ecology*, **78**, 772–788.

Platt, W.J. (1975) The colonization and formation of equilibrium plant species associations on badger disturbances in tallgrass prairie. *Ecological Monographs*, **45**, 285–305.

Platt, W.J. & Weiss, I.M. (1977) Resource partitioning and competition within a guild of fugitive prairie plants. *American Naturalist*, **111**, 479–513.

Platt, W.J. & Weiss, I.M. (1985) An experimental study of competition among fugitive prairie plants. *Ecology*, **66**, 708–720.

Policansky, D. (1981) Sex choice and the size advantage

model in Jack-in-the-pulpit (*Arisaema triphyllum*). *Proceedings of the National Academy of Sciences USA*, **78**, 1306–1308.

Polley, H.W. & Detling, J.K. (1990) Grazing-mediated differentiation in *Agropyron smithii*: evidence from populations with different grazing histories. *Oikos*, **57**, 326–332.

Prairie, Y.T. & Bird, D.F. (1989) Some misconceptions about the spurious correlation problem in the ecological literature. *Oecologia*, **81**, 285–288.

Price, E.A.C., Marshall, C. & Hutchings, M.J. (1992) Studies of growth in the clonal herb *Glechoma hederacea*. I. Patterns of physiological integration. *Journal of Ecology*, **80**, 25–38.

Price, P.W., Westoby, M., Rice, B. *et al.* (1986) Parasite mediation in ecological interactions. *Annual Review of Ecology and Systematics*, **17**, 487–505.

Price, P.W., Westoby, M. & Rice, B. (1988) Parasite-mediated competition: some predictions and tests. *American Naturalist*, **131**, 544–555.

Price, T. & Schluter, D. (1991) On the low heritability of life history traits. *Evolution*, **45**, 853–861.

Primack, R.B. & Antonovics, J. (1981) Experimental ecological genetics in *Plantago* V. Components of seed yield in the ribwort plantain *Plantago lanceolata* L. *Evolution*, **35**, 1069–1079.

Primack, R.B. & Hall, P. (1990) Costs of reproduction in the pink lady's slipper orchid: a four-year experimental study. *American Naturalist*, **136**, 638–656.

Proctor, M. & Yeo, P. (1973) *The Pollination of Flowers*. Collins, London.

Pulliam, H.R. (1989) Sources, sinks and population regulation. *American Naturalist*, **132**, 652–661.

Putz, F.E. (1983) Treefall pits and mounds, buried seeds, and the importance of soil disturbance to pioneer trees on Barro Colorado Island, Panama. *Ecology*, **64**, 1069–1074.

Pyke, G.H. (1991) What does it cost a plant to produce floral nectar? *Nature*, **350**, 58–59.

Rabinowitz, D., Rapp, J.K., Cairns, S. & Mayer, M. (1989) The persistence of rare prairie grasses in Missouri: environmental variation buffered by reproductive output of sparse species. *American Naturalist*, **134**, 525–544.

Radosevich, S.R. & Rousch, M.L. (1990) The role of competition in agriculture. In Grace, J.B. & Tilman, D. (eds) *Perspectives on Plant Competition*, pp. 341–363. Academic Press, San Diego.

Rai, K.N. & Jain, S.K. (1982) Population biology of *Avena*. IX. Gene flow and neighbourhood size in relation to microgeographic variation in *Avena barbata*. *Oecologia*, **53**, 399–405.

Ranker, T.A. (1991) Natural selection and the plastid genome of parasitic angiosperms. *Trends in Ecology and Evolution*, **6**, 205.

Reichman, O.J. (1984) Spatial and temporal variation of seed distributions in Sonoran desert soils. *Journal of Biogeography*, **11**, 1–11.

Reinartz, J.A. (1984) Life history variation of common mullein (*Verbascum thapsus*). I. Latitudinal differences in population dynamics and timing of reproduction. *Journal of Ecology*, **72**, 897–912.

Rejmánek, M., Robinson, G.R. & Rejmankova, E. (1989) Weed–crop competition, experimental designs and models for data analysis. *Weed Science*, **37**, 276–284.

Reznick, D. (1992) Measuring the costs of reproduction. *Trends in Ecology and Evolution*, **7**, 42–45.

Rice, B. & Westoby, M. (1982) Heteroecious rusts as agents of interference competition. *Evolutionary Theory*, **6**, 43–52.

Rice, K.J. (1987) Evidence for the retention of genetic variation in *Erodium* seed dormancy by variable rainfall. *Oecologia*, **72**, 589–596.

Rice, K.J. (1989) Impacts of seed banks on grassland community structure and population dynamics. In Leck, M.A., Parker, V.T. & Simpson, R.L. (eds) *Ecology of Soil Seed Banks*, pp. 211–230. Academic Press, San Diego.

Rice, K.J. & Mack, R.N. (1991) Ecological genetics of *Bromus tectorum* III. The demography of reciprocally sown populations. *Oecologia*, **88**, 91–101.

Rice, W.R. (1983) Parent–offspring pathogen transmission: a selective agent promoting sexual reproduction. *American Naturalist*, **121**, 187–203.

Richards, A.J. (1973) The origin of *Taraxacum* agamospecies. *Botanical Journal of the Linnaean Society*, **66**, 189–211.

Richards, A.J. (1986) *Plant Breeding Systems*. Allen and Unwin, London.

Riera, B. (1985) Importance des buttes de deracinement dans la regeneration forestiere en Guyane Français. *Revue Ecologie (Terre Vie)*, **40**, 321–329.

Riley, J. (1984) A general form of the 'land equivalent ratio'. *Experimental Agriculture*, **20**, 19–29.

Roach, D.A. (1987) Variation in seed and seedling size in *Anthoxanthum odoratum*. *American Midland Naturalist*, **117**, 258–264.

Roberts, H.A. (1970) Viable weed seeds in cultivated soils. *Report of the National Vegetable Research Station*, **1969**, 23–28.

Roberts, H.A. (1981) Seed banks in the soil. *Advances in Applied Biology*, **6**, 1–55.

Roberts, H.A. (1986) Seed persistence in soils and seasonal emergence in plant species from different habitats. *Journal of Applied Ecology*, **23**, 639–656.

Roberts, H.A. & Feast, P.M. (1973) Emergence and longevity of seeds of annual weeds in cultivated and undisturbed soil. *Journal of Applied Ecology*, **10**,

133–143.

Rockwood, L.L. (1985) Seed weight as a function of life form, elevation and life zone in neotropical forests. *Biotropica*, **17**, 32–39.

Rogers, R.W. & Westman, W.E. (1979) Niche differentiation and maintenance of genetic identity in cohabiting *Eucalyptus* species. *Australian Journal of Ecology*, **4**, 429–439.

Roose, M.L. & Gottlieb, L.D. (1976) Genetic and biochemical consequences of polyploidy in *Tragopogon*. *Evolution*, **30**, 818–830.

Rose, M.R. (1991) *Evolutionary Biology of Aging*. Oxford University Press, New York.

Rosewell, J., Shorrocks, B. & Edwards, K. (1990) Competition on a divided and ephemeral resource: testing the assumptions. I. Aggregation. *Journal of Animal Ecology*, **59**, 977–1001.

Ross, M.A. & Harper, J.L. (1972) Occupation of biological space during seedling establishment. *Journal of Ecology*, **60**, 77–88.

Roughgarden, J. (1979) *Theory of Population Genetics and Evolutionary Ecology*. Macmillan, New York.

Runkle, J.R. (1982) Patterns of disturbance in some old-growth mesic forests of Eastern North America. *Ecology*, **63**, 1533–1546.

Russell, P.J., Flowers, T.J. & Hutchings, M.J. (1985) Comparison of niche breadths and overlaps of halophytes on salt marshes of differing diversity. *Vegetatio*, **61**, 171–178.

Saedler, H., Bonas, U., Deumling, B. *et al.* (1983) Transposable elements in plants. In Chaeter, K.F., Cullis, C., Hopwood, A.D.A., Johnson, A.W.B. & Woolhouse, H.W. (eds) *Genetic Rearrangement*. Sinauer, Sunderland, MA.

Sakai, A.K., Karoly, K. & Weller, S.G. (1989) Inbreeding depression in *Schiedea globoza* and *S. salicaria* (Caryophyllaceae), subdioecious and gynodioecious Hawaiian species. *American Journal of Botany*, **76**, 437–444.

Salisbury, E.J. (1942) *The Reproductive Capacity of Plants*. Bell and Sons, London.

Salzman, A.G. & Parker, M.A. (1985) Neighbors ameliorate local salinity stress for a rhizomatous plant in a heterogeneous environment. *Oecologia*, **65**, 273–277.

Samson, D.A. & Werk, K.S. (1986) Size-dependent effects in the analysis of reproductive effort in plants. *American Naturalist*, **127**, 667–680.

Sano, Y., Morishima, H. & Oka, H.-I. (1980) Intermediate perennial–annual populations of *Oryza perennis* found in Thailand and their evolutionary significance. *Botanical Magazine of Tokyo*, **93**, 291–305.

Sarukhán, J. (1974) Studies on plant demography: *Ranunculus repens* L., *R. bulbosus* L. and *R. acris* L.: II. Reproductive strategies and seed population

dynamics. *Journal of Ecology*, **62**, 151–177.

Sarukhán, J. & Harper, J.L. (1973) Studies on plant demography: *Ranunculus repens* L., *R. bulbosus* L. and *R. acris* L.: I. Population flux and survivorship. *Journal of Ecology*, **61**, 675–716.

Sarukhán, J., Piñero, D. & Martínez-Ramos, M. (1985) Plant demography: a community-level interpretation. In White, J. (ed.) *Studies in Plant Demography*, pp. 17–31. Academic Press, London.

Sarvas, R. (1968) Investigations on the flowering and seed crop of *Picea abies*. *Communicationes Instituti Forestalis Fenniae*, **67**, 1–69.

Scaife, M.A. & Jones, D. (1976) The relationship between crop yield (or mean plant weight) of lettuce and plant density, length of growing period, and initial plant weight. *Journal of Agricultural Science*, **86**, 83–91.

Schaal, B.A. (1980) Reproductive capacity and seed size in *Lupinus texensis*. *American Journal of Botany*, **67**, 703–709.

Schaffer, W.M. & Gadgil, M.D. (1975) Selection for optimal life histories in plants. In Cody, M.L. & Diamond, J. (eds) *Ecology and Evolution of Communities*, pp. 142–156. Belknap Press, Cambridge, MA.

Schaffer, W.M. & Schaffer, M.D. (1977) The adaptive significance of variations in reproductive habit in the Agavaceae. In Stonehouse, B. & Perrins C.M. (eds) *Evolutionary Ecology*, pp. 261–276. Macmillan, London.

Schaffer, W.M. & Schaffer, M.D. (1979) The adaptive significance of variations in reproductive habit in the Agavaceae. II. Pollinator foraging behaviour and selection for increased reproductive expenditure. *Ecology*, **60**, 1051–1069.

Schat, H., Ouborg, J. & de Wit, R. (1989) Life history and plant architecture: size-dependent reproductive allocation in annual and biennial *Centaurium* species. *Acta Botanica Neerlandica*, **38**, 183–201.

Scheiner, S.M. & Goodnight, C.J. (1984) The comparison of phenotypic and genetic variation in populations of the grass *Danthonia spicata*. *Evolution*, **38**, 845–855.

Schemske D. (1984) Population structure and local selection in *Impatiens pallida* (Balsamaceae), a selfing annual. *Evolution*, **38**, 817–832.

Schemske, D.W. (1978) Evolution of reproductive characteristics in *Impatiens* (Balsaminaceae): the significance of cleistogamy and chasmogamy. *Ecology*, **59**, 596–613.

Schemske, D.W. (1984) Population structure and local selection in *Impatiens pallida* (Balsaminaceae), a selfing annual. *Evolution*, **38**, 817–832.

Schemske, D.W. & Lande, R. (1985) The evolution of self-fertilization and inbreeding depression in plants. II. Empirical observations. *Evolution*, **39**, 41–52.

Schemske, D.W., Willson, M.F., Melampy, N.M. *et al.* (1978) Flowering ecology of some spring woodland herbs. *Ecology*, **59**, 351−366.

Schlichting, C.D. & Levin, D. (1986) Effects of inbreeding on phenotypic plasticity in cultivated *Phlox*. *Theoretical and Applied Genetics*, **72**, 114−119.

Schlichting, C.D., Stephenson, A.S., Davis, L.E. & Winsor, J.A. (1987) Pollen competition and offspring variance. *Evolutionary Trends in Plants*, **1**, 35−39.

Schmid, B., Puttick, G.M., Burgess, K.H. & Bazzaz, F.A. (1988) Clonal integration and effects of simulated herbivory in old-field perennials. *Oecologia*, **75**, 465−471.

Schmidt, K.P. & Levin, D.A. (1985) The comparative demography of reciprocally sown populations of *Phlox drummondii* Hook I. Survivorship, fecundities and finite rates of increase. *Evolution*, **39**, 396−404.

Schmitt, J. & Antonovics, J. (1986) Experimental studies of the evolutionary significance of sexual reproduction. IV. Effect of neighbor relatedness and aphid infestation on seedling performance. *Evolution*, **40**, 830−836.

Schmitt, J. & Ehrhardt, D.W. (1987) A test of the sib-competition hypothesis for outcrossing advantage in *Impatiens capensis*. *Evolution*, **41**, 579−590.

Schmitt, J. & Gamble, S.E. (1990) The effect of distance from the parental site on offspring performance and inbreeding depression in *Impatiens capensis*: a test of the local adaptation hypothesis. *Evolution*, **44**, 2022−2030.

Schnabel, A. & Hamrick, J.L. (1990) Organization of genetic diversity within and among populations of *Gleditsia triacanthos* (Leguminosae). *American Journal of Botany*, **77**, 1060−1069.

Schoen, D.J. (1982) The breeding system of *Gilia achilleifolia*: variation in floral characteristics and outcrossing rate. *Evolution*, **36**, 352−360.

Schoen, D.J. & Brown, A.H.D. (1991) Intraspecific variation in population gene diversity and effective population size correlates with the mating system in plants. *Proceedings of the National Academy of Sciences USA*, **88**, 4494−4497.

Schoen, D.J. & Latta, R.G. (1989) Spatial autocorrelation of genotypes in populations of *Impatiens pallida* and *Impatiens capensis*. *Heredity*, **63**, 181−189.

Schoen, D.J. & Lloyd, D.G. (1984) The selection of cleistogamy and heteromorphic diaspores. *Biological Journal of the Linnaean Society*, **23**, 303−322.

Schuster, W.S. & Mitton, J.B. (1991) Relatedness within clusters of a bird-dispersed pine and the potential for kin interactions. *Heredity*, **67**, 41−48.

Schwaegerle, K.E. & Bazzaz, F.A. (1987) Differentiation among nine populations of *Phlox*: response to environmental gradients. *Ecology*, **68**, 54−64.

Schwaegerle, K.E. & Schaal, B.A. (1979) Genetic variability and founder effect in the pitcher plant *Sarracenia purpurea* L. *Evolution*, **33**, 1210−1218.

Seavey, S.R. & Bawa, K.S. (1986) Late-acting self-incompatibility in angiosperms. *Botanical Review*, **52**, 195−219.

Selman, M. (1970) The population dynamics of *Avena fatua* (wild oats) in continuous spring barley: desirable frequency of spraying with tri-allate. *Proceedings of the 10th British Weed Control Conference*, **10**, 1176−1188.

Sharitz, R.R. & McCormick, J.F. (1973) Population dynamics of two competing annual plant species. *Ecology*, **54**, 723−740.

Shaw, R.G. (1987) Density dependence in *Salvia lyrata*: an experimental alteration of densities of established plants. *Journal of Ecology*, **75**, 1049.

Shields, W. (1982) *Philopatry, Inbreeding and the Evolution of Sex*. State University of New York Press, Albany.

Shmida, A. & Ellner, S.P. (1984) Coexistence of plants with similar niches. *Vegetatio*, **58**, 29−55.

Shugart, H.H.J. & West, D.C. (1977) Development of an Appalachian deciduous forest succession model and its application to assessment of the impact of chestnut blight. *Journal of Environmental Management*, **5**, 161−180.

Silander, J.A.J. & Antonovics, J. (1982) Analysis of interspecific interactions in a coastal plant community — a perturbation approach. *Nature*, **298**, 557−560.

Silvertown, J. (1980a) Leaf-canopy induced seed dormancy in a grassland flora. *New Phytologist*, **85**, 109−118.

Silvertown, J. (1980b) The evolutionary ecology of mast seeding in trees. *Biological Journal of the Linnaean Society*, **14**, 235−250.

Silvertown, J. (1981) Seed size, lifespan and germination date as co-adapted features of plant life history. *American Naturalist*, **118**, 860−864.

Silvertown, J. (1983a) The distribution of plants in limestone pavement: tests of species interaction and niche separation against null hypotheses. *Journal of Ecology*, **71**, 819−828.

Silvertown, J. (1983b) Why are biennials sometimes not so few? *American Naturalist*, **121**, 448−453.

Silvertown, J. (1984) Phenotypic variety in seed germination behaviour: the ontogeny and evolution of somatic polymorphism in seeds. *American Naturalist*, **124**, 1−16.

Silvertown, J. (1985) Survival, fecundity and growth of wild cucumber *Echinocystis lobata*. *Journal of Ecology*, **73**, 841−849.

Silvertown, J. (1986) 'Biennials': reply to Kelly. *American Naturalist*, **127**, 721−724.

Silvertown, J. (1987a) *Introduction to Plant Population Ecology*. Longman Scientific and Technical, Harlow.

Silvertown, J. (1987b) Possible sexual dimorphism in the double coconut: a reinterpretation of the data of Savidge and Ashton. *Biotropica*, **19**, 282–283.

Silvertown, J. (1987c) The evolution of hermaphroditism. An experimental test of the resource model. *Oecologia*, **72**, 157–159.

Silvertown, J. (1988) The demographic and evolutionary consequences of seed dormancy. In Davy, A.J., Hutchings M.J. & Watkinson, A.R. (eds) *Plant Population Ecology*, pp. 205–219. Blackwell Scientific Publications, Oxford.

Silvertown, J. (1989a) A binary classification of plant life histories and some possibilities for its evolutionary application. *Evolutionary Trends in Plants*, **3**, 87–90.

Silvertown, J. (1989b) The paradox of seed size and adaptation. *Trends in Ecology and Evolution*, **4**, 24–26.

Silvertown, J. (1991a) Modularity, reproductive thresholds and plant population dynamics. *Functional Ecology*, **5**, 577–580.

Silvertown, J. (1991b) Dorothy's dilemma and the unification of plant population biology. *Trends in Ecology and Evolution*, **6**, 346–348.

Silvertown, J. & Dale, P. (1991) Competitive hierarchies and the structure of herbaceous plant communities. *Oikos*, **61**, 441–444.

Silvertown, J., Franco, M., Pisanty, I. & Mendoza, A. (1993) Comparative plant demography: relative importance of life cycle components to the finite rate of increase in woody and herbaceous perennials. *Journal of Ecology*, **81** (in press).

Silvertown, J. & Gordon, D. (1989) A framework for plant behaviour. *Annual Review of Ecology and Systematics*, **20**, 349–366.

Silvertown, J., Holtier, S., Johnson, J. & Dale, P. (1992) Cellular automaton models of interspecific competition for space between clonal plants: the effect of pattern on process. *Journal of Ecology*, **80**, 527–534.

Skellam, J.G. (1951) Random dispersal in theoretical populations. *Biometrika*, **38**, 196–218.

Slade, A.J. & Hutchings, M.J. (1987a) Clonal integration and plasticity in foraging behaviour in *Glechoma hederacea*. *Journal of Ecology*, **75**, 1023–1036.

Slade, A.J. & Hutchings, M.J. (1987b) The effects of nutrient availability on foraging in the clonal herb *Glechoma hederacea*. *Journal of Ecology*, **75**, 95–112.

Slade, A.J. & Hutchings, M.J. (1987c) An analysis of the costs and benefits of physiological integration between ramets in the clonal perennial herb *Glechoma hederacea*. *Oecologia*, **73**, 425–431.

Smith, A.P. & Young, T.P. (1987) Tropical alpine plant ecology. *Annual Review of Ecology and Systematics*, **18**, 137–158.

Smith, C.C., Hamrick, J.L. & Kramer, C.L. (1990) The advantage of mast years for wind pollination. *American Naturalist*, **136**, 154–166.

Snaydon, R.W. & Howe, C.D. (1986) Root and shoot competition between established ryegrass and invading grass seedlings. *Journal of Applied Ecology*, **23**, 667–674.

Snow, A.A. & Whigham, D.F. (1989) Cost of flower and fruit production in *Tipularia discolor*. *Ecology*, **70**, 1286–1293.

Soane, I. & Watkinson, A. (1979) Clonal variation in populations of *Ranunculus repens*. *New Phytologist*, **82**, 557–573.

Sohn, J.J. & Policansky, D. (1977) The costs of reproduction in the mayapple *Podophyllum peltatum* (Berberidaceae). *Ecology*, **58**, 1366–1374.

Solbrig, O.T., Curtis, W.F., Kincaid, D.T. & Newell, S.J. (1988a) Studies on the population biology of the genus *Viola*. VI. The demography of *V. fimbriatula* and *V. lanceolata*. *Journal of Ecology*, **76**, 301–319.

Solbrig, O.T., Sarandon, R. & Bossert, W. (1988b) A density-dependent growth model of a perennial herb, *Viola fimbriatula*. *American Naturalist*, **137**, 385–400.

Spitters, C.J.T. (1983) An alternative approach to the analysis of mixed cropping experiments. 1. Estimation of competition coefficients. *Netherlands Journal of Agricultural Science*, **31**, 1–11.

Stadler, L.J. (1928) Mutations in barley induced by X-rays and radium. *Science*, **68**, 186–187.

Stanton, M.L. (1984) Developmental and genetic sources of seed weight variation in *Raphanus raphanistrum* L. (Brassicaceae). *American Journal of Botany*, **71**, 1090–1098.

Stearns, S. (1987) (ed.) *The Evolution of Sex and its Consequences*. Birkhaüser, Basle.

Stearns, S.C. (1989) Trade-offs in life-history evolution. *Functional Ecology*, **3**, 259–268.

Stephenson, A.G. & Winsor, J.A. (1986) *Lotus corniculatus* regulates offspring quality through selective fruit abortion. *Evolution*, **40**, 453–458.

Sterner, R.W., Ribic, C.A. & Schatz, G.E. (1986) Testing for life historical changes in spatial pattern of four tropical tree species. *Journal of Ecology*, **74**, 621–623.

Stratton, D.A. (1992) Life-cycle components of selection in *Erigeron annuus*: II. Genetic variation. *Evolution*, **46**, 107–120.

Sultan, S.E. (1987) Evolutionary implications of phenotypic plasticity in plants. *Evolutionary Biology*, **21**, 127–178.

Sutherland S. (1986a) Floral sex ratios, fruit set, and resource allocation in plants. *Ecology*, **67**, 991–1001.

Sutherland, S. (1986b) Patterns of fruit-set: what controls fruit–flower ratios in plants? *Evolution*, **40**, 117–128.

Sutherland, S. & Vickery, R.K. (1988) Trade-offs between sexual and asexual reproduction in the genus *Mimulus*. *Oecologia*, **76**, 330–335.

Sutherland, W.J. & Stillman, R.A. (1988) The foraging tactics of plants. *Oikos*, **52**, 239–244.

Symonides, E. (1977) Mortality of seedlings in natural psammophyte populations. *Ekologia Polska*, **25**, 635–651.

Symonides, E. (1983a) Population size regulation as a result of intra-population interactions. II. Effect of density on the survival of individuals of *Erophila verna* (L.) C.A.M. *Ekologia Polska*, **31**, 839–881.

Symonides, E. (1983b) Population size regulation as a result of intra-population interactions. III. Effect of density on the growth rate, morphological diversity and fecundity of *Erophila verna* (L.) CAM individuals. *Ekologia Polska*, **31**, 883–912.

Symonides, E. (1984) Population size regulation as a result of intra-population interactions III. Effect of *Erophila verna* (L.) CAM population density on the abundance of seedlings. Summing-up and conclusions. *Ekologia Polska*, **32**, 557–580.

Symonides, E. (1988) Population dynamics of annual plants. In Davy, A.J., Hutchings, M.J. & Watkinson, A.R. (eds) *Plant Population Ecology*, pp. 221–248. Blackwell Scientific Publications, Oxford.

Symonides, E., Silvertown, J. & Andreasen, V. (1986) Population cycles caused by overcompensating density-dependence in an annual plant. *Oecologia*, **71**, 156–158.

Taggart, J.B., McNally, S.F. & Sharp, P.M. (1990) Genetic variability and differentiation among founder populations of the pitcher plant (*Saracenia purpurea* L.) in Ireland. *Heredity*, **64**, 177–183.

Taylor, D.R. & Aarssen, L.W. (1989) The density-dependence of replacement-series competition experiments. *Journal of Ecology*, **77**, 975–988.

Taylor, D.R. & Aarssen, L.W. (1990) Complex competitive relationships among genotypes of three perennial grasses: implications for species coexistence. *American Naturalist*, **136**, 305–327.

Templeton, A.R. & Levin, D.A. (1979) Evolutionary consequences of seed pools. *American Naturalist*, **114**, 232–249.

ter Borg, S.J. (1979) Some topics in plant population biology. In Werger, M.J.A. (ed.) *The Study of Vegetation*, pp. 13–55. A. Junk, The Hague.

Thoday, J.M. (1953) Components of fitness. *Symposium of the Society of Experimental Biology*, **7**, 96–113.

Thompson, K. (1986) Small-scale heterogeneity in the seed bank of an acidic grassland. *Journal of Ecology*, **74**, 733–738.

Thompson, K. & Grime, J.P. (1979) Seasonal variation in seed banks of herbaceous species in ten contrasting habitats. *Journal of Ecology*, **67**, 893–921.

Thrall, P.H., Pacala, S.W. & Silander J.A. Jr (1989) Oscillatory dynamics in populations of an annual weed species *Abutilon theophrasti*. *Journal of Ecology*, **77**, 1135–1149.

Tiffney, B. (1984) Seed size, dispersal syndromes, and the rise of the angiosperms: evidence and hypotheses. *Annals of the Missouri Botanic Gardens*, **71**, 551–576.

Tilman, D. (1982) *Resource Competition and Community Structure*. Princeton University Press, Princeton, NJ.

Tilman, D. & Wedin, D. (1991) Dynamics of nitrogen competition between successional grasses. *Ecology*, **72**, 1038–1049.

Tissue, D.T. & Nobel, P.S. (1990) Carbon relations of flowering in a semelparous clonal desert perennial. *Ecology*, **71**, 273–281.

Tonsor, S.T. (1989) Relatedness and intraspecific competition in *Plantago lanceolata*. *American Naturalist*, **134**, 897–906.

Torstensson, P. & Telenius, A. (1986) Consequences of differential utilization of meristems in the annual *Spergularia marina* and the perennial *S. media*. *Holarctic Ecology*, **9**, 20–26.

Turkington, R.A. & Harper, J.L. (1979) The growth, distribution and neighbour relationships of *Trifolium repens* in a permanent pasture. *Journal of Ecology*, **67**, 245–254.

Turner, M.D. & Rabinowitz, D. (1983) Factors affecting frequency distributions of plant mass: the absence of dominance and suppression in competing monocultures of *Festuca paradoxa*. *Ecology*, **64**, 469–475.

Turner, M.E., Stephens, J.C. & Anderson, W.W. (1982) Homozygosity and patch structure in plant populations as a result of nearest-neighbor pollination. *Proceedings of the National Academy of Sciences USA*, **79**, 203–207.

Uhl, C. & Clark, K. (1983) Seed ecology of selected Amazon basin successional species. *Botanical Gazette*, **144**, 419–425.

Uma Shaanker, R. & Ganeshaiah, K.N. (1989) Stylar plugging by fertilized ovules in *Kleinhovia hospita* (Sterculeaceae) – a case of vaginal sealing in plants? *Evolutionary Trends in Plants*, **3**, 59–64.

Uyenoyama, M. (1986) Inbreeding and the cost of meiosis: the evolution of selfing in populations practicing biparental inbreeding. *Evolution*, **40**, 388–404.

van Damme J.M.M. (1984) Gynodioecy in *Plantago lanceolata* L. III. Sexual reproduction and the maintenance of male steriles. *Heredity*, **52**, 77–93.

Vandermeer, J. (1984) Plant competition and the yield–density relationship. *Journal of Theoretical Biology*, **109**, 393–399.

Vandermeer, J. (1989) *Ecology of Intercropping*. Cambridge University Press, Cambridge.

van der Reest, P.J. & Rogaar, H. (1988) The effect of earthworm activity on the vertical distribution of plant seeds in newly reclaimed polder soils in the Netherlands. *Pedobiologia*, **31**, 211–218.

van Groenendael, J. (1985) Differences in life histories between two ecotypes of *Plantago lanceolata* L. In White, J. (ed.) *Studies on Plant Demography*, pp. 51–67. Academic Press, London.

van Groenendael, J.M. & Slim, P. (1988) The contrasting dynamics of two populations of *Plantago lanceolata* classified by age and size. *Journal of Ecology*, **76**, 585–589.

Vasek, F.C. (1980) Creosote bush: long-lived clones in the Mohave Desert. *American Journal of Botany*, **67**, 246–255.

Vázquez-Yanes, C. & Orozco-Segovia, A. (1982) Seed germination of a tropical rainforest pioneer tree (*Heliocarpus donellsmithii*) in response to diurnal fluctuations of temperature. *Physiologia Plantarum*, **56**, 295–298.

Vázquez-Yanes, C. & Smith, H. (1982) Phytochrome control of seed germination in the tropical rain forest pioneer trees *Cecropia obtusifolia* and *Piper auritum* and its ecological significance. *New Phytologist*, **92**, 477–485.

Venable, D. (1985) The evolutionary ecology of seed heteromorphism. *American Naturalist*, **126**, 577–595.

Venable, D.L. (1989) Modelling the evolutionary ecology of seed banks. In Leck, M.A., Parker, V.T. & Simpson, R.L. (eds) *Ecology of Soil Seed Banks*, pp. 67–87. Academic Press, San Diego.

Venable, D.L. & Búrquez, A. (1989) Quantitative genetics of size, shape, life-history, and fruit characteristics of the seed-heteromorphic composite *Heterosperma pinnatum*. I. Variation within and among populations. *Evolution* **43**, 113–124.

Venable, D.L. & Lawlor, L. (1980) Delayed germination and dispersal in desert annuals: escape in time and space. *Oecologia*, **46**, 272–282.

Venable, D.L., Búrquez, A., Coral, G., Morales, E. & Espinosa, F. (1987) The ecology of seed heteromorphism in *Heterosperma pinnatum* in Central Mexico. *Ecology*, **68**, 65–76.

Wade, M.J. & Kalisz, S. (1990) The causes of natural selection. *Evolution*, **44**, 1947–1955.

Walker, J. & Peet, R.K. (1983) Composition and species-diversity of pine-wiregrass savannas of the Green Swamp, North Carolina. *Vegetatio*, **55**, 163–179.

Waller, D.M. (1981) Neighbourhood competition in several violet populations. *Oecologia*, **51**, 116–122.

Waller, D.M. (1982) Factors influencing seed weight in *Impatiens capensis* (Balsaminaceae). *American Journal of Botany*, **69**, 1470–1475.

Waller, D.M. (1986) Is there disruptive selection for self-fertilization? *American Naturalist*, **128**, 421–426.

Warner, R.R. & Chesson, P.L. (1985) Coexistence mediated by recruitment fluctuations: a field guide to the storage effect. *American Naturalist*, **125**, 769–787.

Waser, N.M. (1983) The adaptive nature of floral traits: ideas and evidence. In Real, L. (ed.) *Pollination Biology*, pp. 241–285. Academic Press, New York.

Waser, N.M. (1986) Flower constancy: definition, cause and measurement. *American Naturalist*, **127**, 593–603.

Waser, N.M. (1992) Population structure, optimal outbreeding, and assortative mating in angiosperms. In Thornhill, N.W. (ed.) *The Natural History of Inbreeding and Outbreeding: Theoretical and Empirical Perspectives*. University of Chicago Press, Chicago.

Waser, N.M. & Price, M.V. (1983) Optimal and actual outcrossing in plants, and the nature of plant–pollinator interaction. In Jones, C.E. & Little, R.J. (eds) *Handbook of Experimental Pollination Biology*, pp. 341–351. Van Nostrand Reinhold, New York.

Waser, N.M. & Price, M.V. (1985) Reciprocal transplant experiments with *Delphinium nelsonii* (Ranunculaceae): evidence for local adaptation. *American Journal of Botany*, **72**, 1726–1732.

Waser, N.M. & Price, M.V. (1989) Optimal outcrossing in *Ipomopsis aggregata*: seed set and offspring fitness. *Evolution*, **43**, 1097–1109.

Waser, N.M., Price, M.V., Montalvo, A.M. & Gray, R.N. (1987) Female mate choice in a perennial herbaceous wildflower, *Delphinium nelsonii*. *Evolutionary Trends in Plants*, **1**, 29–33.

Waser, N.M. & Thornhill, N.W. (1992) Sex, mating systems, inbreeding, and outbreeding. In Thornhill, N.W. (ed.) *The Natural History of Inbreeding and Outbreeding: Theoretical and Empirical Perspectives*. University of Chicago Press, Chicago.

Watkinson, A.R. (1980) Density-dependence in single-species populations of plants. *Journal of Theoretical Biology*, **83**, 345–357.

Watkinson, A.R. (1981) Interference in pure and mixed populations of *Agrostemma githago* L. *Journal of Applied Ecology*, **18**, 967–976.

Watkinson, A.R. (1984) Yield–density relationships: the influence of resource availability on growth and self-thinning in populations of *Vulpia fasciculata*. *Annals of Botany*, **53**, 469–482.

Watkinson, A.R. (1985) On the abundance of plants along an environmental gradient. *Journal of Ecology*, **73**, 569–578.

Watkinson, A.R. (1986) Plant population dynamics. In Crawley, M.J. (ed.) *Plant Ecology*, pp. 137–184. Blackwell Scientific Publications, Oxford.

Watkinson, A.R. (1990) The population dynamics of

Vulpia fasciculata: a nine-year study. *Journal of Ecology*, **78**, 196–209.

Watkinson, A.R. & Davy, A.J. (1985) Population biology of salt marsh and sand dune annuals. *Vegetatio*, **62**, 487–497.

Watkinson, A.R. & Harper, J.L. (1978) The demography of a sand dune annual *Vulpia fasciculata*: I. The natural regulation of populations. *Journal of Ecology*, **66**, 15–33.

Watkinson, A.R., Lonsdale, W.M. & Andrew, M.H. (1989) Modelling the population dynamics of an annual plant *Sorghum intrans* in the wet–dry tropics. *Journal of Ecology*, **77**, 162–181.

Watson, M.A. (1984) Developmental constraints: effect on population growth and patterns of resource allocation in a clonal plant. *American Naturalist*, **12**, 411–426.

Watson, W.C.R. (1958) *Handbook of the Rubi of Great Britain and Ireland*. Cambridge University Press, Cambridge.

Watt, A.S. (1974) Senescence and rejuvenation in ungrazed chalk grassland (grassland B) in Breckland: the significance of litter and moles. *Journal of Applied Ecology*, **11**, 1157–1171.

Weaver, S.E. & Cavers, P.B. (1979) The effects of date of emergence and emergence order on seedling survival rates in *Rumex crispus* and *R. obtusifolius*. *Canadian Journal of Botany*, **57**, 730–738.

Webb, S.L. (1986) Potential role of passenger pigeons and other vertebrates in the rapid Holocene migrations of nut trees. *Quaternary Research*, **26**, 367–375.

Webb, S.L. (1987) Beech range extension and vegetation history. *Ecology*, **68**, 1993–2005.

Weiner, J. (1986) How competition for light and nutrients affects size variability in *Ipomoea tricolor* populations. *Ecology*, **67**, 1425–1427.

Weiner, J. (1990) Asymmetric competition in plant populations. *Trends in Ecology and Evolution*, **5**, 360–364.

Weiner, J. & Solbrig O.T. (1984) The meaning and measurement of size hierarchies in plant populations. *Oecologia*, **61**, 334–336.

Weiner, J. & Thomas, S.C. (1986) Size variability and competition in plant monocultures. *Oikos*, **47**, 211–222.

Weismann, A. (1893) *The Germ-plasm: a Theory of Heredity*. Walter Scott, London.

Weller, D.E. (1987a) A re-evaluation of the −3/2 power rule of plant self-thinning. *Ecological Monographs*, **57**, 23–43.

Weller, D.E. (1987b) Self-thinning exponent correlated with allometric measures of plant geometry. *Ecology*, **68**, 813–821.

Weller, D.E. (1989) The interspecific size–density relationship among crowded plant stands and its implications for the −3/2 power rule of self thinning. *American Naturalist*, **133**, 20–41.

Weller, D.E. (1990) Will the real self-thinning rule please stand up? — a reply to Osawa & Sugita. *Ecology*, **71**, 1204–1207.

Weller, S.G. & Ornduff, R. (1989) Incompatibility in *Amsinckia grandiflora* (Boraginaceae): distribution of callose plugs and pollen tubes following inter- and intramorph crosses. *American Journal of Botany*, **76**, 277–282.

Wennström, A. & Ericson, L. (1992) Environmental heterogeneity and disease transmission within clones of *Lactuca sibirica*. *Journal of Ecology*, **80**, 71–77.

Werner, P.A. (1975) Predictions of fate from rosette size in teasel (*Dipsacus fullonum* L.). *Oecologia*, **20**, 197–201.

Werner, P.A. & Caswell, H. (1977) Population growth rates and age vs stage-distribution models for teasel (*Dipsacus sylvestris* Huds.). *Ecology*, **58**, 1103–1111.

Werner, P.A. & Platt, W.J. (1976) Ecological relationship of co-occurring goldenrods (*Solidago*: Compositae). *American Naturalist*, **110**, 959–971.

Westoby, M. (1984) The self-thinning rule. *Advances in Ecological Research*, **14**, 167–225.

White, J. (1985) The thinning rule and its application to mixtures of plant populations. In White, J. (ed.) *Studies on Plant Demography*, pp. 291–309. Academic Press, London.

Whitmore, T.C. (1983) Secondary succession from seed in tropical rain forests. *Forestry Abstracts*, **44**, 767–779.

Whitmore, T.C. (1988) The influence of tree population dynamics on forest species composition. In Davy, A.J., Hutchings, M.J. & Watkinson, A.R. (eds) *Plant Population Ecology*, pp. 271–291. Blackwell Scientific Publications, Oxford.

Whitmore, T.C. (1989) Canopy gaps and the two major groups of forest trees. *Ecology*, **70**, 536–538.

Whittaker, R.H. (1969) Evolution of diversity in ecological systems. *Brookhaven Symposia in Biology*, **22**, 178–195.

Widén, B. (1991) Phenotypic selection on flowering phenology in *Senecio integifolius*, a perennial herb. *Oikos*, **61**, 205–215.

Wilcox, M.D. (1983) Inbreeding depression and genetic variances estimated from self- and cross-pollinated families of *Pinus radiata*. *Silvae Genetica*, **32**, 89–96.

Willey, R.W. (1979) Intercropping: its importance and research needs. Part 1. Competition and yield advantage. *Field Crop Abstracts*, **32**, 1–10.

Willey, R.W. & Heath, S.B. (1969) The quantitative relationships between plant population and crop yield. *Advances in Agronomy,* **21**, 281–321.

Williams, G.C. (1975) *Sex and Evolution.* Princeton University Press, Princeton, NJ.

Williamson, M.H. & Brown, K.C. (1986) The analysis and modelling of British invasions. *Philosophical Transactions of the Royal Society of London B,* **314**, 505–522.

Willson, M.F. (1979) Sexual selection in plants. *American Naturalist,* **113**, 777–790.

Willson, M.F. & Burley, N. (1983) *Mate Choice in Plants: Tactics, Mechanisms, and Consequences.* Princeton University Press, Princeton, NJ.

Willson, M.F., Hoppes, W.G., Goldman, D.A., Thomas, P.A., Katusic-Malmbord, P.L. & Bothwell, J.L. (1987) Sibling competition in plants: an experimental study. *American Naturalist,* **129**, 304–311.

Willson, M.F. & Rathcke, B.J. (1974) Adaptive design of the floral display in *AscleDias syriaca* L. *American Midland Naturalist,* **92**, 47–57.

Wilson, J.B. (1988) Shoot competition and root competition. *Journal of Applied Ecology,* **25**, 279–296.

Wilson, J.B. & Levin, D.A. (1986) Some genetic consequences of skewed fecundity distributions in plants. *Theoretical and Applied Genetics,* **73**, 113–121.

Wilson, R.G., Kerr, E.D. & Nelson, L.A. (1985) Potential for using weed seed content in the soil to predict future weed problems. *Weed Science,* **33**, 171–175.

Wilson, S.D. & Tilman, D. (1991) Components of plant competition along an experimental gradient of nitrogen availability. *Ecology,* **72**, 1050–1065.

Winn, A.A. (1985) Effects of seed size and microsite on seedling emergence of *Prunella vulgaris* in four habitats. *Journal of Ecology,* **73**, 831–840.

Wit, C.T. de (1960) On competition. *Verslangen van Landbouwkundige Onderzoekingen,* **66**, 1–82.

Wolf, L.L., Hainsworth, F.R., Mercier, T. & Benjamin, R. (1986) Seed-size variation and pollinator uncertainty in *Ipomopsis aggregata* (Polemoniaceae). *Journal of Ecology,* **74**, 361–371.

Woodell, S.R.J., Mooney, H.A. & Hill, A.J. (1969) The behaviour of *Larrea divaricata* (creosote bush) in response to rainfall in California. *Journal of Ecology,* **57**, 37–44.

Wright, A.J. (1981) The analysis of yield–density relationships in binary mixtures using inverse polynomials. *Journal of Agricultural Science,* **96**, 561–567.

Wright, S. (1943) Isolation by distance. *Genetics,* **28**, 114–138.

Wright, S. (1948) On the roles of directed and random changes in gene frequency in the genetics of populations. *Evolution,* **2**, 279–295.

Wright, S. (1951) The genetical structure of populations. *Annals of Eugenics,* **15**, 323–354.

Wright, S.J. (1982) Competition, differential mortality, and their effect on the spatial pattern of a desert perennial, *Eriogonum inflatum* Torr. and Frem. (Polygonaceae). *Oecologia,* **54**, 266–269.

Wright, S.J. (1983) The dispersion of eggs by a bruchid beetle among Scheelea palm seeds and the effect of distance to the parent palm. *Ecology,* **64**, 1016–1021.

Wulff, R.D. (1986a) Seed size variation in *Desmodium paniculatum.* I. Factors affecting seed size. *Journal of Ecology,* **74**, 87–97.

Wulff, R.D. (1986b) Seed size variation in *Desmodium paniculatum.* III. Effects on reproductive yield and competitive ability. *Journal of Ecology,* **74**, 115–121.

Wyatt, R. (1983) Pollinator–plant interactions and the evolution of breeding systems. In Real, L. (ed.) *Pollination Biology,* pp. 51–95. Academic Press, Orlando.

Yampolsky, E. & Yampolsky, H. (1922) Distribution of sex forms in the phanerogamic flora. *Bibliotheca Genetica,* **3**, 1–62.

Yazdi-Samadi, B., Wu, K.K. & Jain, S.K. (1978) The role of genetic polymorphisms in the outcome of interspecific competition in *Avena. Genetica,* **48**, 151–159.

Yeaton, R.I. (1978) A cyclical relationship between *Larrea tridentata* and *Opuntia leptocaulis* in the northern Chihuahua Desert. *Journal of Ecology,* **66**, 651–656.

Yeh, F.C.-H. & Layton, C. (1979) The organization of genetic variability in central and marginal populations of lodgepole pine, *Pinus contorta* ssp. *latifolia. Canadian Journal of Genetics Cytology,* **21**, 487–503.

Yoda, K., Kira, T., Ogawa, H. & Hozumi, K. (1963) Self-thinning in overcrowded pure stands under cultivated and natural conditions. *Journal of Biology, Osaka City University,* **14**, 107–129.

Young, T.P. (1990) Evolution of semelparity in Mount Kenya lobelias. *Evolutionary Ecology,* **4**, 157–171.

Young, T.P. & Augspurger, C.K. (1991) Ecology and evolution of long-lived semelparous plants. *Trends in Ecology and Evolution,* **6**, 285–289.

Zeide, B. (1985) Tolerance and self-tolerance of trees. *Forest Ecology and Management,* **13**, 149–166.

Zeide, B. (1987) Analysis of the −3/2 power law of self-thinning. *Forest Science,* **33**, 517–537.

Zhang, J. & Maun, M.A. (1990) Effect of sand burial on seed germination, seedling emergence, survival and growth of *Agropyron psammophilum. Canadian Journal of Botany,* **68**, 304–310.

Zuberi, M.I. & Gale, J.S. (1976) Variation in wild populations of *Papaver dubium.* X. Genotype–environment interaction associated with differences in soil. *Heredity,* **36**, 359–368.

Index

Index